AF347514

Polyphenols in Crops, Medicinal and Wild Edible Plants

Polyphenols in Crops, Medicinal and Wild Edible Plants

From Their Metabolism to Their Benefits for Human Health

Special Issue Editors

Marco Landi
Marek Zivcak
Marian Brestic
Oksana Sytar

MDPI • Basel • Beijing • Wuhan • Barcelona • Belgrade • Manchester • Tokyo • Cluj • Tianjin

Special Issue Editors

Marco Landi
Department of Agriculture,
Food and Environment,
University of Pisa
Italy

Marek Zivcak
Department of Plant Physiology,
Slovak University of Agriculture
in Nitra
Slovakia

Marian Brestic
Department of Plant Physiology,
Slovak University of Agriculture
in Nitra
Slovakia

Oksana Sytar
Department of Plant Biology,
Institute Biology and Medicine,
Kiev National University of
Taras Shevchenko
Ukraine

Editorial Office
MDPI
St. Alban-Anlage 66
4052 Basel, Switzerland

This is a reprint of articles from the Special Issue published online in the open access journal *Molecules* (ISSN 1420-3049) (available at: https://www.mdpi.com/journal/molecules/special_issues/polyphenol_crops_plants).

For citation purposes, cite each article independently as indicated on the article page online and as indicated below:

LastName, A.A.; LastName, B.B.; LastName, C.C. Article Title. *Journal Name* **Year**, *Article Number*, Page Range.

ISBN 978-3-03936-116-8 (Pbk)
ISBN 978-3-03936-117-5 (PDF)

Contents

About the Special Issue Editors

Marco Landi is a senior researcher in Plant Biochemistry with the Department of Agriculture, Food and Environment, University of Pisa (Italy). His scientific interests are mainly focused on the structural, biochemical, molecular, and physiological mechanisms through which plants accommodate environmental stress, including flavonoid metabolism. Recent projects have included: (i) photoprotective role of anthocyanins and flavonoids in plants under environmental stress (high light, UV-B, mineral toxicity); (ii) optical models to assess the development of anthocyanin–metal bond exploring the possibility that anthocyanins may additionally function as metal chelators; and (iii) physiological mechanisms adopted by Mediterranean tree species, including fruit tree species, against high levels of tropospheric ozone, drought, salinity, mineral toxicity (e.g., boron, NaCl). Another line of research deals with the evaluation of bioactive compounds in crop species typical of the Mediterranean area when subjected to different pedo-climatic/agronomic/post-harvest factors, with the aim of increasing their nutraceutical values and shelf life. Emphasis has recently included the characterization of the bioactive profile of Mediterranean wild edible species in an attempt to select some of them as new functional foods. Dr. Landi has been a member of SICA since 2014 and the Italian Society of Photobiology since 2015. His scientific activity is documented by more than 100 research or congress papers and 10 book chapters. As Editor-in-Chief of American Journal of Agricultural and Biological Sciences since 2015 and Associate Editor of Photosynthetica and Frontiers in Plant Science since 2016, Dr. Landi has edited several Special Issues in high-ranking journals such as Plants, Molecules, Agronomy, and Frontiers in Plant Science and is the editor of the book "Metal Toxicity in Higher Plants".

Marek Zivcak is an Associate Professor of crop physiology at the Slovak University of Agriculture in Nitra, Slovakia. He is an expert in plant and crop physiology, the biophysics of photosynthesis, and various spectrometric methods for analysis of the photosynthesis in vivo, including different applications of chlorophyll fluorescence methods. Specifically, his research has focused on regulation of photoprotection in crop plants exposed to environmental stress. His recent activities are focused on high-throughput phenotyping techniques in laboratory and field conditions, with special emphasis on leaf optical properties analyses using optical sensors, including non-invasive methods for assessment of major groups of bioactive compounds, such as flavonoids and anthocyanins. Marek Zivcak has a broad experience in managing research projects as well as in the evaluation of the project proposals within different grant schemes, including numerous H2020 projects. He has published over 70 research papers in international peer-reviewed journals; he is a coauthor of one book and 12 book chapters published by renowned publishers. He has edited Special Issues in journals Sensors and Molecules and he is a member of the Editorial Board of Environmental and Experimental Botany.

Marian Brestic is a professor at the Slovak University of Agriculture in Nitra (Slovakia). He has more than 35 years of experience in the field of crop photosynthesis and plant stress physiology. He studies plant tolerance mechanisms to environmental stress and the impact of climate change and drought on the sustainability of agriculture through a better understanding of crop genetic resources to improve the physiological properties and performance of modern varieties. He studies the properties of plant species that are important in terms of human nutrition and disease prevention.

Prof. Brestic has built a workplace with an international reputation and he has been involved in various international research projects and common research papers with EU partners, but also with bilateral projects with partners from Asia and Africa. He published more than 130 scientific papers in WoS journals, 1 book, 3 scientific monographs, and 20 book chapters with partners from 27 countries. He is a co-author of 3 wheat varieties. He is a member of Editorial Boards of Plant Physiology and Biochemistry, BMC Plant Biology, Environmental and Experimental Botany, Plant Physiology and Molecular Biology, Plant Biotechnology Reports, Plant Soil, and Environment. Prof. Brestic has edited several Special Issues in prestigious journals such as Land Degradation and Development, Science of Total Environment, Frontiers in Plant Science, Frontiers in Chemistry, Sensors, Molecules, and International Journal of Genomics.

Oksana Sytar scientific interest is mainly focused on the biochemistry of secondary metabolites and metabolomics, which are important tools in many disciplines, including research on plant resources for food and pharmaceutical use. Special interests include developing methods for obtaining and screening plant secondary metabolites (nathtodianthrones, phenolic acids, catechins, anthocyanins, sulpholipids, and different alliins), which are characterized by health-promoting properties and can be used as functional food components or nutraceuticals. Their main research activity focuses on biodiversity of useful plants, which are a crucial player in the emerging field of functional food and nutrition industry, and sees themselves as especially dedicated to improving the quality of life as well as to ensuring a variability of high quality products for the international food and drugs markets based on plants biodiversity. Their scientific activity is documented by more than 90 research or conference papers and 8 book chapters.

Preface to "Polyphenols in Crops, Medicinal and Wild Edible Plants"

Phenolic compounds from commercial crops, old varieties, medicinal herbs, and wild edible species, including phenolic acids, coumarins, flavonoids, and tannins, may play a crucial role in the prophylaxis of various human diseases. The antioxidant capacity of polyphenols, likely their key prerogative in controlling a plethora of human diseases, varies sensibly depending on their chemical nature, whose complexity has paralleled the evolution of land plants. The aim of this Special Issue was: (i) to describe polyphenols' classification, diversification, and occurrence in the plant kingdom; (ii) to report the effect of external factors on their metabolisms; and (iii) to establish the potential benefits of polyphenols for human pathologies, testing their antioxidant activity with the attempt to exploit the derived secondary metabolites as drug or nutraceutical compounds in fortified foods. This Special Issue of the journal Molecules, entitled "Polyphenols in Crops, Medicinal and Wild Edible Plants: From Their Metabolism to Their Benefits for Human Health" is devoted to studies related to medicinal herbs of different ethnobotanical regions, in an attempt to discover plant resources that can be used for the extraction of targeting polyphenols, leading to the development of new treatments for treating especially complicated and minor diseases. Recent research dealing with polyphenolic chemodiversity and polyphenol-based fingerprint of crops and old varieties, as well as the effect of external factors to polyphenol profile and abundance, are also presented here. The plant kingdom is an open-pit mine of chemical compounds that are still waiting to be explored, a task that can be accomplished in the era of omics sciences.

Marco Landi, Marek Zivcak, Marian Brestic, Oksana Sytar
Special Issue Editors

Article

Thymoquinone Enhances Paclitaxel Anti-Breast Cancer Activity via Inhibiting Tumor-Associated Stem Cells Despite Apparent Mathematical Antagonism

Hanan A. Bashmail [1], Aliaa A. Alamoudi [1], Abdulwahab Noorwali [1], Gehan A. Hegazy [1,2], Ghada M. Ajabnoor [1] and Ahmed M. Al-Abd [3,4,*

[1] Department of Clinical Biochemistry, Faculty of Medicine, King Abdulaziz University, Jeddah 21589, Saudi Arabia; hanan.a.bashmail@gmail.com (H.A.B.); aliaa.alamo@gmail.com (A.A.A.); wal5566@gmail.com (A.N.); gehanhegazy@hotmail.com (G.A.H.); ga_clinbio@yahoo.com (G.M.A.)

[2] Department of Medical Biochemistry, Medical Division, National Research Centre, Giza 12622, Egypt

[3] Department of Pharmaceutical Sciences, College of Pharmacy, Gulf Medical University, Ajman 4184, UAE

[4] Department of Pharmacology, Medical Division, National Research Centre, Giza 12622, Egypt

* Correspondence: ahmedmalabd@pharma.asu.edu.eg; Tel.: +971-(0)56-464-2929

Academic Editor: Marian Brestic
Received: 4 December 2019; Accepted: 17 January 2020; Published: 20 January 2020

Abstract: Thymoquinone (TQ) has shown substantial evidence for its anticancer effects. Using human breast cancer cells, we evaluated the chemomodulatory effect of TQ on paclitaxel (PTX). TQ showed weak cytotoxic properties against MCF-7 and T47D breast cancer cells with IC_{50} values of 64.93 ± 14 µM and 165 ± 2 µM, respectively. Combining TQ with PTX showed apparent antagonism, increasing the IC_{50} values of PTX from 0.2 ± 0.07 µM to 0.7 ± 0.01 µM and from 0.1 ± 0.01 µM to 0.15 ± 0.02 µM in MCF-7 and T47D cells, respectively. Combination index analysis showed antagonism in both cell lines with CI values of 4.6 and 1.6, respectively. However, resistance fractions to PTX within MCF-7 and T47D cells (42.3 ± 1.4% and 41.9 ± 1.1%, respectively) were completely depleted by combination with TQ. TQ minimally affected the cell cycle, with moderate accumulation of cells in the S-phase. However, a significant increase in Pre-G phase cells was observed due to PTX alone and PTX combination with TQ. To dissect this increase in the Pre-G phase, apoptosis, necrosis, and autophagy were assessed by flowcytometry. TQ significantly increased the percent of apoptotic/necrotic cell death in T47D cells after combination with paclitaxel. On the other hand, TQ significantly induced autophagy in MCF-7 cells. Furthermore, TQ was found to significantly decrease breast cancer-associated stem cell clone (CD44+/CD24-cell) in both MCF-7 and T47D cells. This was mirrored by the downregulation of TWIST-1 gene and overexpression of SNAIL-1 and SNAIL-2 genes. TQ therefore possesses potential chemomodulatory effects to PTX when studied in breast cancer cells via enhancing PTX induced cell death including autophagy. In addition, TQ depletes breast cancer-associated stem cells and sensitizes breast cancer cells to PTX killing effects.

Keywords: paclitaxel; thymoquinone; apoptosis; autophagy; tumor-associated stem cells

---

1. Introduction

Over the past three decades, 1355 new drugs were approved for the treatment of malignancies [1,2]. However, there are 18.1 million new cases of cancer, and 9.6 million mortalities due to cancer annually [3]. Breast cancer has the highest incidence, causing the most female mortalities among other malignancies [4]. Breast cancer tissue is a heterogeneous tissue consisting of various cell types, which differ in terms of origin, function, genetic profile, morphology, and sensitivity to therapy [5,6]. Breast cancer stem cells (BCSCs) are a subclone of cancer cells that have gained great attention, and are believed to be responsible for tumor growth and unlimited self-renewal ability. BCSCs possess a remarkable ability to effectively persist after exposure to chemotherapy [7].

Paclitaxel (PTX) is anti-microtubule chemotherapy that has been used successfully for different types of solid tumors including breast cancer for more than 40 years [8–10]. PTX stabilizes tubulin dimmers and suppresses microtubule depolymerization during mitosis, resulting in cell cycle arrest in M-phase; it is also called mitotic catastrophe [9,11]. However, breast cancer patients treated with taxane frequently develop chemotherapeutic resistance [12]. Combination therapy for PTX has been studied by many research teams, including ours, to enhance its anti-tumor activity and protect PTX from tumor resistance [13–16].

Natural compounds are believed to be a promising alternative for many chemotherapeutic remedies in fighting neoplasia [17]. More than 74% of the newly approved anticancer drugs during the past 30 years were of natural origin or inspired by natural product [1,2]. *Nigella sativa* and its constituents are among the most studied medicinal herbs in different health care issues [18]. Thymoquinone (TQ) is the major natural component of *Nigella sativa* seeds; it possesses anti-bacterial, anti-oxidant, anti-allergic, and anti-cancer effects [19–22].

Medicinal plants combined with cancer chemotherapy has gained great attention in recent years, and some studies have demonstrated promising results and outcomes. The main goal of these studies was to reduce the chemotherapeutic resistance associated with conventional chemotherapeutic agents or to protect normal tissues from their toxicity [23]. In our previous publications, thymoquinone was shown to improve the activity of cisplatin and gemcitabine against head and neck squamous cell carcinoma and breast cancer cells in addition to protecting oral epithelial cells from cisplatin-induced apoptosis. Herein, we studied the effect of TQ on the cytotoxicity profile of PTX against breast cancer cells, emphasizing breast-cancer-resistant clones in relation to BCSCs.

2. Results

2.1. The Chemomodulatory Effect of Thymoquinone to PTX within Breast Cancer Cells

A sulfarodamine-B (SRB) assay was used to assess the effect of TQ on the cytotoxic profile of PTX against breast cancer cells by calculating the IC_{50} values and R-fractions of single and combined PTX against MCF-7 and T47D cells. PTX showed a dose-dependent cytotoxic effect. Viability started to drop significantly at a concentration of 0.1 µM with IC_{50} values of 0.2 ± 0.07 µM and 0.1 ± 0.01 µM in MCF-7 and T47D cells, respectively (Figure 1A,B). In contrary, TQ did not exert any cytotoxic activity against either cell line until 30 µM. Higher concentrations of TQ induced a sudden drop in the viability with calculated IC_{50} values of 64.9 ± 14 µM and 165.1 ± 2.8 µM in MCF-7 and T47D cells, respectively (Figure 1A,B). Equitoxic combination (100:1) of TQ with PTX did not further improve the IC_{50} values of PTX against either MCF-7 or T47D cells (0.7 ± 0.01 µM and 0.15 ± 0.02 µM, respectively). Combination index analysis showed that TQ antagonized the cell-killing effect of PTX against MCF-7 and T47D cells, resulting in CI-values of 4.6 and 1.6, respectively (Table 1). Yet, TQ completely abolished the resistance fractions of both MCF-7 and T47D towards PTX from 42.37 ± 1.4% and 41.9 ± 1.1%, respectively, to 0% (Figure 1A,B) (Table 1). These data suggest that TQ does not improve PTX potency against MCF-7 or T47D cells and apparently antagonizes its killing effects. However, TQ significantly abolishes tumor-associated resistant cell clones.

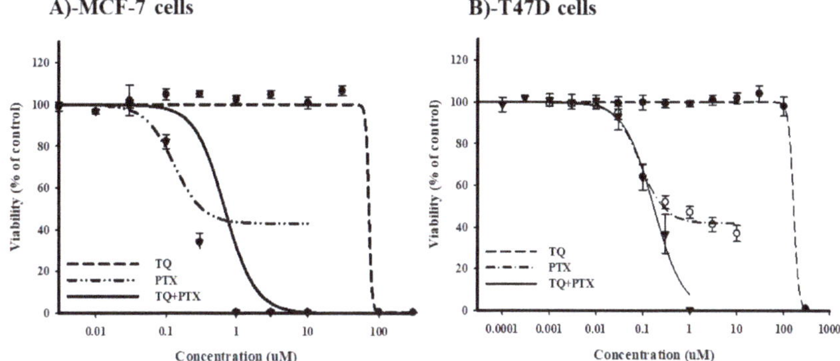

Figure 1. The effect of thymoquinone (TQ) on the dose-response curve of paclitaxel (PTX) in MCF-7 (**A**) and T47D (**B**) breast cancer cell lines. Cells were exposed to the serial dilution of PTX, TQ, or their combination for 72 h. Cell viability was determined using a sulfarodamine-B (SRB) assay, and data are expressed as mean ± SD (*n* = 3).

Table 1. Combination analysis of cell cytotoxicity for TQ, PTX, and their combination against MCF-7 and T47D breast cancer cell lines.

	MCF-7		T47D	
	IC50 (μM)	R-Value (%)	IC50 (μM)	R-Value (%)
PTX	0.2 ± 0.07	42.3 ± 1.4	0.1 ± 0.01	41.9 ± 1.1
TQ	64.9 ± 14.5	1.6 ± 1.3	165.1 ± 2.8	0.1 ± 0.15
PTX+TQ	0.7 ± 0.01	0	0.15 ± 0.02	0
CI-value	Antagonism/4.6		Antagonism/1.6	

2.2. Cell Cycle Distribution Analysis of Breast Cancer Cells

Further assessment for the interaction between TQ and PTX against cell cycle progression was undertaken using DNA content flow cytometry. In MCF-7 cells, PTX significantly arrested the cell cycle at G2/M-phase with a significant increase in the G2/M-phase population from 17.5 ± 2.3% to 71.2 ± 0.8% and from 15.9 ± 2% to 72.1 ± 2.8% after 24 h and 48 h, respectively (Figure 2A,B). TQ alone did not cause any significant change in the cell cycle distribution of MCF-7 cells. However, a combination of TQ with PTX induced a significant increase in the S-phase cell population (from 19.9 ± 0.5% to 23.8 ± 1%) after 24 h (Figure 2A). The cell cycle arrest at G2/M-phase induced by PTX alone and in combination with TQ resulted in cell death; a significant increase of Pre-G phase population was observed from 4.7 ± 1.5% to 27.1 ± 5% and 29.8 ± 4%, respectively, after 24 h (Figure 2C) and from 2.5 ± 0.6% to 17.9 ± 1.6%, 18.9 ± 0.4%, respectively, after 48 h (Figure 2D).

Similar to MCF-7, PTX significantly arrested T47D cells in G2/M-phase with a significant increase in this population from 19.4 ± 1.7% to 62.0 ± 2.9% and from 16.6 ± 1% to 83.3 ± 2.1% after 24 h and 48 h, respectively (Figure 3A,B). After 48 h of exposure, TQ alone and TQ+PTX treatment increased the S-phase T47D cell population from 29.1 ± 1.7% to 38.4 ± 0.2% and from 15.1 ± 1.7% to 28.6 ± 4.1%, respectively (Figure 3B). Interestingly, TQ treatment alone induced significant cell death and increased the Pre-G cell population of T47D cells from 8.5 ± 0.3% to 10.8 ± 0.2% and from 12.6 ± 1.4% to 67.6 ± 5.2% after 24 h and 48 h, respectively (Figure 3C,D). In addition, PTX alone induced a significant increase in the pre-G cell population from 8.5 ± 0.3% to 38.1 ± 5.1% and 12.6 ± 1.45% to 44.5 ± 3.2% after 24 and 48 h, respectively. Combination of PTX with TQ resulted in a significantly higher pre-G cell population compared to PTX treatment alone after 24 and 48 h (69.6 ± 1.1% and 60.4 ± 1.7%, respectively) (Figure 3C,D). Pre-G phase is indicative of cell death. However, it is non-specific and could be programmed cell death (apoptosis or autophagy) or non-programmed cell death (necrosis).

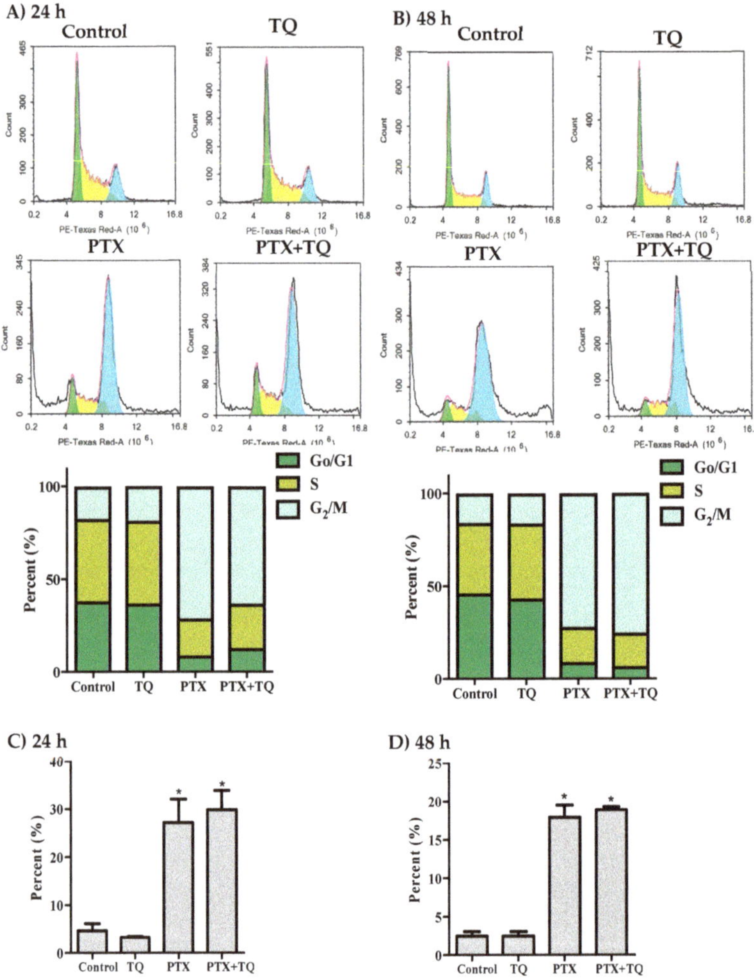

Figure 2. Effect of PTX, TQ, and their combination on the cell cycle distribution of MCF-7 cells. Cells were exposed to PTX, TQ, or their combination for 24 h (**A,C**) or 48 h (**B,D**). Cell cycle distribution was determined using DNA content flowcytometry analysis and different cell phases were plotted as percentage of total events. Sub-G cell population was plotted as percent of total events (**C,D**). Data are presented as mean ± SD; $n = 3$. (*) significantly different from the control group.

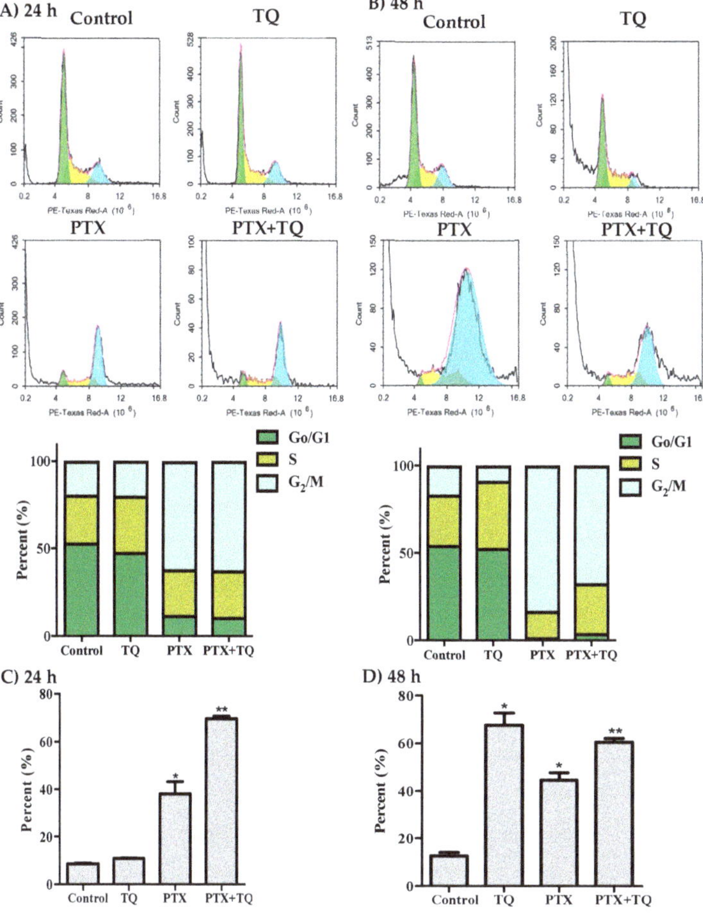

Figure 3. Effect of PTX, TQ, and their combination on the cell cycle distribution of T47D cells. Cells were exposed to PTX, TQ, or their combination for 24 h (**A,C**) or 48 h (**B,D**). Cell cycle distribution was determined using DNA content flowcytometry analysis and different cell phases were plotted as percentage of total events. Sub-G cell population was plotted as percent of total events (**C,D**). Data are presented as mean ± SD; n = 3. (*) significantly different from the control group. (**) significantly different from PTX treatment.

2.3. Apoptosis Assessment

Herein, we investigated the effect of TQ in overcoming MCF-7, T47D cells resistance to PTX by inducing further apoptosis, necrosis, and/or autophagy. T47D cells were exposed to the pre-determined IC_{50} values of PTX, TQ, and their combination for 24 and 48 h rather than 72 h to detect early apoptotic events. Apoptosis/necrosis populations were then determined by Annexin-V/FITC-PI staining coupled with a flowcytometry technique. Both TQ alone and PTX alone induced a significant apoptosis after 24 h of exposure (22.4 ± 3.2% and 10.3 ± 0.8%, respectively) compared to untreated control T47D cells (4.3 ± 0.5%). Yet, PTX combination with TQ induced significantly more apoptosis compared to PTX treatment alone (58.1 ± 2.1%). In addition to apoptosis, TQ, PTX, and their combination induced significant necrotic cell death in T47D by 5.9 ± 0.3%, 7.4 ± 0.7%, and 22.2 ± 0.6%, respectively (compared to 2.3 ± 0.7% necrosis in control cells) (Figure 4A)). Similarly, further exposure (48 h) of T47D cells to PTX alone or TQ alone resulted in more apoptosis (18.4 ± 2.9% and 32.5 ± 2.8%, respectively) compared to untreated control cells (1.9 ± 0.5%). Combination of PTX and TQ did not significantly increase T47D apoptotic cell population compared to TQ treatment alone (31.5 ± 1.3%). Moreover, TQ, PTX, and their combination induced significant necrotic cell death in T47D by 4.0 ± 0.4%, 11.3 ± 0.6%, and 11.0 ± 0.4%, respectively (compared to 2.6 ± 0.1% necrosis in control cells) (Figure 4B). To confirm apoptosis, Western blot analysis was carried out for caspase-3 and PARP proteins. PTX induced the expression of caspase-3 after 24 and 48 h. Further combination of PTX with TQ resulted in more active caspase-3 and more cleavage for its downstream target protein, PARP (Figure 4C,D).

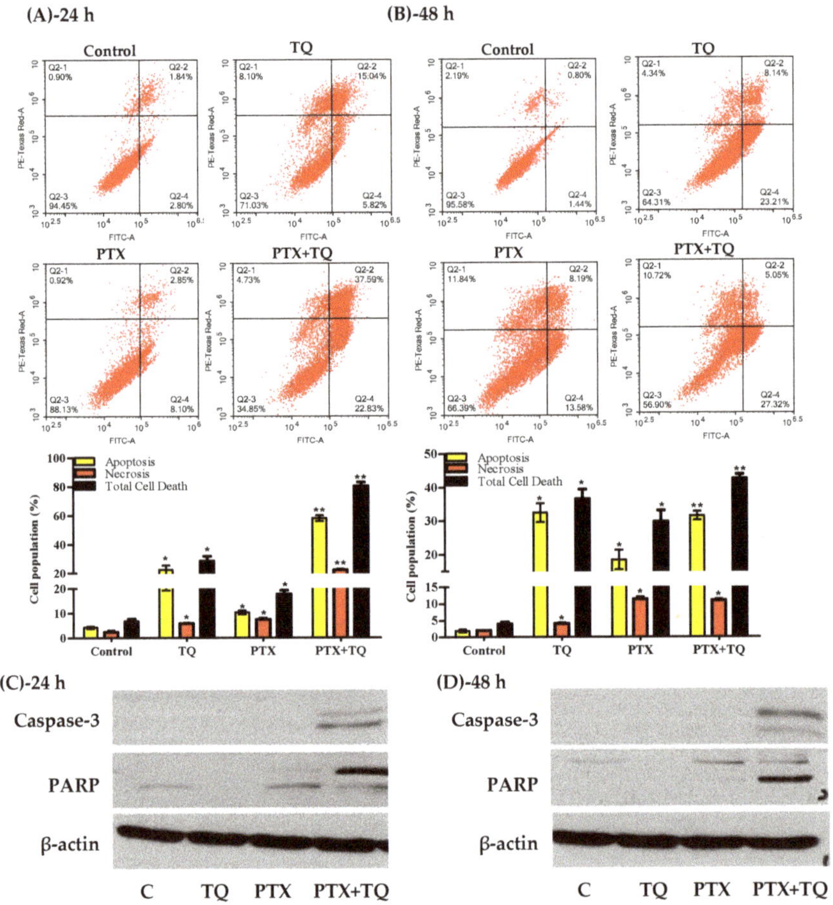

Figure 4. Apoptosis/necrosis assessment in T47D cells after exposure to PTX, TQ, and their combination. Cells were exposed to PTX, TQ, or their combination for 24 h (**A**) and 48 h (**B**). Cells were stained with annexin V-FITC/PI and different cell populations are plotted as a percentage of total events. Western blot analysis for caspase-3 and PARP was assessed for MCF-7 (**C**) and T47D (**D**) cells. Data are presented as mean ± SD; *n* = 3. (*) significantly different from the control group. (**) significantly different from PTX treatment.

2.4. Autophagy Assessment

Besides apoptosis, we were keen to study the effect of PTX, TQ, and their combination on other cell death mechanisms such as the autophagy process. In MCF-7, treatment with PTX, TQ, and the combination of PTX+TQ increased the fluorescent intensity indicative of autophagic cell death by 58.2%, 33.9%, and 49.1%, respectively (Figure 5A). On the other hand, none of the treatments under investigation (PTX, TQ, or their combination) induced any significant change or autophagic cell death in T47D (Figure 5B). CQ (positive control autophagic drug) induced autophagic cell response in MCF-7 and T47D cell lines and increased Cyto-ID fluorescence by 31 and 42%, respectively (Figure 5A,B). For further conformation, two key autophagy genes beclin-1 and LC3-II were assessed by the RT-PCR technique. In MCF-7 cells, treatment with PTX, TQ, and the a combination of PTX+TQ significantly increased the expression of beclin-1 by 4.4, 3.1, and 6.8 folds, respectively, and significantly increased the expression of LC3-II by 3.4, 1.9, and 4.1 folds, respectively. In T47D, only PTX marginally increased the expression of beclin-1 by 1.4 folds (Figure 5C,D)

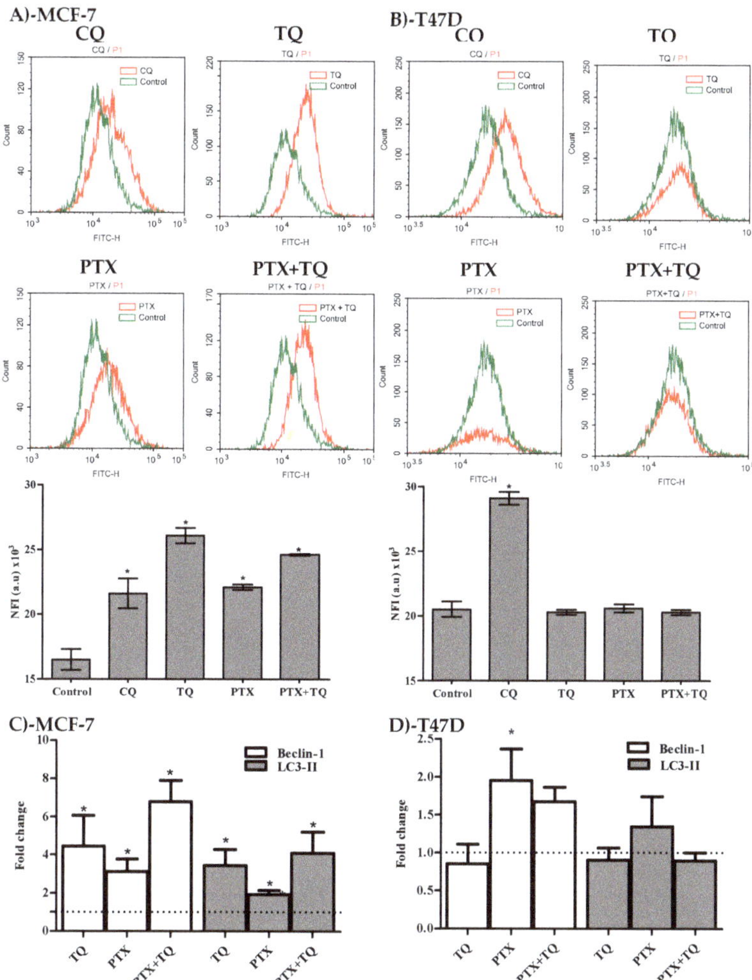

Figure 5. Autophagic cell death assessment in MCF-7 (**A**) and T47D (**B**) cells after exposure to PTX, TQ, and their combination. Cells were exposed to PTX, TQ, or their combination for 24 h, and were stained with a Cyto-ID autophagosome tracker. Net fluorescent intensity (NFI) was plotted and compared to the basal fluorescence of the control group. Gene expression fold changes for beclin-I and LC3-II were assessed for MCF-7 (**C**) and T47D cells (**D**). Data are presented as mean ± SD; *n* = 3. (*) significantly different from the control group.

2.5. Breast Cancer-Associated Stem Cell (CD44+/CD24- Cell Clone) Detection

Furthermore, we assessed the breast cancer-associated stem cell clone (CD44+/CD24-) and endothelial mesenchymal transition gene expression in relation to treatment with TQ, PTX, and their combination. TQ alone induced a significant decrease in the CD44+/CD24- stem cell clone by 12.4 ± 0.8%. However, PTX treatment reduced CD44+/CD24- cell clone by only 7.6 ± 0.1%. Interestingly, the combination of PTX with TQ significantly abolished the tumor-associated stem cell clone (CD44+/CD24-) by 32.3 ± 0.08% (Figure 6A). In addition, TQ significantly decreased the T47D associated stem cell clone (CD44+/CD24) by 19.9 ± 0.8%, while PTX caused a 9.9 ± 0.2% decrease in the tumor-associated stem cell clone. Yet, the combination of PTX with TQ further decreased the tumor-associated stem cell clone by 23.9 ± 1.6% (Figure 6B).

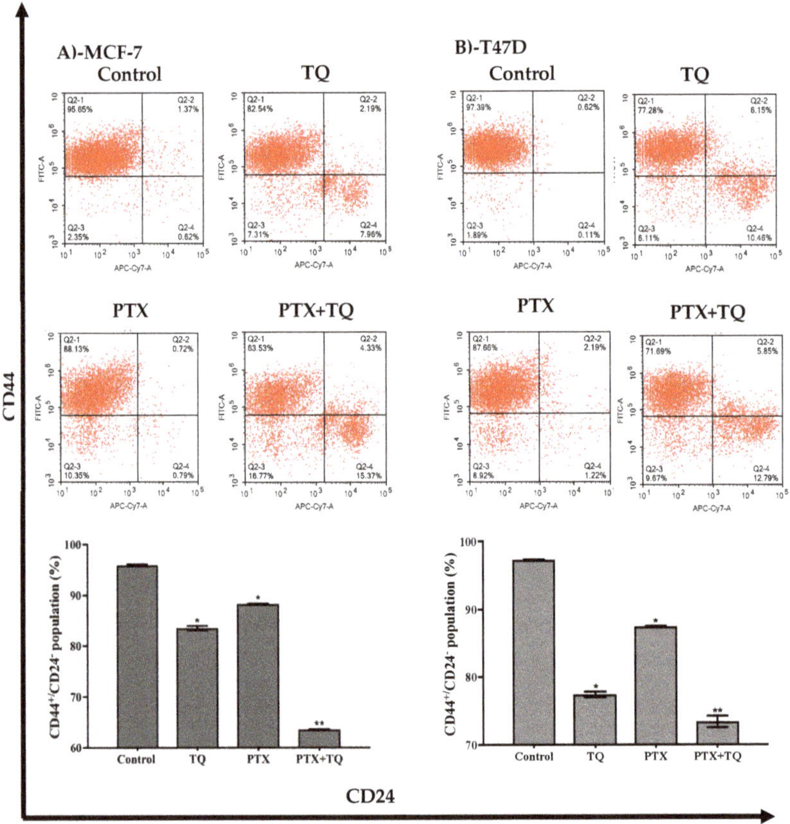

Figure 6. Effect of PTX, TQ, and their combination on the expression of CD44 and CD24 stem cell markers. MCF-7 (**A**) and T47D (**B**) cells were exposed to PTX, TQ, or their combination for 24 h. Expression levels of CD44 and CD24 were assessed using flowcytometry and plotted as percentage of total events. Data are presented as mean ± SD; $n = 3$. (*) significantly different from the control group. (**) significantly different from PTX treatment.

2.6. EMT Genes Expression Assessment

Further assessment for the expression of key EMT genes (ZEB-2, TWIST-1, SNAIL-1, and SNAIL-2) after treatment with TQ and PTX was undertaken using the RT-PCR technique. No significant changes in the expression of ZEB-2 due to the treatment of TQ or PTX could be detected. TQ induced a significant increase in the expression level of both SNAIL-1 and SNAL-2 compared to the untreated control by 3.8 and 3.7 folds, respectively. On the other hand, TQ significantly downregulates TWIST-1 expression to 18% of the control expression level. PTX did not induce any significant change in the expression level of the SNAIL-1, SNAIL-2, or TWIST-1 genes. Taken together, TQ efficiently diminishes tumor-associated stem cells (Figure 7).

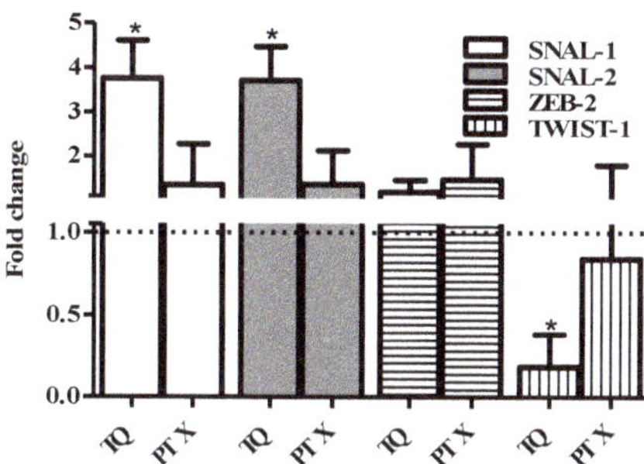

Figure 7. Effect of PTX and TQ on the expression of EMT-related genes in MCF-7 cells. Cells were exposed PTX or TQ for 24 h. Total RNA was extracted and subjected to RT-qPCR to measure gene expression. Data were plotted using the 2-ΔΔCt method (expression normalized to the housekeeping gene GAPDH). Fold expression and significance was calculated relative to control untreated cells (dotted line). Data are presented as mean ± SD; n = 3. (*) significantly different from the control group.

3. Discussion and Conclusion

Breast cancer remains the most common malignancy in females and a leading cause of death worldwide [3,24]. PTX is a cornerstone and commonly used chemotherapeutic drug for the treatment of breast cancer [9,25]. Despite its promising initial clinical response, it might be discontinued due to the emergence of resistance and toxicities [26,27]. TQ is a major active component of the *Nigella sativa* plant, which is commonly used for different medicinal purposes [18,28]. Herein, we are testing the hypothesis that a combination of PTX with TQ can decrease the breast cancer cell resistance to PTX.

According to our data, TQ alone showed significantly weaker cytotoxic/antiproliferative effects compared to PTX. Apparently, TQ combined with PTX resulted in decreased PTX potency in the form of a slight increase in its IC_{50} values. It was interesting to discover that TQ significantly abolished the resistance fractions of both breast cancer cell lines to PTX (R-fractions were above 40% in both cell lines). In one of our previous publications, it was found that curcuminoid-based synthetic compounds increased the IC_{50} of PTX against colorectal cancer cells but significantly decreased the resistance fractions of these cells towards PTX [16]. Herein, we have a similar scenario with the natural and safe compound, TQ. Cumulative evidence within the literature reports the safety and potency of TQ in enhancing many chemotherapeutic compounds against different tumor cell lines [29–32].

It is well known that PTX causes cell cycle arrest in the G2/M phase [33]. According to our finding in the current study, TQ induced accumulation of cells in S-phase. This could explain the apparent antagonism between PTX and TQ. However, this combination resulted in an increase in the pre-G cell population. Yet, TQ pushed quiescent stem cells to proliferate and become sensitized to PTX treatment via entering S-phase temporally and then entering G2/M-phase [34].

The elevated Pre-G cell population due to PTX treatment alone or in combination with TQ is indicative of cell death. Apoptotic, necrotic, and autophagic cell death induced by PTX, TQ, and their combination were examined to test the above hypothesis. Our observations showed that TQ significantly increased apoptosis in T47D cells by more than 4 folds and 16 fold after 24 h and 48 h, respectively. Many previous publications from our team and from others have highlighted the apoptotic effects of TQ against several cancer cells [31,32,35,36]. The enhancement effect of TQ towards PTX against T47D could be explained by the supra increase of apoptosis; PTX in combination with

TQ increased the apoptosis by more than 4 folds compared to PTX alone. It is worth mentioning that combination treatment induced marked elevation rates of necrosis compared to PTX alone. It was previously reported by Canatan and colleagues that the combination of PTX with TQ increased the expression of several genes involved in apoptosis in triple-negative breast cancer [37]. On the other hand, MCF-7 does not undergo normal apoptosis due to a lack of caspase-3 expression [38]. In the current work, caspase-3 was not detected by Western blotting in MCF-7 cells. Yet, it was suggested in our previous publications that autophagy represents an alternative cell death pathway in MCF-7 cells [32]. Herein, TQ induced autophagic pro-death effects in MCF-7 cells. In other words, PTX and TQ increased the concentration of caspase-3 enzyme in T47D cells with a prominent elevation of cleaved PARP concentration, which is indicative of apoptosis-dependent cell death. In MCF-7 (caspase-3 deficient cell line), TQ and PTX forced the cells to proceed in autophagy-dependent cell death.

Breast cancer stem cells play an important role in resisting chemotherapeutic treatments, and targeting or depleting this clone could effectively sensitize cancer cells to drugs [39]. Overall, our results provide further evidence to the ant-resistance effect exhibited by TQ in combination with PTX through studying their effect against breast cancer-associated stem cells (CD44+/CD24-) [40]. To the best of our knowledge, this is the first study demonstrating the effect of TQ in depleting BCSCs. Herein, TQ remarkably decreased the percentage of the CD44+/CD24- cell clone of MCF-7 and T47D, while PTX caused a slight decrease in the CD44+/CD24- cell clone. Interestingly, TQ in combination with PTX significantly decreased the aforementioned clone far more than TQ or PTX treatment alone. Resistance to chemotherapy and poor cancer patient prognosis was previously attributed to the failure of chemotherapy in depleting this stem cell clone [41]. Finally, the TQ effect in depleting tumor-associated stem cells was confirmed via studying the expression levels of key EMT genes. BCSCs possess high expression levels of mesenchymal markers such as TWIST-1 and a low expression of epithelial markers [42]. Herein, TQ significantly downregulates EMT-regulatory protein (TWIST-1) expression. Previously, it was found that TQ downregulates TWIST-1 expression in other tumor types, resulting in improved efficacy [43]. Other key EMT genes, the SNAIL gene family, are known to promote the migration and invasion of cancer cells [44]. Further details showed that the SNAIL family increased the sensitivity to anti-tubulin drugs such as PTX through the downregulation of bIII and bIVa-tubulin [45]. Thus, the observed increased expression of SNAIL1 and SNAIL2 due to TQ treatment, in the current study, provides PTX with a wider target to induce mitotic catastrophe, resulting in ultimately reduced resistance.

In conclusion, TQ proved and is still proving to have a potential effect in chemosensitizing several tumor types such as breast cancer to many chemotherapeutic agents such as PTX. Several aspects of TQ in decreasing breast cancer cell resistance to PTX are shown in this study. Mechanisms underlying this effect include cell cycle synchronization followed by cell death via apoptosis, necrosis, and autophagy. In addition, TQ interacts with EMT properties resulting in diminishing tumor-associated resistant stem cell fraction. Further in vivo studies that translate these molecular observations into therapeutic values is highly recommended.

4. Materials and Methods

4.1. Drugs and Chemicals

Thymoquinone (TQ), paclitaxel (PTX), and sulfarodamine-B (SRB) were all obtained from Sigma-Aldrich Chemical Co. Cell culture media, fetal bovine serum, and trypsin were obtained from Gibco™, Thermo Fisher Scientific.

4.2. Cell Culture

MCF-7 and T47D, were obtained from Nawah Scientific (Mokkatam, Cairo, Egypt). Cells were maintained in full DMEM media with heat-inactivated fetal bovine serum (10% *v/v*), streptomycin

(100 µg/mL), and penicillin (100 units/mL). Cells were kept in a humidified, 5% (*v/v*) CO2 atmosphere at 37 °C.

4.3. Cell Viability Assay

An SRB assay was used to evaluate the cytotoxicity effect of TQ, PTX, and/or their combination against MCF7 and T47D as previously described. Cells were seeded at 3000–5000 cells/well and treated with a serial concentration of TQ (0.01–300 µM), PTX (0.001–10 µM), and their combination for 72 h. Following that, cells were fixed by adding TCA (10% *w/v*) to each well and incubated for 1 h at 4 °C. After washing, a 0.4% SRB staining solution (*w/v*) was added, and the following steps and incubations followed what was previously described. The absorbance was measured at 540 nm with an ELISA microplate reader and calculated as the percent viability of control cells (cells exposed to drug-free media). DMSO concentrations were less than 0.1% in all treatment conditions.

4.4. Data Analysis

The dose–response curves of TQ, PTX, and their combination were analyzed using the E_{max} model [46] according to the following formula

$$\% \text{ Cell viability} = (100 - R) \times \left(1 - \frac{[D]^m}{K_d{}^m + [D]^m}\right) + R \tag{1}$$

The CI-value was calculated from the following formula:

Combination index (CI) was calculated from the formula:

$$CI = \frac{IC50 \text{ of drug}(x) \text{ combination}}{IC50 \text{ of drug}(x) \text{ alone}} + \frac{IC50 \text{ of drug}(y) \text{ combination}}{IC50 \text{ of drug}(y) \text{ alone}} \tag{2}$$

The nature of the drug interaction was defined according to Chou and Talalay as synergism if CI < 0.8, as antagonism if CI > 1.2, and as additive if CI ranges from 0.8 to 1.2 [47].

4.5. Cell Cycle Analysis by Flow Cytometry

To evaluate the effect of the drugs on cell cycle distribution, both cell lines were treated by the pre-determined IC_{50} values of TQ, PTX, or both drugs in combination, for 24 or 48 h. An additional drug-free medium-treated group acted as a control group. Post-treatment, cells were trypsinized, collected, and washed with ice-cold PBS and re-suspended in 0.5 mL of PBS. To ensure fixation of cells, 2 mL of 60% ice-cold ethanol were added on the cells while vortexing. Cells were incubated at 4 °C for 1 h. Prior to analysis, cells were washed and re-suspended in 1 mL of PBS containing 50 µg/mL RNAase A and 10 µg/mL propidium iodide (PI). Cells were incubated for 20 min in the dark at 37 °C and analyzed for DNA content using flow cytometry analysis FL2 (λex/em 535/617 nm). In all of the following flow cytometer analysis, 12,000 events were acquired, and NovoExpress™ software was used for analysis.

4.6. Analysis of Cell Apoptosis by Flow Cytometry

An Annexin V-FITC apoptosis detection kit (Abcam Inc., Cambridge Science Park, Cambridge, UK) was used to determine the effect of drugs on apoptosis and necrosis. Briefly, the cells were treated by the pre-determined IC_{50} values of either TQ, PTX, or both drugs combined for 24 h. A drug-free media-treated group was used as control. Cells were collected and washed twice with PBS and incubated in a dark place with 0.5 mL of Annexin V-FITC/PI solution for 30 min at room temperature according to the manufacturer's protocol. FITC and PI fluorescent signals were then analyzed using FL1 and FL2 signal detector, respectively (λex/em 488/530 nm for FITC and λex/em 535/617 nm for PI). Positive FITC and/or PI cells were quantified by quadrant analysis.

4.7. Analysis of Cell Autophagy by Flow Cytometry

To further confirm the cell death mechanism induced by the drugs, autophagic cell death was quantitatively analyzed using a Cyto-ID Autophagy Detection Kit (Abcam Inc., Cambridge Science Park, Cambridge, UK). In brief, cells were treated for 24 h by the IC_{50} values of the test compounds (single or combined treatments). Chloroquine treatment (10 µM) was used as a positive control, while a drug-free medium was used as a negative control. Cells were then washed twice with PBS and stained with Cyto-ID Green in the dark at 37 °C for 30 min according to the manufacturer's protocol. After staining, cells were analyzed for Cyto-ID differential green/orange fluorescent signals using an FL2 signal detector (λex/em 535/617 nm). Mean net fluorescent intensities (NFI) were quantified.

4.8. Stem Cell Detection by Flow Cytometry

For assessing the effects of TQ, PTX, and their combination against the breast cancer-associated stem cell clone (CD44+/CD24-), cells underwent FITC-labeled anti-CD44 and APC/Cy7-labeled anti-CD24 antibody (Abcam Inc. Cambridge Science Park, Cambridge, UK) staining and flowcytometry assessment. Briefly, cells were treated with the predetermined IC_{50} values of test drugs (single or combined treatments), or with drug-free media as a control group, for 24 h. After treatment, cells were trypsinized and washed with ice-cold PBS supplemented with 10% FCS. Cells were stained with the conjugated anti-CD44 and anti-CD24 antibodies and kept in the dark at room temperature for 30 min. Cells were then washed three times with ice-cold PBS containing 10% FCS. Cells were then analyzed for FITC and APC/CY7 fluorescent signals using FL1 and FL2 signal detector, respectively (λex/em 488/530 nm for FITC and λex/em 535/617 nm for APC/CY7).

4.9. EMT Gene Expression Analysis

Real-time polymerase chain reaction (PCR) was performed to assess the expression of CDH1, CDH2, SNAL1, SNAL2, ZEB2, and TWIST1 genes after treatment with the pre-determined IC_{50} values of TQ and PTX. After 24 h of treatment, RNA was extracted using mirVana™ RNA isolation kit (Invitrogen, Carlsbad, CA, USA). The RNA and purity were confirmed (A260/280>2.0) using a DeNovix DS-11™ microvolume spectrophotometer (Thermo Fisher Scientific, Wilmington, DE, USA). Subsequently, the total RNA samples of all treatments were reverse-transcribed to construct a cDNA library using the SuperScript™ Master Mix kit (Invitrogen, Carlsbad, CA, USA). The cDNA were then subjected to quantitative real-time PCR reactions using Custom TaqMan® Gene Expression Assay (Applied Biosystems, Foster City, CA, USA). GAPDH was used as a housekeeping gene; normalized fold changes for all genes of interest were calculated using the following formula: 2-$\Delta\Delta$Cq.

4.10. Autophagy Gene Expression Analysis

Real-time polymerase chain reaction (PCR) was performed on the cDNA prepared in the previous experiment to assess the expression of beclin-1 and LC3-II autophagy genes after treatment with the pre-determined IC_{50} values of TQ and PTX. The beclin-1 forward primer was 3'-GGCTGAGAGACTGGATCAGG-5'; the backward primer was 5'- CTGCGTCTGGGCATAACG-3'; the LC3-II forward primer was 3'-GAGAAGCAGCTTCCTGTTCTGG-5'; the backward primer was 5'-GTGT CCGTTCACCAACAGGAAG-3'. The housekeeping β-actin gene was as a reference gene with a forward primer of 3'-GAGAGGCGGCTAAGGTGTTT-5' and a backward primer of 5'-TGGTGTAGACGGGGATGACA-3'.

4.11. Western Blot Analysis and Detection of Apoptosis Related Signals

Apoptosis proteins, caspase-3, and PARP were assessed within cell lysate after treatment with TQ, PTX, and their combination to confirm cell death propagation via apoptosis. Briefly, cells were treated with the pre-determined IC_{50} values of PTX and TQ for 24 h and 48 h. Cell lysates were extracted using an RIPA-buffer and electrophoresed using SDS-PAGE (10%) and then transferred to PVDF-membrane.

Caspase-3 and cleaved PARP-1 proteins were detected using rabbit monoclonal anti-active caspase-3 and rabbit monoclonal anti-PARP (Abcam Inc., Cambridge Science Park, Cambridge, UK). Bands were visualized using HRP-conjugated anti-rabbit secondary antibodies (Abcam Inc., Cambridge Science Park, Cambridge, UK).

4.12. Statistical Analysis

Data are presented as mean ± SD using Prism® for Windows, ver. 5.00 (GraphPad Software Inc., La Jolla, CA, USA). To assess significance, analysis of variance (ANOVA) with an LSD post hoc test was used with SPSS® for Windows, version 17.0.0. A cut off value of $p < 0.05$ was used for significance.

Author Contributions: Conceptualization, A.M.A. Data curation, H.A.B.; Formal analysis, H.A.B., A.A.A., and A.M.A.; Funding acquisition, A.M.A.; Methodology, A.M.A.; Project administration, A.M.A.; Resources, A.A.A. and A.N.; Supervision, A.A.A., A.N., G.A.H., and G.M.A.A.; Validation, G.A.H.; Writing – original draft, H.A.B.; Writing – review & editing, G.A.H., G.M.A.A., and A.M.A.

Funding: This work was supported by the Deanship of Scientific Research (DSR), King Abdulaziz University, Jeddah, Saudi Arabia [grant numbers 303/166/1436].

Conflicts of Interest: The authors declare that there is no conflict of interest. The funders had no role in the design of the study; in the collection, analyses, or interpretation of data; in the writing of the manuscript; or in the decision to publish the results.

Abbreviations

The following abbreviations are used in this manuscript:

BCSC	Breast cancer stem cell
CQ	Chloroquine
PTX	Paclitaxel
SRB	Sulfarodamine-B
TQ	Thymoquinone

References

1. Clardy, J.; Walsh, C. Lessons from natural molecules. *Nature* **2004**, *432*, 829. [CrossRef] [PubMed]
2. Newman, D.J.; Cragg, G.M. Natural products as sources of new drugs over the 30 years from 1981 to 2010. *J. Nat. Prod.* **2012**, *75*, 311–335. [CrossRef] [PubMed]
3. Bray, F.; Ferlay, J.; Soerjomataram, I.; Siegel, R.L.; Torre, L.A.; Jemal, A. Global cancer statistics 2018. GLOBOCAN estimates of incidence and mortality worldwide for 36 cancers in 185 countries. *CA Cancer J. Clin.* **2018**, *68*, 394–424. [CrossRef] [PubMed]
4. Torre, L.A.; Islami, F.; Siegel, R.L.; Ward, E.M.; Jemal, A. Global cancer in women, burden and trends. *Cancer Epidemiol. Biomarkers Prev.* **2017**, *26*, 444–457. [CrossRef] [PubMed]
5. Ellsworth, R.E.; Blackburn, H.L.; Shriver, C.D.; Soon-Shiong, P.; Ellsworth, D.L. Molecular heterogeneity in breast cancer, State of the science and implications for patient care. *Semin. Cell. Dev. Biol.* **2017**, *64*, 65–72. [CrossRef] [PubMed]
6. Prasetyanti, P.R.; Medema, J.P. Intra-tumor heterogeneity from a cancer stem cell perspective. *Mol. Cancer* **2017**, *16*, 41. [CrossRef]
7. Aponte, P.M.; Caicedo, A. Stemness in cancer, Stem cells, cancer stem cells, and their microenvironment. *Stem Cells Int.* **2017**, *2017*, 5619472. [CrossRef]
8. Pazdur, R.; Kudelka, A.P.; Kavanagh, J.J.; Cohen, P.R.; Raber, M.N. The taxoids, paclitaxel (Taxol) and docetaxel (Taxotere). *Cancer Treat Rev.* **1993**, *19*, 351–386. [CrossRef]
9. Perez, E.A. Paclitaxel in Breast Cancer. *Oncologist* **1998**, *3*, 373–389.
10. Gudena, V.; Montero, A.J.; Glück, S. Gemcitabine and taxanes in metastatic breast cancer, a systematic review. *Ther. Clin. Risk Manag.* **2008**, *4*, 1157–1164.

11. Klimaszewska-Wisniewska, A.; Halas-Wisniewska, M.; Tadrowski, T.; Gagat, M.; Grzanka, D.; Grzanka, A. Paclitaxel and the dietary flavonoid fisetin, a synergistic combination that induces mitotic catastrophe and autophagic cell death in A549 non-small cell lung cancer cells. *Cancer Cell. Int.* **2016**, *16*, 1. [CrossRef] [PubMed]

12. Rivera, E.; Gomez, H. Chemotherapy resistance in metastatic breast cancer, the evolving role of ixabepilone. *Breast Cancer Res.* **2010**, *12* (Suppl. 2). [CrossRef] [PubMed]

13. Toppmeyer, D.; Seidman, A.D.; Pollak, M.; Russell, C.; Tkaczuk, K.; Verma, S.; Overmoyer, B.; Garg, V.; Ette, E.; Harding, M.W.; et al. Safety and efficacy of the multidrug resistance inhibitor Incel (biricodar, VX-710) in combination with paclitaxel for advanced breast cancer refractory to paclitaxel. *Clin. Cancer Res.* **2002**, *8*, 670–678. [PubMed]

14. Gill, C.; Walsh, S.E.; Morrissey, C.; Fitzpatrick, J.M.; Watson, R.W. Resveratrol sensitizes androgen independent prostate cancer cells to death-receptor mediated apoptosis through multiple mechanisms. *Prostate* **2007**, *67*, 1641–1653. [CrossRef]

15. Tolba, M.F.; Esmat, A.; Al-Abd, A.M.; Azab, S.S.; Khalifa, A.E.; Mosli, H.A.; Abdel-Rahman, S.Z.; Abdel-Naim, A.B. Caffeic acid phenethyl ester synergistically enhances docetaxel and paclitaxel cytotoxicity in prostate cancer cells. *IUBMB Life* **2013**, *65*. [CrossRef]

16. El-Araby, M.E.; Omar, A.M.; Khayat, M.T.; Assiri, H.A.; Al-Abd, A.M. Molecular mimics of classic p-glycoprotein inhibitors as multidrug resistance suppressors and their synergistic effect on paclitaxel. *PLoS ONE* **2017**, *12*. [CrossRef]

17. Dastjerdi, M.N.; Mehdiabady, E.M.; Iranpour, F.G.; Bahramian, H. Effect of Thymoquinone on P53 Gene Expression and Consequence Apoptosis in Breast Cancer Cell Line. *Int. J. Prev. Med.* **2016**, *7*, 66. [CrossRef]

18. Ahmad, A.; Husain, A.; Mujeeb, M.; Alam Khan, S.; Najmi, A.K.; Siddique, N.A.; Damanhouri, Z.A.; Anwar, F.; Kishore, K. A review on therapeutic potential of Nigella sativa, A miracle herb. *Asian Pac. J. Trop Biomed.* **2013**, *3*, 337–352. [CrossRef]

19. Kassab, R.B.; El-Hennamy, R.E. The role of thymoquinone as a potent antioxidant in ameliorating the neurotoxic effect of sodium arsenate in female rat. *Egypt J. Basic Appl. Sci.* **2017**, *4*, 160–167. [CrossRef]

20. Randhawa, M.A.; Alenazy, A.K.; Alrowaili, M.G.; Basha, J. An active principle ofNigella sativaL., thymoquinone, showing significant antimicrobial activity against anaerobic bacteria. *J. Intercult. Ethnopharmacol.* **2017**, *6*, 97–101. [CrossRef]

21. Abd El Aziz, A.E.; El Sayed, N.S.; Mahran, L.G. Anti-asthmatic and Anti-allergic effects of Thymoquinone on Airway-Induced Hypersensitivity in Experimental Animals. *J. Appl. Pharm. Sci.* **2011**, *1*, 109–117.

22. Mostofa, A.G.M.; Hossain, M.K.; Basak, D.; Bin Sayeed, M.S. Thymoquinone as a Potential Adjuvant Therapy for Cancer Treatment, Evidence from Preclinical Studies. *Front. Pharmacol.* **2017**, *8*, 295. [CrossRef] [PubMed]

23. Mokhtari, R.B.; Homayouni, T.S.; Baluch, N.; Morgatskaya, E.; Kumar, S.; Das, B.; Yeger, H. Combination therapy in combating cancer. *Oncotarget* **2017**, *8*, 38022–38043. [CrossRef] [PubMed]

24. Ghoncheh, M.; Pournamdar, Z.; Salehiniya, H. Incidence and Mortality and Epidemiology of Breast Cancer in the World. *Asian Pac. J. Cancer Prev.* **2016**, *17*, 43–46. [CrossRef]

25. Miura, D.; Yoneyama, K.; Furuhata, Y.; Shimizu, K. Paclitaxel Enhances Antibody-dependent Cell-mediated Cytotoxicity of Trastuzumab by Rapid Recruitment of Natural Killer Cells in HER2-positive Breast Cancer. *J. Nippon. Med. Sch.* **2014**, *81*, 211–220. [CrossRef]

26. Dorman, S.N.; Baranova, K.; Knoll, J.H.; Urquhart, B.L.; Mariani, G.; Carcangiu, M.L.; Rogan, P.K. Genomic signatures for paclitaxel and gemcitabine resistance in breast cancer derived by machine learning. *Mol. Oncol.* **2016**, *10*, 85–100. [CrossRef]

27. Barbuti, A.M.; Chen, Z.-S. Paclitaxel Through the Ages of Anticancer Therapy, Exploring Its Role in Chemoresistance and Radiation Therapy. *Cancers (Basel)* **2015**, *7*, 2360–2371. [CrossRef]

28. Khader, M.; Eckl, P.M. Thymoquinone, an emerging natural drug with a wide range of medical applications. *Iran J. Basic Med. Sci.* **2014**, *17*, 950–957.

29. Jafri, S.H.; Glass, J.; Shi, R.; Zhang, S.; Prince, M.; Kleiner-Hancock, H. Thymoquinone and cisplatin as a therapeutic combination in lung cancer, In vitro and in vivo. *J. Exp. Clin. Cancer Res.* **2010**, *29*, 87. [CrossRef]

30. Zhang, L.; Bai, Y.; Yang, Y. Thymoquinone chemosensitizes colon cancer cells through inhibition of NF-κB. *Oncol. Lett.* **2016**, *12*, 2840–2845. [CrossRef]

31. Alaufi, O.M.; Noorwali, A.; Zahran, F.; Al-Abd, A.M.; Al-Attas, S. Cytotoxicity of thymoquinone alone or in combination with cisplatin (CDDP) against oral squamous cell carcinoma in vitro. *Sci. Rep.* **2017**, *7*. [CrossRef] [PubMed]

32. Bashmail, H.A.; AlAmoudi, A.A.; Noorwali, A.; Hegazy, G.A.; Ajabnoor, G.; Choudhry, H.; Al-Abd, A.M. Thymoquinone synergizes gemcitabine anti-breast cancer activity via modulating its apoptotic and autophagic activities. *Sci. Rep.* **2018**, *8*, 11674. [CrossRef] [PubMed]

33. Han, W.B.; Lu, Y.H.; Zhang, A.H.; Zhang, G.F.; Mei, Y.N.; Jiang, N.; Lei, X.; Song, Y.C.; Ng, S.W.; Tan, R.X. Curvulamine, a new antibacterial alkaloid incorporating two undescribed units from a Curvularia species. *Org. Lett.* **2014**, *16*, 5366–5369. [CrossRef] [PubMed]

34. Takeishi, S.; Nakayama, K.I. To wake up cancer stem cells, or to let them sleep, that is the question. *Cancer Sci.* **2016**, *107*, 875–881. [CrossRef] [PubMed]

35. Alobaedi, O.H.; Talib, W.H.; Basheti, I.A. Antitumor effect of thymoquinone combined with resveratrol on mice transplanted with breast cancer. *Asian Pac. J. Trop Med.* **2017**, *10*, 400–408. [CrossRef] [PubMed]

36. Ganji-Harsini, S.; Khazaei, M.; Rashidi, Z.; Ghanbari, A. Thymoquinone Could Increase The Efficacy of Tamoxifen Induced Apoptosis in Human Breast Cancer Cells, An In Vitro Study. *Cell J.* **2016**, *18*, 245–254. [CrossRef]

37. Şakalar, Ç.; İzgi, K.; İskender, B.; Sezen, S.; Aksu, H.; Çakır, M.; Kurt, B.; Turan, A.; Canatan, H. The combination of thymoquinone and paclitaxel shows anti-tumor activity through the interplay with apoptosis network in triple-negative breast cancer. *Tumor Biol.* **2016**, *37*, 4467–4477. [CrossRef]

38. Jänicke, R.U. MCF-7 breast carcinoma cells do not express caspase-3. *Breast Cancer Res. Treat* **2009**, *117*, 219–221. [CrossRef]

39. Zhao, J. Cancer stem cells and chemoresistance, The smartest survives the raid. *Pharmacol. Ther.* **2016**, *160*, 145–158. [CrossRef]

40. Horimoto, Y.; Arakawa, A.; Sasahara, N.; Tanabe, M.; Sai, S.; Himuro, T.; Saito, M. Combination of Cancer Stem Cell Markers CD44 and CD24 Is Superior to ALDH1 as a Prognostic Indicator in Breast Cancer Patients with Distant Metastases. *PLoS ONE* **2016**, *11*, e0165253. [CrossRef]

41. Alfarouk, K.O.; Stock, C.-M.; Taylor, S.; Walsh, M.; Muddathir, A.K.; Verduzco, D.; Bashir, A.H.H.; Mohammed, O.Y.; O ElHassan, G.; Harguindey, S.; et al. Resistance to cancer chemotherapy, failure in drug response from ADME to P-gp. *Cancer Cell Int.* **2015**, *15*, 71. [CrossRef] [PubMed]

42. Spaeth, E.L.; Labaff, A.M.; Toole, B.P.; Klopp, A.; Andreeff, M.; Marini, F.C. Mesenchymal CD44 expression contributes to the acquisition of an activated fibroblast phenotype via TWIST activation in the tumor microenvironment. *Cancer Res.* **2013**, *73*, 5347–5359. [CrossRef] [PubMed]

43. Khan, A.; Tania, M.; Wei, C.; Mei, Z.; Fu, S.; Cheng, J.; Xu, J.; Fu, J. Thymoquinone inhibits cancer metastasis by downregulating TWIST1 expression to reduce epithelial to mesenchymal transition. *Oncotarget* **2015**, *6*, 19580. [CrossRef] [PubMed]

44. Brabletz, T.; Jung, A.; Spaderna, S.; Hlubek, F.; Kirchner, T. Migrating cancer stem cells—An integrated concept of malignant tumour progression. *Nat. Rev. Cancer* **2005**, *5*, 744. [CrossRef] [PubMed]

45. Tamura, D.; Arao, T.; Nagai, T.; Kaneda, H.; Aomatsu, K.; Fujita, Y.; Matsumoto, K.; De Velasco, M.A.; Kato, H.; Hayashi, H.; et al. Slug increases sensitivity to tubulin-binding agents via the downregulation of βIII and βIVa-tubulin in lung cancer cells. *Cancer Med.* **2013**, *2*, 144–154. [CrossRef] [PubMed]

46. Skehan, P.; Scudiero, M.; Vistica, D.; Bokesch, H.; Kenney, S.; Storeng, R.; Monks, A.; McMahon, J.; Warren, J.T.; Boyd, M.R. New colorimetric cytotoxicity assay for anticancer-drug screening. *J. Natl. Cancer Inst.* **1990**, *82*, 1107–1112. [CrossRef] [PubMed]

47. Chou, T.C.; Talalay, P. Generalized equations for the analysis of inhibitions of Michaelis-Menten and higher-order kinetic systems with two or more mutually exclusive and nonexclusive inhibitors. *Eur. J. Biochem.* **1981**, *115*, 207–216. [CrossRef]

Sample Availability: Samples of the used compounds are available from the authors.

Article

Changes in Content of Polyphenols and Ascorbic Acid in Leaves of White Cabbage after Pest Infestation

Zuzana Kovalikova [1,*], Jan Kubes [2], Milan Skalicky [2,*], Nikola Kuchtickova [1], Lucie Maskova [1], Jiri Tuma [1], Pavla Vachova [2] and Vaclav Hejnak [2]

[1] Department of Biology, Faculty of Science, University of Hradec Kralove, Rokitanskeho 62, 500 03 Hradec Kralove, Czech Republic
[2] Department of Botany and Plant Physiology, Faculty of Agrobiology, Food and Natural Resources, Czech University of Life Sciences Prague, 165 00 Prague, Czech Republic
* Correspondence: zuzana.kovalikova@uhk.cz (Z.K.); skalicky@af.czu.cz (M.S.); Tel.: +420-224-382-520 (M.S.)

Received: 1 June 2019; Accepted: 17 July 2019; Published: 18 July 2019

Abstract: Crops, such as white cabbage (*Brassica oleracea* L. var. *capitata* (L.) f. *alba*), are often infested by herbivorous insects that consume the leaves directly or lay eggs with subsequent injury by caterpillars. The plants can produce various defensive metabolites or free radicals that repel the insects to avert further damage. To study the production and effects of these compounds, large white cabbage butterflies, *Pieris brassicae* and flea beetles, *Phyllotreta nemorum*, were captured in a cabbage field and applied to plants cultivated in the lab. After insect infestation, leaves were collected and UV/Vis spectrophotometry and HPLC used to determine the content of stress molecules (superoxide), primary metabolites (amino acids), and secondary metabolites (phenolic acids and flavonoids). The highest level of superoxide was measured in plants exposed to fifty flea beetles. These plants also manifested a higher content of phenylalanine, a substrate for the synthesis of phenolic compounds, and in activation of total phenolics and flavonoid production. The levels of specific phenolic acids and flavonoids had higher variability when the dominant increase was in the flavonoid, quercetin. The leaves after flea beetle attack also showed an increase in ascorbic acid which is an important nutrient of cabbage.

Keywords: *Brassica oleracea*; *Pieris brassicae*; *Phyllotreta* sp., phenolics; ascorbic acid

---

1. Introduction

Brassica crops, commonly known as crucifers, are grown worldwide for food and as animal feed and represent a significant economic value due to their nutritional, medicinal, bioindustrial, biocontrol, and crop rotation properties [1]. White cabbage (*Brassica oleracea* var. *capitata*) is a widely-cultivated crucifer vegetable with a high nutritive value due to its richness in active phytochemicals, such as vitamins C and E, carotenoids, minerals, dietary fiber, glucosinolates, phenolic acids, flavonoids, and anthocyanins [2]. White cabbage is extensively cultivated throughout the world, and the crop is often severely damaged by herbivorous insects such as the cabbage white butterfly, *Pieris brassicae*, and flea beetles which are the most common pests of the Brassicaceae family.

The large white butterfly (*Pieris brassicae* L.; Lepidoptera: Pieridae) has a life cycle lasting 45 days from egg to adult and there may be two to three broods per year with the first caterpillars hatching in spring (April to June), and the second during summer (July to August). Adults drink nectar of various plant species, while larvae exclusively feed on crucifers. Severe caterpillar infestations can destroy entire plants not only because of their feeding but also their feces. Cruciferous plants produce defensive compounds to repel insect attackers, but some larvae are able to safely accumulate the poisonous compounds in their bodies, which provides them protection from being eaten by some bird species. Because of this phenomenon, the cabbage butterfly has been used as a model species in the field or laboratory to study insect pest biology [1]. Flea beetles, primarily *Phyllotreta nemorum* and *P. undulata* (Coleoptera: Chrysomelidae), are the most common species on *B. oleracea* plants. They feed on cotyledons, young developing leaves, and stems of seedlings, leading to loss of photosynthetic capability and often to plant death. Feeding starts at the first two weeks after beetle emergence, and produces a shot-hole appearance and necrosis. Injury from larvae feeding on secondary roots hairs, however, caused a negligible effect on plant survival [3].

Plants have developed various defensive strategies against herbivorous insects. Apart from structural features like trichomes, thorns, or waxy leaf coatings, plants can reconfigure their metabolism to produce and accumulate specific chemical compounds that repel or even kill insects. These plant defense compounds act both as constitutive substances to repel herbivores through direct toxicity, or antifeeding properties by lowering the digestibility of plant tissues (e.g., by lignification), and as inducible substances produced in response to direct damage by herbivores. In addition, they may also play roles as antioxidants or as volatile attractants for predators [2,4]. In addition to the well-studied glucosinolate–myrosinase products, there are also a number of volatile compounds, such as lectins, phytoalexins, and phytoanticipins [5]. Phenolic compounds are common secondary metabolites in vascular plants. They exhibit great structural diversity, embodying a variety of functions in plant–herbivore interactions. They play important roles in pollination and oviposition, in host plant recognition by phytophagous insects, as feeding repellents, and in insect pest management [6]. Among the phenolics derived from phenylalanine are simple phenylpropanoids such as caffeic and ferulic acid, phenylpropanoid lactones (coumarins), and benzoic acid derivatives such as vanillin and salicylic acid [7]. The most common flavonoids in *Brassica* crops are quercetin, kaempferol, and isorhamnetin, commonly found as *O*-glycosides, mainly conjugated to glucose. They are also commonly acylated by different hydroxycinnamic acids [2].

Most of the published work has focused on the glucosinolate–myrosinase system, which is the best-studied chemical defense in crucifers [8,9]. Most research dealing with changes in phenolic metabolism has focused mainly on the absorption and sequestration of phenols and flavonoids in the bodies of *P. brassicae* caterpillars [10,11]. There have been no published reports about changes in phenolic metabolism after *Phyllotreta* attack. Therefore, the aim of this research was to study the changes in the accumulation of the main phenolic compounds in *Brassica oleracea* following attack by common insect pests. We focused on the herbivory of adult flea beetles (*Phyllotreta nemorum*) and the larvae of the large white cabbage butterfly (*Pieris brassicae*), specifically the direct feeding of 2nd instar larvae, the effects of oviposition, and the subsequent feeding of hatched caterpillars.

2. Results and Discussion

Plants have been generating complex defense mechanisms against various herbivorous insect feeding strategies over the entire long period of their evolution. Here, we studied the effects of adult flea beetle predation in the lab at two different levels, 50 or 100 insects per exposure (FB_{50}, FB_{100}), the feeding of 2nd instar larvae (WBC), or oviposition with subsequent feeding of hatched caterpillars of the large white cabbage butterfly (WBA). As soon as herbivore feeding starts on a plant, several defense signals are induced, leading to different defense responses. The injury by chewing of insects causes an immediate burst of reactive oxygen species (ROS), mainly hydrogen peroxide and superoxide radical, giving rise to both local and systemic responses [7]. In our study, the concentration of superoxide radical was elevated in nearly all plants exposed to insect predation. The presence of flea beetles, especially at lower numbers (FB_{50}), as well as exposure to cabbage butterfly oviposition and larvae, WB_A and WB_C, significantly increased the superoxide level in infested plants compared with controls (Figure 1a). Evidence from the literature, shows that the characteristics of the oxidative burst differ according to the type of plant and insect pest. For example, a strong accumulation of H_2O_2 was observed within 3 h of aphid infestation in wheat [12]. Additionally, histochemical staining of *Arabidopsis* leaves showed an accumulation of H_2O_2 72 h after oviposition by *P. brassicae* [13].

In contrast, no ROS accumulation was observed in *Arabidopsis* plants up to 48 h after attack by the phloem feeding aphid, *Brevicoryne brassicae*. No evidence of ROS was found for up to 21 days after feeding of the sweet potato whitefly, *Bemisia tabaci*, however. A number of genes associated with oxidative stress, ROS scavengers, such as ascorbate peroxidase or catalase, were reported to be upregulated, which suggests that increased levels of ROS were not essential for triggering enhanced expression of oxidative defense genes, and secondary signaling pathways may be involved [14,15].

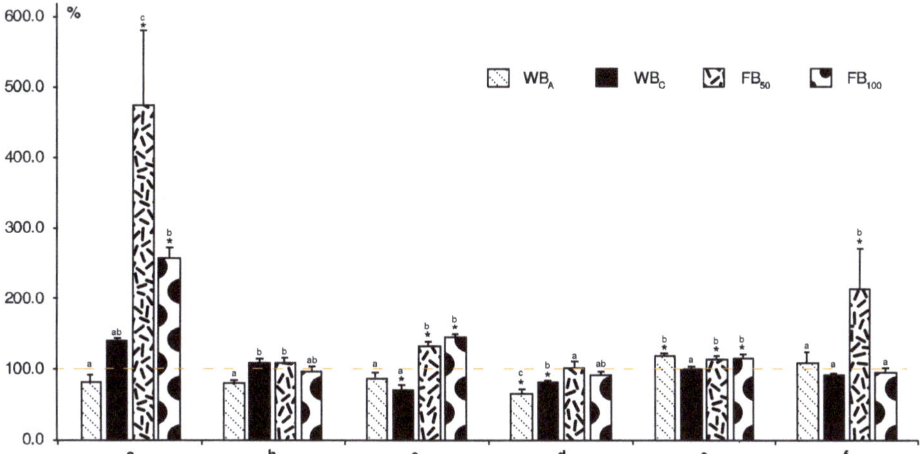

Figure 1. Effect of insect predation on production of stress compounds, and primary and secondary metabolites in white cabbage: (**a**) Superoxide, (**b**) total soluble proteins, (**c**) phenylalanine, (**d**) tyrosine, (**e**) total phenols content, and (**f**) total flavonoid content. WB$_A$, adult cabbage butterflies (egg laying and hatching larvae), WB$_C$ 2nd instar larvae, FB$_{50}$ fifty flea beetles per box, and FB$_{100}$, one hundred flea beetles per box. All bar values were recalculated relative to the compound content in untreated samples taken as 100% (dashed line). Data are means of three repeats ± SE. The asterisk (*) represents a significant difference between insect-damaged plants and controls, and different letters between the values of one compound. *p* < 0.05 by Fisher's least significant difference (LSD) test.

One characteristic plant response to insect attack is elevated protein content as a result of the induction of plant enzymes and nonenzymatic proteins involved in plant defense. Here, the content of total soluble proteins in injured leaves was not measurably affected by any of the used treatments (Figure 1b). In previous studies, contrary results were found in kale plants attacked by *P. brassicae* [16] and in corn plants infested with *Spodoptera frugiperda* [17], where a significant increase in total protein was measured. In general, phenolic compounds play a major role in host plant resistance to herbivores, including insects. Phenolics are synthesized in plants via the shikimic acid pathway. Phenylalanine ammonia-lyase (PAL) is the key enzyme catalyzing deamination of the aromatic amino acid, phenylalanine (Phe), to *t*-cinnamic acid, which participates in further reactions by conjugation with coenzyme A or hydroxylation to *p*-coumaric acid. Alternatively, *p*-coumaric acid is produced directly from tyrosine (Tyr), the hydroxyl derivative of Phe, via tyrosine ammonia-lyase, an analogue of PAL (Figure 2) [2,6,7]. The activity of PAL was strongly elevated in *Chrysanthemum* during the early period (0.5 to 6 h) after aphid infestation [18] and in kale after *P. brassicae* herbivory [16], and the enhanced PAL activity was correlated with elevated concentration of phenols. In our study, the activity of specific enzymes was not analyzed, but the levels of enzyme precursors were measured. The amount of Phe significantly rose only in plants attacked by flea beetles while the level of Tyr stayed unchanged. In contrast, herbivory by cabbage butterfly larvae resulted in a significant decrease in Phe and Tyr (Figure 1c,d).

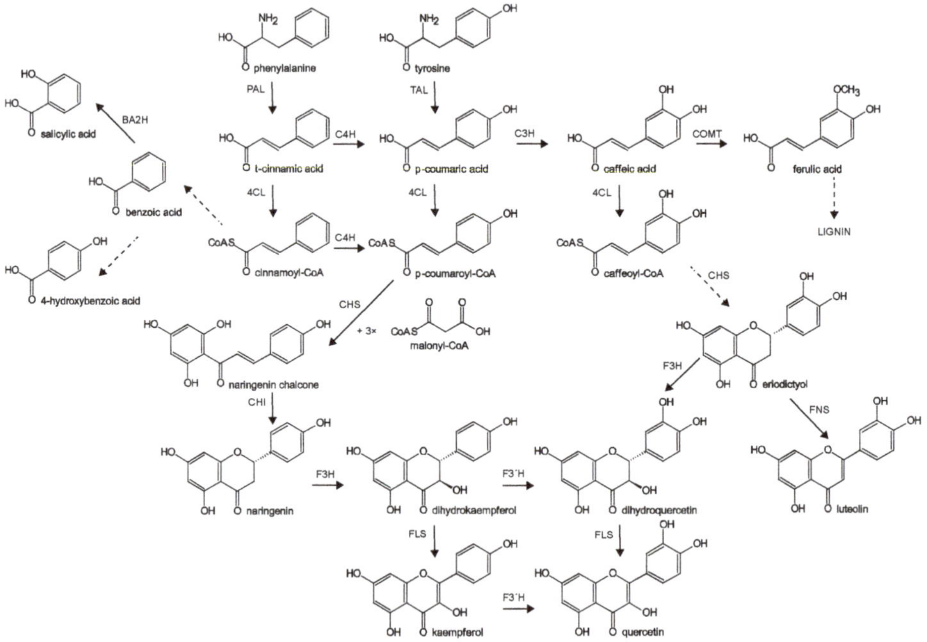

Figure 2. Pathways of phenolic acid and flavonoid metabolism [19–22]. PAL—phenylalanine ammonia-lyase; TAL—tyrosine ammonia-lyase; C4H—cinnamate 4-hydroxylase; 4CL—4-coumarate-CoA ligase; C3H—*p*-coumarate 3-hydroxylase; COMT—caffeic acid 3-O-methyltransferase; BA2H—benzoic acid 2-hydroxylase; CHS—chalcone synthase; CHI—chalcone isomerase; F3H—flavanone 3-hydroxylase; FLS—flavonol synthase; F3'H—flavonoid 3'-hydroxylase; FNS—flavone synthase.

Phenols possessing antioxidant activity are known to be produced by stressed plants. They may neutralize ROS directly or through enzymatic reactions. Their antioxidant activity depends on their chemical structure, the position and increased number of hydroxyl groups in the molecule leading to higher antioxidant activity. Even when phenolic compounds are oxidized by polyphenol oxidase or peroxidase to quinones, they can still be effective in defense reactions against herbivores [23,24]. On the other hand, glycosylation, the addition of a sugar moiety, results in lowering of this antioxidant activity [2]. In our study, plants exposed to flea beetles (FB_{50} and FB_{100}) and hatching caterpillars (WB_A) showed a significantly higher content of total soluble phenols in comparison with controls; but treatments did not differ between themselves. The WBC plants did not differ from controls (Figure 1e). In the literature, the described changes in the content of total soluble phenols are not consistent. Some studies reported higher phenols content [25,26] while others noted a decrease in their level [23]. Analysis of total phenols content is not only specific for molecules like phenolic acids or flavonoids, but other types of compounds can also be detected. However, it is still the most widely used method, especially in stress-related research, and values of particular metabolites are then usually determined in detail using more accurate methods, such as liquid chromatography or mass spectrometry.

Another commonly used analytical procedure in stressed plants is the evaluation of flavonoid content via spectrophotometric assay based on the formation of an aluminum chloride complex (sometimes expressed as $AlCl_3$-reacted flavonols). Here, the content was significantly higher only in cabbage plants exposed to 50 flea beetles, FB_{50} (Figure 1f). Total flavonoid content includes amount of various molecules that are synthesized and further metabolized by subsequent reactions. Figure 2 shows various pathways leading to three individually analyzed aglycones—kaempferol, quercetin, and luteolin. In this study, the preceding flavonoid substrates for these compounds or their glycosides and other metabolites were not analyzed. Overall, the highest content was monitored for quercetin. Here, a significant increase compared to control was found for WB_C and FB_{50} plants. Surprisingly, hatching caterpillars (WBA) did not change the amount of quercetin. Total levels of kaempferol and luteolin were considerably lower in comparison with quercetin. Only flea beetle infestation resulted in significant enhancements in both compounds (Figure 3).

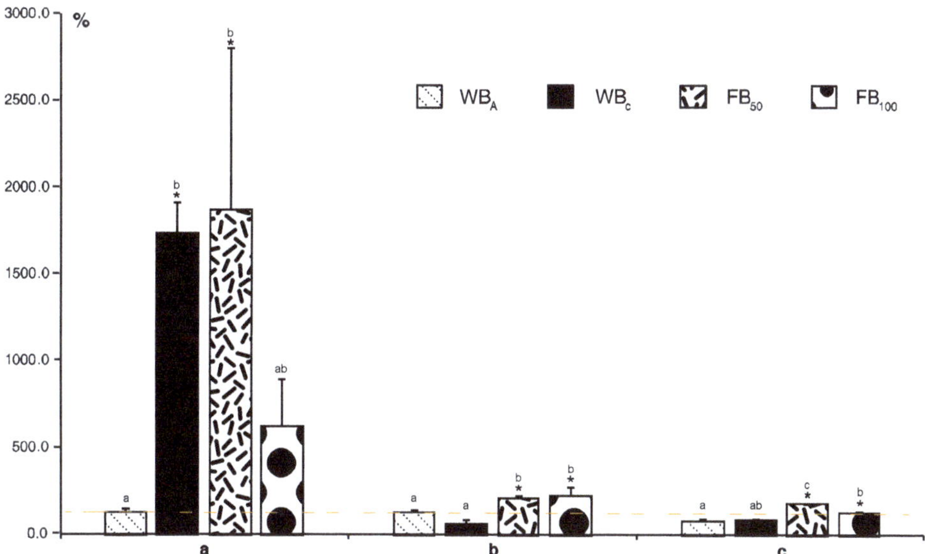

Figure 3. Effect of insect herbivory on flavonoid content in white cabbage. (**a**) Quercetin; (**b**) luteolin; (**c**) kaempferol. All bar values for experimental levels of compounds were recalculated relative to the content in untreated samples taken as 100% (dashed line). WB_A, adult cabbage butterflies (egg laying and hatching larvae), WB_C 2nd instar larvae, FB_{50} fifty flea beetles per box, and FB_{100}, one hundred flea beetles per box. Data are means of three repeats $\pm$ SE. The asterisk (*) represents a significant difference between treated samples and controls, and different letters between the values of one compound. $p < 0.05$ by Fisher's least significant difference (LSD) test.

Beyond their well-known antioxidant properties, flavonoids also play an important role in insect–plant interactions. Studies of a variety of flavonoids have demonstrated their feeding deterrent and stimulant activity. The significant inhibition of feeding by flavone and dihydroquercetin, and the ability of apigenin or isorhamnetin to stimulate feeding suggest that the degree of predation of flea beetles on crucifers may depend on flavonoid structure. The most abundant cabbage flavonoids, quercetin, and kaempferol, showed only slightly elevated deterrent properties. However, a mixture of these flavonoids may influence feeding preferences [27]. There is evidence of selective uptake of flavonoids by *P. brassicae* larvae from their food sources (*B. napa*, *B. oleracea*) and their subsequent bioconversion to provide beneficial functions, such as protection from harmful UV radiation [10,11].

With regard to the phenolic acids participating in flavonoid biosynthesis, only *t*-cinnamic acid was strongly increased in plants subjected to herbivory by second instar larvae (WB_C). Surprisingly, other insect exposures caused significantly lower values in comparison with controls (Figure 4a). The content of other phenolic acids, *p*-coumaric and caffeic, rose slightly after flea beetle attack, FB_{50}, and both FB_{50} and FB_{100}, respectively. The other two treatments led to a decrease in phenolic acid concentration, more so for caffeic acid (Figure 4b,c). These phenolic acids also serve as substrates for other compounds like ferulic or chlorogenic acid; but no correlation between herbivory and their content was observed.

Moreover, their content dropped in all tested plants, with the exception of ferulic acid in plants exposed to one hundred flea beetles, FB_{100}, where a significant increase was seen (Figure 4d,e). One possible explanation for this may be their involvement in lignin biosynthesis.

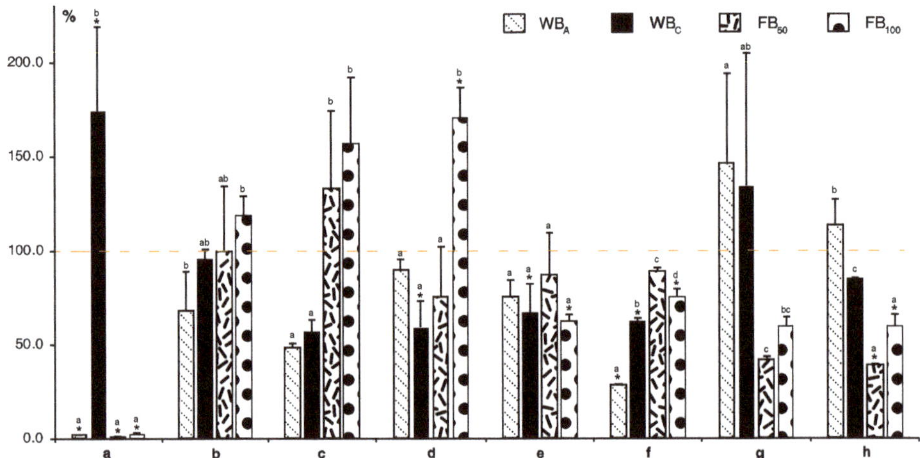

Figure 4. Effect of insect treatment on phenolic acids content in white cabbage. (**a**) *t*-cinnamic acid; (**b**) *p*-coumaric acid; (**c**) caffeic acid; (**d**) ferulic acid; (**e**) chlorogenic acid; (**f**) benzoic acid; (**g**) salicylic acid; (**h**) 4-hydroxy benzoic acid. All bar values were recalculated relative to the compound content in untreated samples taken as 100%. Data are means of three repeats ± SE. The asterisk (*) represents a significant difference between treated samples and controls, and different letters between the values of one compound. $p < 0.05$ by Fisher's least significant difference (LSD) test.

The phenolic acids, *t*-cinnamate and *p*-coumarate are also metabolized by β-oxidation to benzoate and 4-hydroxybenzoate, respectively. Benzoate can be further hydroxylated to salicylic acid which can serve as a signal molecule. The measured results, however, showed that insect-stressed plants did not have a significantly higher content of any of these compounds compared to controls (Figure 4f–h), but there were differences between individual exposures. Moreover, markedly lower amounts of salicylic and 4-hydroxybenzoic acid (Figure 4g,h) were detected in plants after flea beetle predation. Phenolics are known to play an antioxidant role in plant defense systems as a backup to the primary ascorbate-dependent detoxification system. In our study, the results showed a significant increase of ascorbic acid content after the flea beetles attack, while cabbage butterfly larvae herbivory showed only a weak response (Table 1). Few studies directly quantifying ascorbic acid content after damage of both chewing and sap feeding herbivores showed oxidation and loss of ascorbate in host plants [28,29]. In addition, enzyme activities and the transcript abundance of ascorbate peroxidase or oxidase, enzymes catalyzing generation of monodehydroascorbate and dehydroascorbate, were differentially modulated. Oviposition of *P. brassicae* downregulated a transcript encoding dehydroascorbate reductase [13], whereas larval feeding by a closely-related species, *P. rapae*, upregulated this transcript [30]. Surprisingly, the elevation of ascorbate content in plants by supplying them with its precursor enhanced the expansion rates of aphid colonies. One possible interpretation is that the excess ascorbate was utilized by aphids to enhance their metabolism. An alternative explanation is that an increased leaf ascorbate concentration decreased the lifetime of ROS signals in the plants and thus altered the balance of redox signaling pathways [31].

Oviposition and subsequent larval feeding represent a serious threat to crop plants. Thus, host plants have evolved direct defenses against egg laying (necrotic zones at the oviposition site) and indirect defenses such as the emission of volatiles to attract egg parasitoids or the synthesis of toxic or antifeeding compounds [32]. Geiselhardt et al. 2013 [8] showed that prior egg deposition of *P. brassicae* on *Arabidopsis* plants negatively influenced the feeding, growth, and survival of larvae but the concentrations of the major antiherbivory glucosinolates were not significantly increased by oviposition.

Table 1. The content of ascorbic acid ($mg \cdot g^{-1}$ FW) in extracts from white cabbage after predation by cabbage butterfly larvae or flea beetles.

Control	WB_A	WB_C	FB_{50}	FB_{100}
0.034 ± 0.003 [a]	0.052 ± 0.004 [a]	0.054 ± 0.013 [a]	0.116 ± 0.006 [b]	0.155 ± 0.033 [b]

WB_A, exposure to adult cabbage butterflies (egg laying and hatching larvae), WB_C 2nd instar larvae, FB_{50} fifty flea beetles per box, and FB_{100}, one hundred flea beetles per box. Data are means of three repeats $\pm$ SE. Different letters between the values of each sample show significant difference at $p < 0.05$ by Fisher's protected least significant test (LSD).

Treatment with the phenolic acids, *p*-coumaric, ferulic, salicylic, and protocatechuic, stimulated oviposition of *P. brassicae*, with the highest egg number after *p*-coumaric acid treatment. Here, the highest weight of feeding caterpillars was observed [33]. An antagonistic effect of other phenolic acids (syringic, coumaric, cinnamic, and vanillic acid) was found in castor plants. Their content rose after infestation with the castor semilooper, *Achaea janata*, or the tobacco cutworm, *Spodoptera litura*, and subsequent treatment of plants with these acids resulted in altered larval feeding preferences, where vanillic and cinnamic acids acted as repellents, and syringic and coumaric acids as attractants. The same preferences were also found for oviposition [34]. In our study, we determined only the changes in phenolic acid content. The oviposition of *P. brassicae* and subsequent larval hatching (WB_A—exposure to adult cabbage butterflies laying eggs and larval hatching) led to a decrease of almost all monitored acids, significantly in the case of cinnamic and benzoic acid. A considerable increase in salicylic acid was monitored in extracts of cabbage plants subjected to oviposition and larval predation (Figure 4g). Its elevated level may point to a major role of salicylate in egg-induced plant responses. Oviposition as well as applications of extracts of *P. brassicae* eggs led to a strong accumulation of salicylic acid in *Arabidopsis* leaves [35]. The expression of several salicylic acid-responsive genes (PR1) was enhanced in leaf tissue beneath and close to *P. brassicae* eggs [13]. However, salicylic acid can be metabolized in *A. thaliana* to methyl salicylate. This volatile compound was not analyzed in this study and it could explain the difference between the production of benzoic acid metabolites and salicylates. Methyl salicylate acts as an attractant for the wasp, *Cotesia rubecula*, that parasitizes *Pieris rapa* caterpillars and its concentration was increased in the presence of herbivores [36]. The authors also stated that transcription of PAL was higher and they connected this fact with the induction of methyl salicylate biosynthesis. A similar mechanism could be involved in white cabbage, but this must be proven by further research.

3. Materials and Methods

3.1. Cultivation

Cabbage (*Brassica oleracea* var. *capitata* f. *alba*) seeds were sown in propagators using a mixture of multipurpose compost, garden soil, and perlite in a ratio of 2:2:1. Two weeks later, the seedlings were repotted into pots measuring $5 \times 5 \times 5$ cm and left to grow for 25 days. Each 8-cell tray was placed in a 29 L transparent plastic box with a lid, and a thin layer of perlite was sprinkled on the box bottom. All boxes were equipped with two lateral ventilation windows measuring 10×15 cm, covered by metal nets, and a watering hose. During the whole experiment, the plants were cultivated in a Phytotron growth room (Weiss Technik) under the following conditions: 15-h photoperiod, temperature 22 °C (day)/17 °C (night), and humidity 60% (day)/70% (night). The plants were watered with an equal mixture of tap water and demineralized water.

Flea beetles (*Phyllotreta nemorum*) and large white cabbage butterflies (*Pieris brassicae*) (both adults and caterpillars) were captured on a cabbage field near Bolehost (50.2131900N, 16.0778428E, Czech Republic, 260 m a.s.l.), using aspirators for collecting the beetles, and entomological net bags for butterflies. On the same day, the insects were added into the cultivation boxes in the following way (treatments are bold):

- C—control (untreated plants)
- WB$_A$—boxes with adult butterflies (five butterflies per box) to determine the influence of oviposition and feeding by hatched caterpillars
- WB$_C$—boxes with caterpillars (six caterpillars per box) in 2nd instar
- FB$_{50}$—boxes with 50 flea beetles per box
- FB$_{100}$—boxes with 100 flea beetles per box.

The experiments were conducted as follows: The controls and the plants exposed to 2nd instar caterpillars and beetles were analyzed after four days; the plants exposed to adult butterflies were analyzed after eight days, after the females had laid eggs and new caterpillars had hatched. Plants were withdrawn from the boxes, leaves were gently cleansed with a brush and used for analysis.

3.2. Quantification of Stress-Related Compounds and Ascorbic Acid

Fresh leaves (0.1 g) were homogenized with 50 mM potassium phosphate buffer (pH 7.0) and centrifuged for 15 min at 14,000 rpm and 4 °C. The content of superoxide radical was determined by measuring nitrite formation from hydroxylamine (530 nm) [37]. A volume of 30 µL of supernatant and bovine serum albumin as the standard (595 nm) were used for determination of total soluble proteins [38]. All measurements were done with a Cintra spectrophotometer (Cintra 101, Dandenong, Australia).

The ascorbic acid assay was carried out as previously described by [39]. The leaves were kept at −18 °C between the harvest and the analysis. For each sample, 0.5 g of frozen leaves were homogenized with K_2HPO_4-PBS buffer (pH 2.5). The suspension was heated in a water bath to 75 °C for 45 min, cooled, and centrifuged (3000× g). The analysis was performed using an HPLC system (Agilent 1260 Series; Santa Clara, California, United States) with Kinetex C18 column (150 × 4.6 mm, 5 µm). The mobile phase was composed of a mixture of 97% 0.01 M K_2HPO_4-PBS (pH 2.5) and 3% methanol and was at isocratic elution with a flow rate of 1.0 mL min^{-1}. The detection was at 210 nm and the concentration was calculated by comparison with a standard calibration curve.

3.3. Quantification of Phenols, Flavonoids, and Phenolic Acids

Extracts were prepared from fresh leaves using 80% methanol (*w/v* 0.2 g 2 mL^{-1}). Total soluble phenols were estimated using the Folin–Ciocalteu method with gallic acid as a standard (750 nm). Total flavonoids ($AlCl_3$-reacted flavonols) were quantified using $AlCl_3$ as a reagent and quercetin was used as a standard (420 nm) [37]. All measurements were done with a Cintra spectrophotometer (Cintra 101, Dandenong, Australia). The content of phenolic acids and selected flavonoids was determined by UHPLC on a 2.1 × 50 mm, 1.8 µm Zorbax RRHD Eclipse plus C18 column (Agilent) with a 6470 Series Triple Quadrupole mass spectrometer (Agilent) (electrospray ionization—negative ion mode) as detector. Eluents: (A) 0.05% formic acid in water and (B) 0.05% formic acid in acetonitrile were used in the following gradient program: 0–1 min (5% B), 2.0–4.0 min (20% B), 8.0–9.5 min (70% B), 10.0–11.0 min (5% B). The MS source conditions were as follows: Gas temperature 350 °C, gas flow 9 L min^{-1}, nebulizer 35 psi, sheath gas temperature 380 °C, sheath gas flow 12 L min^{-1}, capillary 2500 V, and nozzle voltage 0 V. Selected MRM transitions were followed for each compound: 4-hydroxybenzoic acid (137.0 => 108.0, 92.0), benzoic acid (121.0 => 77.1), caffeic acid (179.0 => 135.0, 107.0), chlorogenic acid (353.1 => 191.0, 127.0), *t*-cinnamic acid (147.1 => 103.0, 77.0), ferulic acid (193.1 => 134.1, 178.0), kaempferol (185.1 => 1185.0, 239.0), luteolin (285.1 => 133.0, 151.0), *p*-coumaric acid (163.1 => 119.0, 104.9), quercetin (301.0 => 151.0, 179.0), salicylic acid (137.0 => 93.0, 65.0).

3.4. Quantification of Amino Acids

0.1 g of powdered dry leaves (80 °C for 24 h) was mixed with 1.5 mL of 70% ethanol containing 10 mM norvaline (as an internal standard). The suspension was blended for 15 min using a MultiReax shaker (Heidolph, Schwabach, Germany) and subsequently heated to 120 °C for 10 min. After cooling and centrifugation for 5 min at 4000 rpm, the supernatants were stored in Eppendorf tubes. The pellet was resuspended with 70% ethanol and shaken for another 15 min and again centrifuged. This process was repeated twice. The collected supernatants were centrifuged at 13,000 rpm for 3 min and heated to 60 °C under a N_2 atmosphere (NDK 200-2, Hangzhou MIU Instruments Co., Ltd., Hangzhou, China) until they evaporated. Prior to analysis, the dry extract was dissolved in a mixture of mobile phases, 98% A and 2% B. The analysis was performed using an HPLC system (Agilent 1260 Series; Santa Clara, California, USA) with ZORBAX Eclipse Plus RRHT C18 column (50 × 4.6 mm, 1.8 μm) heated to 40 °C. Derivatization reagents (borate buffer 5061-3339, OPA Reagent 5061-3337) and amino acid standards 1 nmol μL^{-1} (5061–3330) were purchased from Agilent. Mobile phases, A (mixture of 10 mM Na_2HPO_4 and 10 mM $Na_2B_4O_7$, pH 8.2) and B (acetonitrile/methanol/water 45/45/10) were run at a flowrate of 2.0 mL min^{-1} under the following gradient program: 0 min 2% B; 0.2 min 2% B; 7.67 min 57% B; 7.77 min 100% B; 8.3 min 100% B; 8.4 min 2% B; 9.0 min 2% B. The DAD (UV) detection was set at 338 and 390 nm for 0–6.1 min; 262 and 324 for 6.1–9 min and FLD detection was set at Ex 340 nm/Em 455 nm [40].

3.5. Data Processing

The values for the concentrations of the various compounds in control samples that were taken as 100% (Figures 1, 3 and 4) are given in Tables S1–S3 in the Supplementary Materials section. The concentrations of the experimental samples were related to the 100% values. Each tested group was only compared with the compound content in the corresponding subculture. A mixed-model procedure, with a repeated statement for each parameter, was used to analyze the data set. Data from each measurement was tested separately. Fisher's LSD test ($p < 0.05$) was used to determine significant differences. All statistical tests presented in this study were performed using a Statistica 13 (StatSoft Inc., Tulsa, OK, USA) software package. Principal component analysis (PCA); for establishing the effects of treatments on amounts of metabolite compounds was analyzed by the CANOCO 5 [41] software package.

4. Conclusions

In this paper, we showed that infestation of white cabbage by white cabbage butterflies or flea beetles caused changes in metabolism of stress compounds for both insect species. The results showed that these species had different effects on superoxide levels in predated leaves. The exposure to oviposition by butterflies and subsequent feeding by newly hatched caterpillars did not manifest in higher total phenolic content in comparison to predation by flea beetles. Despite the increased total flavonoid content in the case of the lower number of flea beetles, there was no clear prediction of which species could affect these secondary metabolites in principal component analysis (PCA) (Figure 5). The primary and secondary compounds like phenolic acids and flavonoids may play an important role in the defense against biotic stressors. Further detailed analysis of changes in individual phenolic metabolites are needed to explain their role in the defense response.

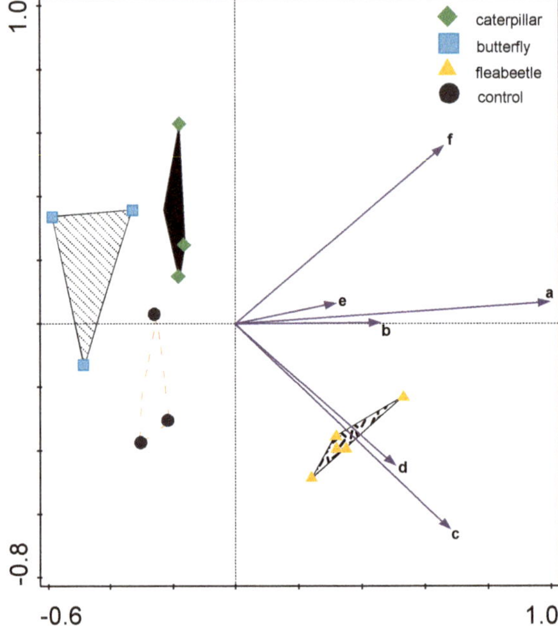

Figure 5. The principal component analysis (PCA) includes Figure 1 compounds and the Euclidean distances between particular samples. The total variation was 10.6. The explained variation on first two axes was 86.15%.

Supplementary Materials: The following are available online, Table S1: The content of observed stress compounds, primary and secondary metabolites in control samples of white cabbage extract. Table S2: The content of observed phenolic acids in control samples of white cabbage extract. Table S3: The content of observed flavonoids in control samples of white cabbage extract.

Author Contributions: Conceptualization, Z.K. and J.T.; Data curation, M.S. and P.V.; Formal analysis, M.S. and P.V.; Investigation, J.K., N.K. and L.M.; Methodology, Z.K., J.K. and J.T.; Software, M.S. and P.V.; Supervision, J.T. and V.H.; Visualization, J.K. and M.S.; Writing—original draft, Z.K. and J.K.; Writing—review & editing, J.K. and M.S.

Funding: This research was funded by the Ministry of Education, Youth and Sports of the Czech Republic grant number [S grant of MSMT CR] and by Specific Research Project of Faculty of Science, University of Hradec Kralove, No. 2111, 2018.

Acknowledgments: The authors are thankful to Matej Semerak for providing the insect sampling and valuable assistance during chemical analyses, and to the company, Agrospol Bolehost, for permission to collect beetles and butterflies.

Conflicts of Interest: The authors declare no conflicts of interest. The founding sponsors had no role in the design of the study; in the collection, analyses, or interpretation of data; in the writing of the manuscript, and in the decision to publish the results.

References

1. Ahuja, I.; Rohloff, J.; Bones, A.M. Defence Mechanisms of Brassicaceae: Implications for Plant-Insect Interactions and Potential for Integrated Pest Management. In *Sustainable Agriculture Volume 2*; Lichtfouse, E., Hamelin, M., Navarrete, M., Debaeke, P., Eds.; Springer: Dordrecht, The Netherlands, 2011; pp. 623–670. ISBN 9789048126651.
2. Cartea, M.E.; Francisco, M.; Soengas, P.; Velasco, P. Phenolic Compounds in Brassica Vegetables. *Molecules* **2010**, *16*, 251–280. [CrossRef] [PubMed]

3. Metspalu, L.; Kruus, E.; Ploomi, A.; Williams, I.H.; Hiiesaar, K.; Jõgar, K.; Veromann, E.; Mänd, M. Flea beetle (Chrysomelidae: Alticinae) species composition and abundance in different cruciferous oilseed crops and the potential for a trap crop system. *Acta Agric. Scand. Sect. B—Soil Plant Sci.* **2014**, *64*, 572–582. [CrossRef]

4. War, A.R.; Paulraj, M.G.; Ahmad, T.; Buhroo, A.A.; Hussain, B.; Ignacimuthu, S.; Sharma, H.C. Mechanisms of plant defense against insect herbivores. *Plant Signal. Behav.* **2012**, *7*, 1306–1320. [CrossRef] [PubMed]

5. Kumar, S. Plant secondary metabolites (PSMs) of Brassicaceae and their role in plant defense against insect herbivores—A review. *J. Appl. Nat. Sci.* **2017**, *9*, 508–519. [CrossRef]

6. Harborne, J.B.; Grayer, R.J. *The Flavonoids Advances in Research Since 1986*; Routledge: Abingdon, UK, 2017; ISBN 9780203736692.

7. Fürstenberg-Hägg, J.; Zagrobelny, M.; Bak, S. Plant Defense against Insect Herbivores. *Int. J. Mol. Sci.* **2013**, *14*, 10242–10297. [CrossRef] [PubMed]

8. Geiselhardt, S.; Yoneya, K.; Blenn, B.; Drechsler, N.; Gershenzon, J.; Kunze, R.; Hilker, M. Egg Laying of Cabbage White Butterfly (*Pieris brassicae*) on Arabidopsis thaliana Affects Subsequent Performance of the Larvae. *PLoS ONE* **2013**, *8*, e59661. [CrossRef] [PubMed]

9. Kos, M.; Houshyani, B.; Wietsma, R.; Kabouw, P.; Vet, L.E.M.; van Loon, J.J.A.; Dicke, M. Effects of glucosinolates on a generalist and specialist leaf-chewing herbivore and an associated parasitoid. *Phytochemistry* **2012**, *77*, 162–170. [CrossRef] [PubMed]

10. Ferreres, F.; Fernandes, F.; Pereira, D.M.; Pereira, J.A.; Valentão, P.; Andrade, P.B. Phenolics Metabolism in Insects: *Pieris brassicae—Brassica oleracea* var. costata Ecological Duo. *J. Agric. Food Chem.* **2009**, *57*, 9035–9043. [CrossRef]

11. Ferreres, F.; Valentão, P.; Pereira, J.A.; Bento, A.; Noites, A.; Seabra, R.M.; Andrade, P.B. HPLC-DAD-MS/MS-ESI Screening of Phenolic Compounds in *Pieris brassicae* L. Reared on *Brassica rapa* var. rapa L. *J. Agric. Food Chem.* **2008**, *56*, 844–853. [CrossRef]

12. Moloi, M.J.; van der Westhuizen, A.J. The reactive oxygen species are involved in resistance responses of wheat to the Russian wheat aphid. *J. Plant Physiol.* **2006**, *163*, 1118–1125. [CrossRef]

13. Little, D.; Gouhier-Darimont, C.; Bruessow, F.; Reymond, P. Oviposition by Pierid Butterflies Triggers Defense Responses in Arabidopsis. *Plant Physiol.* **2006**, *143*, 784–800. [CrossRef]

14. Kunierczyk, A.; Winge, P.; Jrstad, T.S.; Troczyska, J.; Rossiter, J.T.; Bones, A.M.; Kuśnierczyk, A.; Jørstad, T.S.; Troczyńska, J. Towards global understanding of plant defence against aphids timing and dynamics of early Arabidopsis defence responses to cabbage aphid (*Brevicoryne brassicae*) attack. *Plant Cell Environ.* **2008**, *31*, 1097–1115. [CrossRef]

15. Kempema, L.A.; Cui, X.; Holzer, F.M.; Walling, L.L. Arabidopsis Transcriptome Changes in Response to Phloem-Feeding Silverleaf Whitefly Nymphs. Similarities and Distinctions in Responses to Aphids. *Plant Physiol.* **2006**, *143*, 849–865. [CrossRef]

16. Ibrahim, S.; Mir, G.M.; Rouf, A.; War, A.R.; Hussain, B. Herbivore and phytohormone induced defensive response in kale against cabbage butterfly, *Pieris brassicae* Linn. *J. Asia-Pac. Entomol.* **2018**, *21*, 367–373. [CrossRef]

17. Chen, Y.; Ni, X.; Buntin, G.D. Physiological, Nutritional, and Biochemical Bases of Corn Resistance to Foliage-Feeding Fall Armyworm. *J. Chem. Ecol.* **2009**, *35*, 297–306. [CrossRef]

18. He, J.; Chen, F.; Chen, S.; Lv, G.; Deng, Y.; Fang, W.; Liu, Z.; Guan, Z.; He, C. Chrysanthemum leaf epidermal surface morphology and antioxidant and defense enzyme activity in response to aphid infestation. *J. Plant Physiol.* **2011**, *168*, 687–693. [CrossRef]

19. Dixon, R.A.; Paiva, N.L. Stress-Induced Phenylpropanoid Metabolism. *Plant Cell* **1995**, *7*, 1085–1097. [CrossRef]

20. Tsai, C.J.; Harding, S.A.; Tschaplinski, T.J.; Lindroth, R.L.; Yuan, Y. Genome-wide analysis of the structural genes regulating defense phenylpropanoid metabolism in Populus. *New Phytol.* **2006**, *172*, 47–62. [CrossRef]

21. Morreel, K.; Goeminne, G.; Storme, V.; Sterck, L.; Ralph, J.; Coppieters, W.; Breyne, P.; Steenackers, M.; Georges, M.; Messens, E.; et al. Genetical metabolomics of flavonoid biosynthesis in Populus: A case study. *Plant J.* **2006**, *47*, 224–237. [CrossRef]

22. Leon, J.; Shulaev, V.; Yalpani, N.; Lawton, M.A.; Raskin, I. Benzoic acid 2-hydroxylase, a soluble oxygenase from tobacco, catalyzes salicylic acid biosynthesis. *Proc. Natl. Acad. Sci. USA* **1995**, *92*, 10413–10417. [CrossRef]

23. Khattab, H. The Defense Mechanism of Cabbage Plant Against Phloem-Sucking Aphid (*Brevicoryne brassicae* L.). *Aust. J. Basic Appl. Sci.* **2007**, *1*, 56–62.

24. Bhonwong, A.; Stout, M.J.; Attajarusit, J.; Tantasawat, P. Defensive role of tomato polyphenol oxidases against cotton bollworm helicoverpa armigera and beet armyworm spodoptera exigua. *J. Chem. Ecol.* **2009**. [CrossRef]

25. Kumar, S.; Singh, Y.P.; Singh, S.P.; Singh, R. Physical and biochemical aspects of host plant resistance to mustard aphid, Lipaphis erysimi (Kaltenbach) in rapeseed-mustard. *Arthropod-Plant Interact.* **2017**, *11*, 551–559. [CrossRef]

26. Palial, S.; Kumar, S.; Sharma, S. Biochemical changes in the Brassica juncea-fruticulosa introgression lines after Lipaphis erysimi (Kaltenbach) infestation. *Phytoparasitica* **2018**, *46*, 499–509. [CrossRef]

27. Onyilagha, J.C.; Gruber, M.Y.; Hallett, R.H.; Holowachuk, J.; Buckner, A.; Soroka, J.J. Constitutive flavonoids deter flea beetle insect feeding in *Camelina sativa* L. *Biochem. Syst. Ecol.* **2012**, *42*, 128–133. [CrossRef]

28. Bi, J.L.; Murphy, J.B.; Felton, G.W. Antinutritive and Oxidative Components as Mechanisms of Induced Resistance in Cotton to Helicoverpa zea. *J. Chem. Ecol.* **1997**, *23*, 97–117. [CrossRef]

29. Jiang, Y. Oxidative interactions between the spotted alfalfa aphid (Therioaphis trifolii maculata) (Homoptera: Aphididae) and the host plant Medicago sativa. *Bull. Entomol. Res.* **1996**, *86*, 533–540. [CrossRef]

30. Broekgaarden, C.; Poelman, E.H.; Steenhuis, G.; Voorrips, R.E.; Dicke, M.; Vosman, B. Genotypic variation in genome-wide transcription profiles induced by insect feeding: *Brassica oleracea*—Pieris rapae interactions. *BMC Genom.* **2007**, *8*, 239. [CrossRef]

31. Goggin, F.L.; Avila, C.A.; Lorence, A. Vitamin C content in plants is modified by insects and influences susceptibility to herbivory. *BioEssays* **2010**, *32*, 777–790. [CrossRef]

32. Hilker, M.; Fatouros, N.E. Plant Responses to Insect Egg Deposition. *Annu. Rev. Entomol.* **2015**, *60*, 493–515. [CrossRef]

33. Walker, K.S.; Bray, J.L.; Lehman, M.E.; Lentz-Ronning, A.J. Effects of host plant phenolic acids and nutrient status on oviposition and feeding of the cabbage white butterfly, Pieris rapae. *Bios* **2014**, *85*, 95–101. [CrossRef]

34. Usha Rani, P.; Pratyusha, S. Role of castor plant phenolics on performance of its two herbivores and their impact on egg parasitoid behaviour. *BioControl* **2014**, *59*, 513–524. [CrossRef]

35. Bruessow, F.; Gouhier-Darimont, C.; Buchala, A.; Metraux, J.P.; Reymond, P. Insect eggs suppress plant defence against chewing herbivores. *Plant J.* **2010**, *62*, 876–885. [CrossRef]

36. Van Poecke, R.M.P.; Posthumus, M.A.; Dicke, M. Herbivore-induced volatile production by Arabidopsis thaliana leads to attraction of the parasitoid Cotesia rubecula: Chemical, behavioral, and gene-expression analysis. *J. Chem. Ecol.* **2001**, *27*, 1911–1928. [CrossRef]

37. Dučaiová, Z.; Sajko, M.; Mihaličová, S.; Repčák, M. Dynamics of accumulation of coumarin-related compounds in leaves of Matricaria chamomilla after methyl jasmonate elicitation. *Plant Growth Regul.* **2016**, *79*, 81–94. [CrossRef]

38. Bradford, M.M. A rapid and sensitive method for the quantitation of microgram quantities of protein utilizing the principle of protein-dye binding. *Anal. Biochem.* **1976**, *72*, 248–254. [CrossRef]

39. Simek, J.; Kovalikova, Z.; Dohnal, V.; Tuma, J. Accumulation of cadmium in potential hyperaccumulators Chlorophytum comosum and Callisia fragrans and role of organic acids under stress conditions. *Environ. Sci. Pollut. Res.* **2018**, *25*, 28129–28139. [CrossRef]

40. Sajko, M.; Kovalíková-Dučaiová, Z.; Paľove-Balang, P.; Repčák, M. Physiological Responses of Matricaria chamomilla to Potassium Nitrate Supply and Foliar Application of Ethephon. *J. Plant Growth Regul.* **2018**, *37*, 360–369. [CrossRef]

41. Smilauer, P.; Lepš, J. *Multivariate Analysis of Ecological Data Using CANOCO 5*; Cambridge University Press: Cambridge, UK, 2014; ISBN 9781139627061.

Sample Availability: Samples of the compounds are available from the authors of Department of Biology.

 *molecules*

Article

Polyphenolic Profiling, Quantitative Assessment and Biological Activities of Tunisian Native *Mentha rotundifolia* (L.) Huds.

Imen Ben Haj Yahia [1,†], Yosr Zaouali [1,†], Maria Letizia Ciavatta [2], Alessia Ligresti [2], Rym Jaouadi [1], Mohamed Boussaid [1] and Adele Cutignano [2,*]

[1] Department of Biology, National Institute of Applied Science and Technology, B.P. 676, 1080 Tunis Cedex, Tunisia

[2] Institute of Biomolecular Chemistry (ICB), National Research Council (CNR), 80078 Pozzuoli (NA), Italy

* Correspondence: adele.cutignano@icb.cnr.it; Tel.: +39-081-8675313; Fax: +39-081-8041770

† These authors contributed equally to this work.

Academic Editors: Marian Brestic, Marek Zivcak, Oksana Sytar and Marco Landi
Received: 31 May 2019; Accepted: 24 June 2019; Published: 26 June 2019

Abstract: Phenolic profiling of ten plant samples of *Mentha rotundifolia* (L.) Huds. collected from different bioclimatic areas of Tunisia, was for the first time carried out by using a fast ultra-high-performance liquid chromatography (UHPLC)-high resolution tandem mass spectrometry (HRMS/MS) method on a Q Exactive platform equipped with an electrospray ionization (ESI) source. An intraspecific, interpopulation variability was evidenced and a total of 17 polyphenolic metabolites were identified and quantified by using the UHPLC-HRESIMS/MS method, here validated for specificity, linearity, limit of detection, limit of quantitation, accuracy and precision. The quantitative method resulted sensitive at the nM level and reliable for rapid polyphenol quantification in vegetal matrices. The metabolomic study allowed us to identify a new compound, named salvianolic acid W, which was isolated and characterized mainly by NMR and MS analysis. A statistical correlation of the phenolic composition with antioxidant and anti-acetylcholinesterase activities was provided.

Keywords: *Mentha rotundifolia*; *Lamiaceae*; UHPLC-MS; polyphenolics; salvianolic acid W; antioxidant activity; anti-acetylcholinesterase activity

---

1. Introduction

The genus *Mentha*, encompassing about 40 among species and recognized hybrids distributed worldwide is one of the most important genera of the *Lamiaceae* family [1]. *Mentha* species are a well-known source of terpene-rich essential oils used in traditional medicine as well as in flavoring, beverage, culinary and for cosmetic applications [2,3]. Furthermore, like other members of the *Lamiaceae* family, mint extracts contain a wealth of compounds, collectively named polyphenols, which include phenolic acids, flavones and flavanols, in a free form or as glycoconjugates, mainly responsible of the antioxidant properties of the plant. The interest towards natural polyphenolic compounds is increasing during the past years also in view of a possible technological use in the food industry as a safer alternative to synthetic molecules such as butylated hydroxyanisole (BHA) and butylated hydroxytoluene (BHT) [4]. The antioxidant potential of plant extracts and pure compounds is still the major factor in characterizing plants and, in general, nutritional health food by virtue of their bioactive components [5]. However, along with antioxidant properties, polyphenols exhibit other diverse biological activities, such as anti-inflammatory [6], anticancer [7,8], anti-atherosclerotic [9] as well as they contribute to maintaining the balance of gut microbiota [10,11]. Interestingly, *Mentha* extracts showed anti-acetylcholinesterase activity and the reversible inhibition of AChE activity has been proposed for the treatment of various diseases, including gastrointestinal disorders and Alzheimer's disease [12,13]. In Tunisia, mint is represented by few species and namely *Mentha rotundifolia* L., *M. longifolia* (L.) Huds., *M. spicata* (*M. viridis*) L., *M. aquatica* L. and *M. pulegium* L. *M. rotundifolia* (L.) Huds. is a hybrid between *M. longifolia* (L.) and *M. suaveolens* Ehrh and has been considered as a synonym of *M. suaveolens* Ehrh [14]. Chemical studies on this species are scarce and mainly directed toward essential oils composition and bioactivity [15]. Furthermore, previous studies have demonstrated that medicinal plants growing wild in diverse environments may show differences in chemical constituents [16].

As far as we know there are no previous studies on the intraspecific, interpopulation chemical variability of Tunisian *M. rotundifolia* methanolic extracts, so the aim of the present study was to:

(i)   Identify by both mass-spectrometric and NMR approaches polyphenolics from ten plant populations of Tunisian *M. rotundifolia* (L.) Huds growing wild in two bioclimatic zones;

(ii) Quantify metanolic extracts by a validated analytical strategy;

(iii) Evaluate the antioxidant and anti-acetylcholinesterase activities of extracts towards their chemical composition.

2. Results and Discussion

Despite that the genus *Mentha* has been largely studied for both essential oil and phenolic content, chemical characterization of polyphenols from *M. rotundifolia* has been documented in very few reports but never from Tunisian populations [17,18]. In view of exploring the chemistry and the bioactivity potential of *M. rotundifolia* from Tunisia, we carried out a collection of plant samples from different sites. The list of the *Mentha* populations labeled as MROT-1 to 10 used for the present study is reported in Supporting Material (Table S1). Extraction yields (mg extract/g dried material) ranged from 8.12% (MROT-3) to 22.54% (MROT-9; Table 1). The amount of total phenolic content (TPC) of plant extracts varied significantly ($p < 0.05$) from 5.70 (MROT-3) to 57.11 mg GAE /g DM (MROT-1). Total flavonoids (TFC) ranged from 5.12 mg ER/g DM to 24.11 mg ER/g DM (Table 1). On average, these results are in line with those reported for the same species from Algeria [19,20]. Prompted by these results showing differences in phenolic amounts among extracts, we decided to investigate their phenolic composition to get deeper insight into specific components of the polyphenolic mixture.

Table 1. Extraction yield, total polyphenol content, total flavonoid content, antioxidant activity (DPPH, β-carotene and ferric reducing power activity—FRAP) and AChE inhibition of extracts from *Mentha rotundifolia*.

	MROT-1	MROT-2	MROT-3	MROT-4	MROT-5	MROT-6	MROT-7	MROT-8	MROT-9	MROT-10
Extraction Yield (%)	17.79 ± 2.16 [c]	14.83 ± 0.47 [d]	8.12 ± 0.57 [f]	14.84 ± 1.37 [d]	13.37 ± 2.63 [e]	13.92 ± 1.33 [e]	19.26 ± 3.62 [b]	17.15 ± 0.03 [c]	22.54 ± 2.73 [a]	19.54 ± 0.81 [b]
Polyphenols (mg GAE/g DW)	57.11 ± 1.53 [a]	39.47 ± 0.28 [b]	5.70 ± 0.39 [f]	30.09 ± 1.19 [cd]	33.17 ± 3.96 [c]	25.76 ± 0.49 [ed]	33.30 ± 3.05 [c]	22.61 ± 0.07 [e]	24.80 ± 1.44 [ed]	20.55 ± 1.35 [e]
Flavonoids (mg ER/g DW)	24.11 ± 0.33 [a]	21.65 ± 0.09 [a]	5.12 ± 0.29 [d]	23.66 ± 2.24 [a]	23.73 ± 2.34 [a]	16.28 ± 1.96 [b]	20.55 ± 0.02 [a]	10.92 ± 0.01 [c]	20.20 ± 0.36 [a]	11.79 ± 0.41 [c]
DPPH (IC50 µg/mL)	15.16 ± 0.16 [e]	57.44 ± 3.87 [c]	83.50 ± 0.95 [b]	21.40 ± 0.85 [e]	54.40 ± 0.40 [c]	62.26 ± 9.64 [c]	40.66 ± 0.66 [d]	96.66 ± 3.33 [a]	75.50 ± 3.40 [b]	86.50 ± 0.97 [ab]
β-carotene (IC50 µg/mL)	106.66 ± 6.66 [f]	115.08 ± 9.34 [f]	816.66 ± 11.78 [b]	902.50 ± 22.74 [a]	936.25 ± 0.69 [a]	110.93 ± 10.96 [f]	796.66 ± 26.66 [b]	196.66 ± 20.27 [e]	745.55 ± 4.37 [c]	458.33 ± 1.82 [d]
FRAP (µmol eqFe^{2+}/g)	574.03 ± 0.98 [a]	154.54 ± 9.81 [h]	62.02 ± 0.95 [j]	497.43 ± 21.31 [b]	251.15 ± 2.22 [f]	115.06 ± 13.77 [i]	372.11 ± 0.01 [d]	289.65 ± 1.12 [e]	209.94 ± 11.43 [g]	445.41 ± 12.67 [c]
AChE (IC50 mg/mL)	0.70 ± 0.03 [d]	0.24 ± 0.01 [fg]	0.21 ± 0.01 [g]	0.27 ± 0.01 [f]	0.21 ± 0.01 [g]	0.83 ± 0.02 [c]	0.87 ± 0.03 [c]	0.93 ± 0.01 [b]	0.54 ± 0.02 [e]	2.16 ± 0.02 [a]

Values were means of three replications ± SD. Extraction yield (%): (mg extract × 100)/g dry material. DPPH scavenging activity: IC50 values for the positive control Trolox: 3.77 ± 0.12 µg/mL; β-carotene bleaching inhibition: IC50 values for the positive control BHT: 29.40 ± 0.11 µg/mL; FRAP potential: µmol $FeSO_4.7H_2O$/g extract; AChE activity: IC50 values for the positive control Donepezil: 18.0 ± 0.1 µg/mL. Values in lines followed by the same letter are not significantly different (Duncan's test, $p < 0.05$).

A small aliquot of the raw extract prepared from each population was used for a preliminary untargeted LC–MS screening. A novel ultra-high-performance liquid chromatography (UHPLC)- high resolution electrospray ionization tandem mass spectrometry (HRESIMS/MS) method was developed for the determination of the phenolic composition of the plant material. The chromatographic separation relies on a column based on Core-Shell technology (Kinetex) packed with C18-phase particles of 2.6 µm, assuring high performance, comparable to sub-2 µm particles of an UHPLC column, with significantly lower back pressure. The resolution among the various compounds detected in the extracts was optimized by using a H_2O 0.1% FA/ACN 0.1% FA gradient. The total run time of 13 min (including 3 min for re-equilibration) was far shorter than common HPLC methods adopted in the literature for polyphenol analysis, and assured a complete elution of less polar components to avoid interference in the successive runs. The LC method was improved by the coupling with a high resolution hybrid Quadrupole-Orbitrap (Q Exactive) mass spectrometer. The mass spectrometry (MS) method workflow comprised a full MS scan followed by a set of data dependent scans with a fragmentation energy applied to gather untargeted tandem mass data along with accurate mass measurements in a single analytical run. ESI source was operating in negative ion mode and an acquisition range of 100–800 *m/z* was selected. A careful inspection of LC–MS data for each extract revealed distinct fingerprints. Furthermore, differences were evident in the relative amount of polyphenols, some of which were detectable as isomeric species (Figure 1).

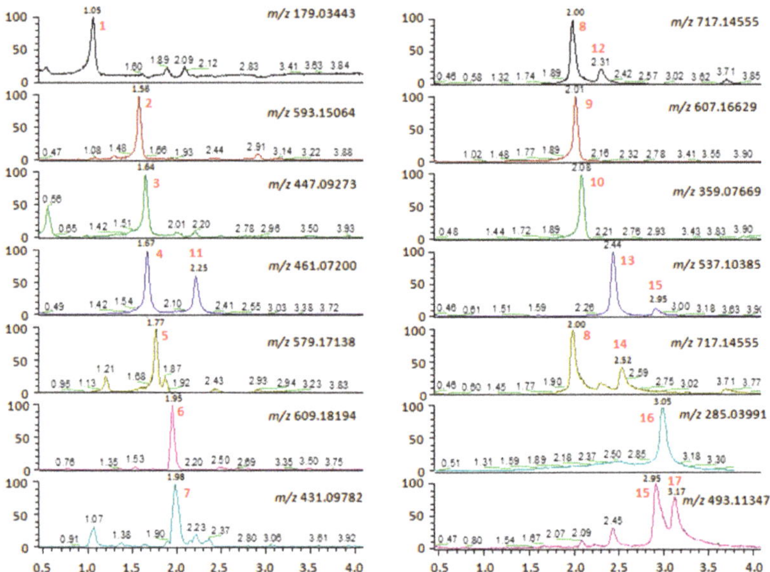

Figure 1. Extracted ion chromatograms (XICs) from UHPLC-HRESI-MS traces of polyphenols identified in the different populations of *M. rotundifolia*.

Merging literature and experimental data, a few phenolic metabolites were tentatively listed and a pool of standard reference compounds was subsequently run to confirm molecular species identification by matching experimental data. This initial screening allowed to detect the regular presence of rosmarinic acid (peak 10), which indeed is considered as a biomarker for this genus; other polyphenolic metabolites included caffeic acid (peak 1), luteolin-7-rutinoside (peak 2), luteolin-7-glucoside (peak 3), hesperidin (peak 6), apigenin-7-glucoside (peak 7) diosmin (peak 9) and luteolin (peak 16). Isomeric species were detected in all populations analyzed, and specifically two isomers of luteolin-glucuronide at [M − H]⁻ *m/z* 461.07200, three isomers of salvianolic acid B at [M − H]⁻ *m/z* 717.14555, and two isomeric compounds at [M − H]⁻ *m/z* 537.10385 (Figure 1). By co-processing with commercially available standards, luteolin-7-glucuronide (peak 4) was identified as the peak eluting at t_R =1.67 min whereas its isomer, exhibiting the same fragment at m/z 285 and eluting at 2.25 min in our LC–MS condition, was purified by subsequent chromatographic steps (LH-20 and RP18 HPLC) and easily identified as luteolin-3'-glucuronide (peak 11) by NMR data interpretation (Figures S1–S5, Supporting Information). Three species, i.e., peaks 8, 12 and 14 eluting at t_R = 2.00, 2.30 and 2.52 min, respectively, exhibited [M − H]⁻ at m/z 717.14716 (Figure 1). Peak 8 eluting at t_R = 2.00 min was isolated by HPLC and identified as salvianolic acid L by interpretation of both NMR data (Figures S6–S10, Supporting Information) and MS/MS spectra (Figure 2a). The isomer eluting at t_R = 2.30 min (peak 12) was a minor component of the extracts, not isolated during this work and here generally indicated as isosalvianolic acid B (Figure 2b). The peak 14 eluting later (Figure 2c) was identified by both retention time and MS/MS analysis as salvianolic acid B by comparison with commercial standard (Figure 2d).

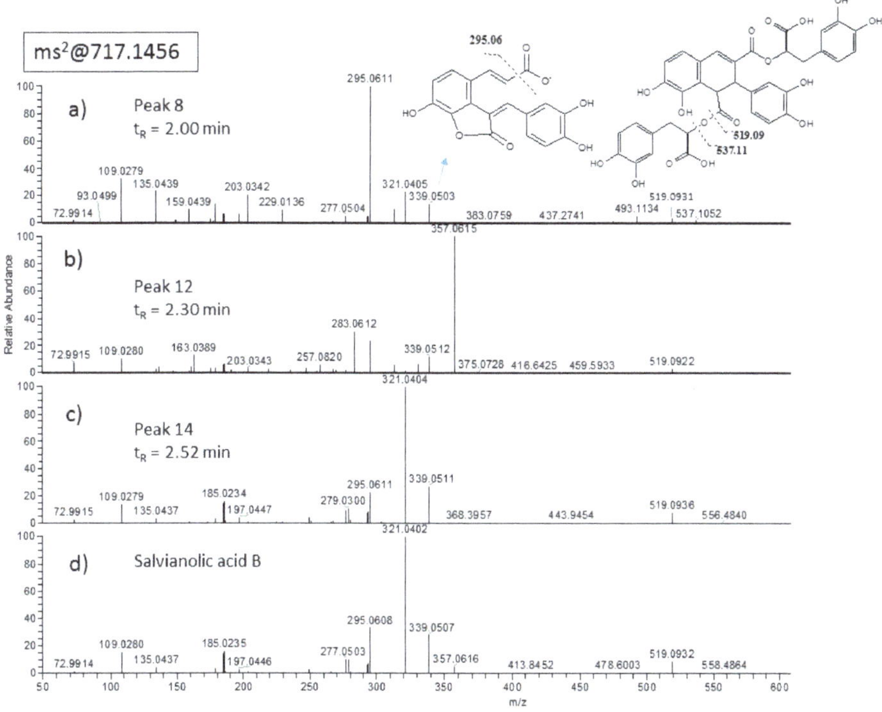

Figure 2. ESI-MS/MS spectra of isomeric compounds at m/z 717. (**a**) Peak 8, at t_R = 2.00 min; (**b**) peak 12 at t_R = 2.30 min; (**c**) peak 14 at t_R = 2.52 min in comparison with (**d**) salvianolic acid B pure standard eluting at t_R = 2.52 min.

A compound with molecular ion at m/z 493.11462 eluting at t_R = 3.17 min (Figure 3a, peak 17) did not correspond to the reference compound salvianolic acid A (Figure 3b), which indeed eluted at t_R = 2.86 min. Therefore, it was labeled as isosalvianolic acid A. Furthermore, a metabolite at m/z 579.17224 originally identified in our analysis as naringin resulted to be its structural isomer, exhibiting different retention time but identical fragmentation when compared to naringin standard and therefore was indicated as isonaringin (peak 5, t_R = 1.77 min).

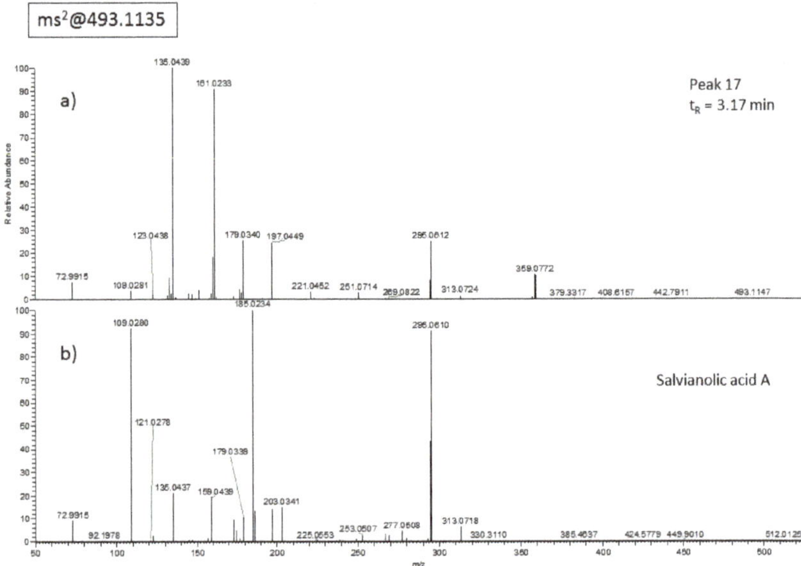

Figure 3. ESI-MS/MS spectra of compound at *m/z* 493. (**a**) Peak 17 at t_R = 3.17 min compared with (**b**) salvianolic acid A pure standard eluting at t_R = 2.86 min.

Finally, two peaks eluted at t_R = 2.44 (peak 13) and 2.95 min (peak 15) exhibiting the same molecular ion at *m/z* 537.10352 and fragmentation pattern (Figure 4). The compound eluting at t_R = 2.95 min (Figure 4a) was isolated by HPLC and fully characterized by a combination of spectroscopic and spectrometric methods, resulting in a new diastereoisomer of salvianolic acid J, here named salvianolic acid W (**15**, Figure 5) and chemically described below. The isomeric compound **13** degraded during HPLC isolation; therefore it remains uncharacterized and is here indicated as isosalvianolic acid W.

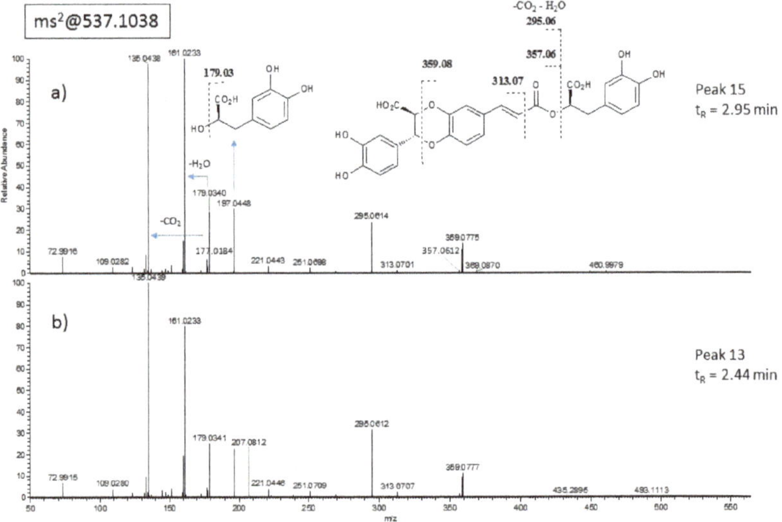

Figure 4. ESI-MS/MS spectra of compound at *m/z* 537 exhibiting the same fragmentation pattern. (**a**) Peak 15 at t_R = 2.95 min and (**b**) peak 13 at t_R = 2.44 min.

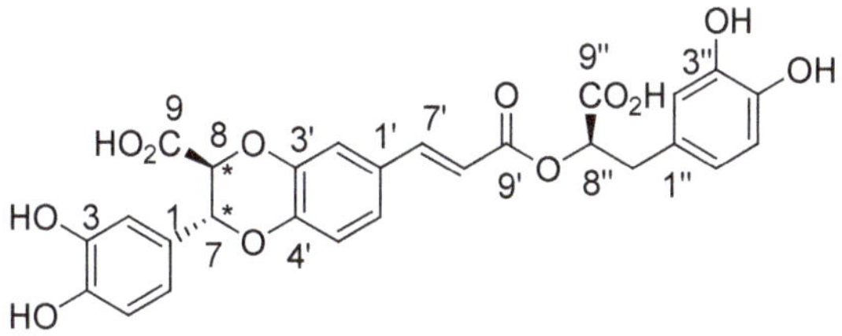

salvianolic acid W (**15**)

Figure 5. Structure of novel salvianolic acid W (**15**) detected in all Tunisian *M. rotundifolia* populations analyzed.

Compound **15** (Figure 5) was isolated as an optically active yellowish amorphous solid with $[\alpha]_D$ 10.0 (c = 0.1, MeOH), displaying UV absorption at 209, 219, 289 and 324 nm. Its molecular ion $[M - H]^-$ measured at *m/z* 537.1038 by HR-ESIMS indicated the molecular formula $C_{27}H_{22}O_{12}$, and was accompanied even in mild ionization conditions by an in-source fragment ion at *m/z* 493.1152 due to the loss of CO_2, suggesting the presence of a COOH group in the structure. ^{1}H and ^{1}H,^{1}H-COSY NMR data (600 MHz, CD$_3$OD; Table 2) revealed three sets of ABX coupled systems, i.e., δ 6.88 (1H, d, 1.8, H-2), 6.75 (1H, d, 8.1, H-5) and 6.79 (1H, dd overlapped, H-6); 7.16 (1H, d, 1.8, H-2'), 6.96 (1H, d, 8.4, H-5') and 7.12 (1H, dd, 8.4, 1.8, H-6'); 6.80 (1H, d, overlapped, H-2"), 6.70 (1H, d, 8.1, H-5") and 6.66 (1H, dd, 8.1, 1.8, H-6"), which were attributed by combining HSQC and HMBC data to three catechol residues. In the olefinic region was present a *trans* coupling proton system at δ 7.57 (1H, d, 16.0, H-7') and 6.38 (1H, d, 16.0, H-8') conjugated to a carboxyl functionality at 168.8 ppm (C-9'), which indicated the occurrence of a caffeoyl-unit. On the other hand, an oxygenated α-carboxyl methine proton appeared at δ 5.11 (dd, 10.0, 3.2, H-8") coupled to the methylene protons at δ 2.95 (dd, 10.0, 14.3, H-7"a) and 3.12 (dd, 3.2, 14.3, H-7"b). HMBC correlations with a carboxyl group at 176.9 ppm (C-9") and the aromatic carbons at 131.3 (C-1"), 117.1 (C-2") and 121.7 (C-6") ppm, supported the presence of a (3,4-dihydroxyphenyl)-lactic moiety. Completed the proton resonances an isolated spin system constituted by two oxymethine at δ 5.16 (1H, d, 5.5, H-7) and 4.52 (1H, d, 5.5, H-8) attributed to a 7,8-disubstituted benzodioxane subunit. Diagnostic HMBC correlations between H-7 and the aromatic carbons 129.9 (C-1), 115.6 (C-2) and 120.2 ppm (C-6) and between H-8 and the quaternary carbons at 174.5 (C-9) and 146.8 ppm (C-3') allowed to place the phenyl ring and the carboxyl function on carbons C-7 and C-8, respectively. The planar structure resembled the one described for salvianolic acid J [21]. In particular, the junction of the benzene and the dioxane rings was secured by diagnostic hetero-correlations observed for H-7 and H-6' with C-4' (145.8 ppm) thus excluding the other possible junctional-isomer reported as salvianolic acid P [22]. However, the coupling constant of 5.5 Hz between H-7 and H-8 suggested a *trans*-orientation instead of *cis* as in salvianolic acid J for these two protons on the dioxane cycle [23,24].

The pattern of fragmentation of molecular ion observed in ESIMS/MS spectra (Figure 4a) was in agreement with the structure depicted. In fact, diagnostic fragments at *m/z* 359.0775 and 357.0612 were compatible with the breakage of the dioxane ring and of the ester bond, generating the (dehydro)-caffeoyl acid fragments at *m/z* 177.0184 and 179.0340, respectively. Absolute stereochemistry of carbon C-8" was assumed as *R* in accordance with rosmarinic acid stereochemistry reported so far. Hence, compound **15** resulted in a novel diastereoisomer of salvianolic acid J, here named salvianolic acid W (Figure 5).

On the whole, 17 polyphenolics were identified (Figure 6) and among these eight were recurrent in all the examined extracts, i.e., the phenolic acids caffeic acid (**1**), salvianolic acid L (**8**) rosmarinic acid (**10**), isosalvianolic acid A (17) and novel salvianolic acid W (**15**) together with flavonoids represented by luteolin-7-glucuronide (**4**), luteolin-3'-glucuronide (**11**) and luteolin (**16**). Furthermore, luteolin-7-rutinoside (**2**), luteolin-7-glucoside (**3**), isonaringin (**5**), hesperidin (**6**), apigenin-7-glucoside (**7**) and salvianolic acid B (**14**) were for the first time reported in *M. rotundifolia*.

Table 2. NMR data of salvianolic acid W (**15**; 600 MHz).

C		^{1}H, δ, m, *J* (Hz) [a]	^{13}C, ppm [a]	^{1}H, δ, m, *J* (Hz) [b]	^{13}C, ppm [b]
1	C	-	129.9	-	129.5
2	CH	6.88, d, 1.8	115.6	6.98, d, 1.1	115.2
3	C	-	146.2	-	144.0 [g]
4	C	-	146.8 [e]	-	145.4
5	CH	6.75, d, 8.1	116.2	6.85 [d]	116.7
6	CH	6.79 [c]	120.2	6.84 [d]	120.1
7	CH	5.16, d, 5.5	78.2	5.28, d. 4.9	77.0
8	CH	4.52, d, 5.5	81.1	4.69, d, 4.9	79.2
9	C	-	174.5	-	174.5
1′	C	-	129.3	-	128.6
2′	CH	7.16, d, 1.8	117.5	7.20, d, 1.7	117.3
3′	C	-	146.8 [e]	-	146.0 [h]
4′	C	-	145.8 [f]	-	144.1
5′	CH	6.96, d, 8.4	118.4	7.0, d, 8.8	117.8
6′	CH	7.12, dd, 8.4, 1.8	123.1	7.19, dd, 8.8, 1.7	123.6
7′	CH	7.57, d, 16.0	146.3	7.57, d, 16.0	146.0 [h]
8′	CH	6.38, d, 16.0	117.2	6.39, d, 16.0	116.0
9′	C	-	168.8	-	169.2
1″	C	-	131.3	-	131.0
2″	CH	6.80 [c]	117.1	6.85 [d]	119.6
3″	C	-	145.8 [f]	-	144.9
4″	C	-	145.1	-	144.0 [g]
5″	CH	6.70, d, 8.1	116.3	6.80, d, 8.1	116.4
6″	CH	6.66, dd, 8.1, 1.8	121.7	6.74, dd, 8.1, 1.8	121.7
7″	CH$_2$	2.95, dd, 14.3, 10.0 3.12, dd, 14.3, 3.2	38.9	2.96, dd, 14.3, 10.0 3.13, dd, 14.3, 3.4	37.9
8″	CH	5.11, dd, 10.0, 3.2	77.9	5.04, dd, 10.0, 3.4	77.3
9″	C	-	176.9	-	177.5

[a] CD$_3$OD, [b] CD$_3$OD:D$_2$O 1:1, [c–h] overlapped signal.

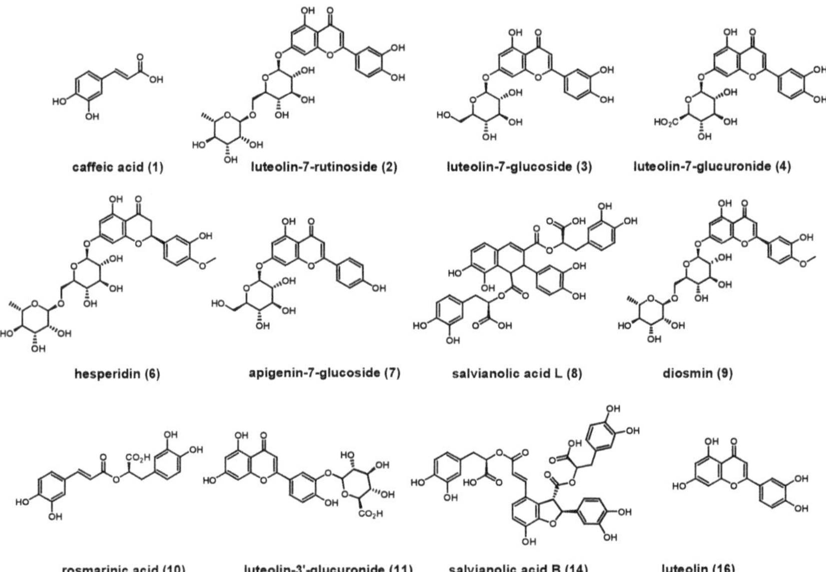

Figure 6. Structures of known compounds identified in *M. rotundifolia*.

To quantify the polyphenols, we set up a quantitative method by using a pool of commercially available reference standards. Sinapic acid, which resulted absent in all natural samples analyzed in the untargeted qualitative pre-screening, was chosen as internal standard (IS). The LC–MS method was validated according to ICH Q2 (R1) guidelines. It is suitable for quantitative measurement of polyphenolic compounds in *Mentha* and in other similar matrices. Validation parameters were reported in Table 3. The linearity of the method was evaluated by analyzing for each standard seven calibration points in triplicate over the nominal range. All standard curves showed good linearity with R^2 values within 0.9966–0.9995. Limit of detection (LOD) and limit of quantitation (LOQ) varied greatly among the different molecular species, ranging from 4.7/14.4 ng/mL, respectively, for diosmin to 75.5/229.0 ng/mL, respectively, for salvianolic acid B. Recovery was established on sinapic acid and was 62.4% ± 20.2%. The precision, expressed as the relative standard deviation (RSD; %), met the acceptance criteria, being always below 15% for all calibration points in both intra- and inter-assay measurements. Accuracy values were all within 100% ± 15% range. Quantitative measurement of polyphenolic levels was achieved by internal standard approach and results are reported in Table 4. Differences in occurrence and concentration (mg/g extract) of the identified compounds among the ten populations were observed. However, polyphenolic acids represented by caffeic acid (**1**), rosmarinic acid (**10**), salvianolic acid L (**8**), salvianolic acid W (**15**), isosalvianolic acid A (**17**) and flavonoids represented by luteolin-7-glucuronide (**4**), luteolin-3'-glucuronide (**11**) and luteolin (**16**), were always detected in all samples. Therefore, these components could be considered as polyphenolic biomarkers for this Tunisian species. Other polyphenolics occurred at different concentrations in various samples, some of them are almost ubiquitous (e.g., luteolin-7-rutinoside (**2**) and luteolin-7-glucoside (**3**)) others are scattered (e.g., salvianolic acid B (**14**), hesperidin (**6**), isonaringin (**5**) and diosmin (**9**)).

The population MROT-1 and -2 from the two localities Tamra and Oued maaden exhibited the highest contents of polyphenols, with major compounds represented by rosmarinic acid (**10**), salvianolic acid L (**8**), luteolin-7-glucuronide (**4**) and luteolin-3'-glucuronide (**11**). Isosalvianolic acid B (**12**) and isosalvianolic acid W (**13**) were also abundant and at the highest concentration in MROT-1 compared to the other populations. MROT-7 exhibited the highest amount of caffeic acid (**1**) per g of dry extract (1.38 mg/g) and MROT-3 exhibited the uppermost content in luteolin (**16**; 4.91 mg/g) and diosmin (**9**; 3.95 mg/g). The highest amount of hesperidin (**6**; 10.0 mg/g) was detected in MROT-4.

The extraction yield and chemical variability evidenced in the polyphenolic composition of populations of *M. rotundifolia* may be traced back to the influence of various climatic and edaphic conditions on the biosynthetic pathways responsible for the production of compounds related to the adaptive strategy of plants against environmental constraints [25]. Changes in environmental conditions may occur also over short distances thus supporting the differences we observed among populations even from the same bioclimatic zone (lower humid or upper semi-arid).

Phenolic acids and flavonoids are associated with potent antioxidant activity as well as with a plethora of beneficial effects on the human health. We evaluated the antioxidant activity of Tunisian mint extracts in vitro by three different tests, namely free radical scavenging assay (DPPH assay), β-carotene bleaching inhibition and ferric ion reducing antioxidant power (FRAP). Results of antioxidant assays are reported in Table 1. Antiradical (DPPH) activity expressed as IC50 (μg/mL) varied significantly among the populations tested, from 15.16 (MROT-1) to 96.66 μg/mL (MROT-8). Three populations showed antioxidant potential (MROT-1, 4 and -7) of 15.16, 21.40 and 40.66 μg/mL, respectively, which resulted significantly in comparison with the standard (Trolox = 3.77 μg/mL). Previous studies on the same species reported similar DPPH potential [17,19,26]. As for the β-carotene bleaching assay, again MROT-1 along with MROT-2 and -6 showed the highest inhibition with an IC50 of 106.66, 115.08 and 110.93 μg/mL, respectively. Literature data on *M. rotundifolia* extracts in this test are quite discordant, varying from very low to extremely high activity [17,27]. However, data on *M. rotundifolia* are often difficult to compare due to differences in adopted methodologies [28]. The FRAP values also ranged remarkably from 62.02 μmol Fe^{2+}/g in MROT-3 to 574.03 μmol Fe^{2+}/g in MROT-1. This latter population definitely resulted the most promising for antioxidant potential.

Table 3. Validation parameters for the quantitative UHPLC MS method on polyphenolic standards.

Standard Compound	t_R	$[M - H]^-$ m/z	Error (ppm)	MS/MS m/z	R^2	Working Range (μg/mL)	LOD (ng/mL)	LOQ (ng/mL)	QC (μg/mL)	Precision (% RSD)		Accuracy (%)	
										Intra-Day	Inter-Day	Intra-Day	Inter-Day
Caffeic acid	1.03	179.03443	1.90	135.04	0.9976	0.250–10	43.7	132.6	0.250	10.8	9.1	98.4	96.7
									1	7.6	6.4	100.7	109.8
									10	4.9	6.5	94.1	106.4
Luteolin-7-rutinoside	1.56	593.15064	2.26	285.04	0.9977	0.125–10	29.8	90.4	0.125	7.8	7.7	94.0	96.2
									1	5.4	6.4	100.0	101.8
									10	2.9	4.4	100.9	99.5
Luteolin-7-glucoside	1.64	447.09273	1.81	285.04	0.9960	0.125–5	17.0	35.5	0.125	2.8	4.7	96.3	99.4
									1	2.4	2.4	115.0	116.0
									5	2.5	2.7	105.5	106.6
Luteolin-7-glucuronide	1.67	461.07200	1.84	285.04	0.9983	0.125–5	39.8	120.7	0.125	10.8	9.6	96.3	94.8
									1	7.9	8.0	98.8	107.6
									5	4.4	7.0	94.1	108.4
Naringin	1.85	579.17138	1.48	151.00	0.9981	0.125–10	6.1	18.3	0.125	3.9	5.8	90.3	94.0
									1	4.3	4.1	99.7	101.4
									10	6.2	6.9	97.8	101.3
Hesperidin	1.95	609.18194	0.61	301.07	0.9973	0.125–5	3.3	10.1	0.125	5.6	7.0	85.5	89.5
									1	4.4	4.6	104.7	107.0
									5	3.9	6.1	101.3	104.5
Apigenin-7-glucoside	1.99	431.09782	0.74	268.04	0.9984	0.125–5	25.1	76.2	0.125	2.0	3.3	87.4	88.6
									1	2.4	2.1	108.6	109.1
									5	2.6	3.3	107.8	109.5
Diosmin	2.01	607.16629	1.35	299.06	0.9993	0.125–5	4.7	14.4	0.125	3.2	5.9	90.1	94.0
									1	4.7	4.8	106.9	109.4
									5	3.1	4.5	104.1	107.0
Rosmarinic acid	2.08	359.07669	1.86	161.02	0.9978	0.125–10	6.3	19.0	0.125	1.2	5.2	89.1	92.7
									1	2.7	2.6	104.3	105.6
									10	3.9	4.2	101.7	103.4
Salvianolic acid B	2.49	717.14555	2.24	321.04	0.9989	0.250–10	75.5	229.0	0.250	6.3	10.3	100.3	99.8
									1	12.7	14.3	93.4	93.5
									10	12.2	14.9	100.8	101.9
Salvianolic acid A	2.86	493.11347	2.33	185.02	0.9977	0.125–10	11.3	34.2	0.125	9.4	13.8	94.0	95.3
									1	11.8	14.1	86.9	89.8
									10	7.3	14.7	103.4	105.8
Luteolin	3.03	285.03991	1.96	133.03	0.9966	0.250–5	47.0	142.6	0.250	12.2	11.0	100.1	95.2
									1	9.6	7.5	105.1	113.9
									5	5.9	5.9	94.4	106.6

Table 4. Absolute quantitative composition (mg/g extract) of polyphenols in ten *M. rotundifolia* populations from different geographical areas. Values are means of three replicates ± SD.

Peak	Compound	t_R	$[M - H]^-$ m/z	MS/MS m/z	MROT-1	MROT-2	MROT-3	MROT-4	MROT-5	MROT-6	MROT-7	MROT-8	MROT-9	MROT-10
1	Caffeic acid	1.05	179.03443	135.04	1.21 ± 0.01	1.16 ± 0.10	0.13 ± 0.05	0.60 ± 0.09	0.96 ± 0.14	0.94 ± 0.01	1.38 ± 0.16	0.85 ± 0.14	0.70 ± 0.12	0.38 ± 0.02
2	Luteolin-7-rutinoside	1.56	593.15064	285.04	2.67 ± 0.68	6.87 ± 0.69	-	0.79 ± 0.10	1.84 ± 0.15	1.86 ± 0.21	8.02 ± 0.47	2.05 ± 0.03	1.62 ± 0.26	1.36 ± 0.02
3	Luteolin-7-glucoside	1.64	447.09273	285.04	3.84 ± 0.49	4.60 ± 0.24	-	0.53 ± 0.06	2.39 ± 0.30	1.62 ± 0.02	3.66 ± 0.78	1.23 ± 0.20	0.64 ± 0.03	0.57 ± 0.01
4	Luteolin-7-glucuronide	1.67	461.07200	285.04	26.34 ± 7.88	22.71 ± 2.10	1.91 ± 0.63	5.16 ± 0.94	8.73 ± 0.40	15.26 ± 1.91	9.71 ± 3.49	4.39 ± 0.74	1.75 ± 0.17	1.81 ± 0.16
5	Isonaringin	1.77	579.17138	271.06	0.25 ± 0.03	-	-	0.48 ± 0.14	-	-	0.67 ± 0.02	-	-	-
6	Hesperidin	1.95	609.18194	301.07	2.05 ± 0.15	-	9.54 ± 2.29	10.00 ± 3.46	3.58 ± 0.93	-	-	-	-	-
7	Apigenin-7-glucoside	1.98	431.09782	268.04	0.70 ± 0.03	0.67 ± 0.03	-	-	0.23 ± 0.02	0.21 ± 0.01	-	-	-	-
8	Salvianolic acid L	2.00	717.14555	295.06	276.44 ± 91.65	103.72 ± 2.17	2.02 ± 0.89	23.70 ± 4.69	66.31 ± 4.06	74.05 ± 8.68	29.92 ± 5.02	14.82 ± 3.44	6.27 ± 1.03	10.86 ± 0.51
9	Diosmin	2.02	607.16629	299.06	-	1.70 ± 0.68	3.95 ± 0.37	-	-	-	-	-	-	-
10	Rosmarinic acid	2.08	359.07669	161.02	116.15 ± 6.57	65.65 ± 0.39	6.53 ± 1.46	29.00 ± 0.99	68.99 ± 11.16	87.89 ± 8.52	54.51 ± 9.72	37.68 ± 4.82	50.21 ± 6.35	23.97 ± 1.15
11	Luteolin-3'-glucuronide	2.25	461.07200	285.04	42.59 ± 15.06	37.43 ± 1.05	3.50 ± 0.34	8.40 ± 1.32	15.62 ± 0.82	32.44 ± 3.36	16.59 ± 5.55	7.61 ± 1.28	3.61 ± 0.31	4.31 ± 0.28
12	Isosalvianolic acid B	2.30	717.14555	357.06	61.10 ± 24.35	-	-	-	-	21.75 ± 0.70	-	-	2.35 ± 0.19
13	Isosalvianolic acid W	2.44	537.10385	135.04	28.23 ± 8.89	3.24 ± 0.01	-	2.27 ± 0.53	9.95 ± 0.57	7.80 ± 0.22	5.19 ± 2.25	2.68 ± 0.47	3.98 ± 0.37	3.37 ± 0.0.01
14	Salvianolic acid B	2.52	717.14555	321.04	-	10.07 ± 2.97	1.32 ± 0.37	2.06 ± 0.08	5.22 ± 0.92	-	0.27 ± 0.01	0.50 ± 0.04	1.55 ± 0.01	-
15	Salvianolic acid W	2.95	537.10385	161.02	71.34 ± 15.14	10.28 ± 0.97	2.05 ± 0.43	8.83 ± 0.66	25.91 ± 0.38	21.89 ± 1.04	5.99 ± 1.99	1.92 ± 0.22	2.57 ± 0.24	2.62 ± 0.27
16	Luteolin	3.05	285.03991	133.03	2.25 ± 0.03	4.78 ± 0.44	4.91 ± 0.19	4.01 ± 0.33	3.01 ± 0.60	0.76 ± 0.25	3.68 ± 0.27	1.53 ± 0.12	0.10 ± 0.02	0.36 ± 0.04
17	Isosalvianolic acid A	3.17	493.11347	135.04	18.27 ± 3.30	2.96 ± 0.11	1.24 ± 0.17	2.98 ± 0.09	7.12 ± 0.97	3.74 ± 0.01	2.78 ± 0.90	1.31 ± 0.17	1.83 ± 0.48	0.92 ± 0.04

Polyphenols have also emerged as possible candidates for the treatment of gastrointestinal and neurodegenerative disease by virtue of their reversible inhibitory effects on AChE [29,30]. Few *Mentha* species have been tested for the inhibition of AChE activity, i.e., *M. spicata, M. pulegium* and *M. piperita* showing an IC50 in the range 0.72–1.93 mg/mL [12]. This activity has been associated to rosmarinic acid or other phenolic acids [12,31,32] or to flavonoids, which exhibited similar or even more potent activity [18,30,33]. The ten *M. rotundifolia* extracts exhibited moderate anticholinesterase activities in comparison with the positive control (Donepezil: 18.0 ± 0.1 µg/mL) (Table 1). The highest values attributed to MROT-2, -3, -4 and -5 were around 0.2 mg/mL, which were higher than those reported in the literature for other *Mentha* species.

The correlation analysis (Table 5) showed that phenolic acids represented by caffeic acid ($r = -0.48$, $p < 0.05$), rosmarinic acid ($r = -0.47$, $p < 0.05$), isosalvianolic acid B ($r = -0.41$, $p < 0.05$), salvianolic acid L ($r = -0.56$, $p < 0.01$), salvianolic acid W ($r = -0.58$, $p < 0.01$), isosalvianolic acid A ($r = -0.61$, $p < 0.01$) and isosalvianolic acid W ($r = -0.51$; $p < 0.01$) and flavonoid pool constituted by luteolin ($r = -0.37$, $p < 0.05$), luteolin-7-glucoside ($r = -0.45$, $p < 0.01$), luteolin-7-glucuronide ($r = -0.51$, $p < 0.01$), luteolin-3'-glucuronide ($r = -0.45$, $p < 0.01$), apigenin-7-glucoside ($r = -0.40$, $p < 0.05$) and isonaringin ($r = -0.65$, $p < 0.01$) were negatively correlated to DPPH.

Table 5. Correlation coefficients between identified compounds and biological activities.

	DPPH	β-carotene	FRAP	AChE
Polyphenols	-0.70 **	-0.36 *	0.45 *	-0.21 ns
Flavonoids	-0.73 **	0.14 ns	0.34 *	-0.43 *
Phenolic acids				
Caffeic acid	-0.48 *	-0.32 *	0.14 ns	-0.22 ns
Salvianolic acid L	-0.56 **	-0.49 *	0.30 *	-0.15 ns
Rosmarinic acid	-0.47 *	-0.49 *	0.07 ns	-0.16 ns
Isosalvianolic acid B	-0.41 *	-0.51 **	0.31 *	0.06 ns
Isosalvianolic acid W	-0.51 **	-0.30 *	$0,41$ *	-0.03 ns
Salvianolic acid B	-0.12 ns	-0.03 ns	-0.30 *	-0.49 **
Salvianolic acid W	-0.58 **	-0.31 *	0.34 *	-0.16 ns
Isosalvianolic acid A	-0.61 **	-0.21 ns	0.41 *	-0.19 ns
Flavonoids				
Luteolin	-0.37 *	0.29 ns	-0.12 ns	-0.65 **
Luteolin-7-rutinoside	-0.25 ns	-0.26 ns	0.002 ns	-0.07 ns
Luteolin-7-glucoside	-0.45 *	-0.39 *	0.07 ns	-0.23 ns
Luteolin-7-glucuronide	-0.51 **	-0.61 **	0.05 ns	-0.24 ns
Luteolin-3'-glucuronide	-0.45 *	-0.65 **	-0.02 ns	-0.20 ns
Apigenin-7-glucoside	-0.40 *	-0.58 **	0.03 ns	-0.23 ns
Diosmin	0.26 ns	0.11 ns	-0.54 **	-0.36 *
Isonaringin	-0.65 **	0.30 *	0.53 **	-0.09 ns
Hesperidin	-0.25 ns	0.56 **	-0.004 ns	-0.48 *

*, ** = significant at $p < 0.05$ and $p < 0.01$, respectively; ns = not significant.

A positive correlation was found between isosalvianolic acid B ($r = 0.31$, $p < 0.05$), salvianolic acid L ($r = 0.30$, $p < 0.01$), salvianolic acid W ($r = 0.34$, $p < 0.01$), isosalvianolic acid A ($r = 0.41$, $p < 0.05$), isosalvianolic acid W ($r = 0.41$, $p < 0.05$) and FRAP. Rosmarinic acid ($r = -0.49$, $p < 0.05$), isosalvianolic acid B ($r = -0.51$, $p < 0.01$), salvianolic acid L ($r = -0.49$, $p < 0.05$), salvianolic acid W ($r = -0.31$, $p < 0.05$), isosalvianolic acid W ($r = -0.30$, $p < 0.05$), luteolin-7-glucoside ($r = -0.39$, $p < 0.05$), luteolin-7-glucuronide ($r = -0.61$, $p < 0.01$), luteolin-3'-glucuronide ($r = -0.65$, $p < 0.01$) and apigenin-7-glucoside ($r = -0.58$, $p < 0.05$) were also found to be negatively correlated to the β-carotene bleaching activity. The population MROT-1, exhibiting the highest amounts in phenolic acids and in the majority of flavonoids, showed the best antiradical, β-carotene bleaching and ferric reducing capacities. Based on the above statistical analysis, this higher antioxidant activity seems to be related to its richness in phenolic compounds and not to a specific composition.

Differently from antioxidant activity, anti-cholinesterase activity appeared associated to specific compounds. Salvianolic acid B ($r = -0.30$, $p < 0.05$), hesperidin ($r = -0.48$, $p < 0.05$), luteolin ($r = -0.65$, $p < 0.01$) and diosmin ($r = -0.54$, $p < 0.1$) were negatively correlated to the anticholinesterase activity (Table 5). The populations MROT-2, -3, -4 and MROT-5 exhibiting the highest anti-AChE activities, were the richest in at least one of these compounds. Katalinic et al. reported that a high AChE inhibition potency is attributed to luteolin [34]. Hesperidin was previously found to exhibit cholinesterase inhibition [35] while pure diosmin was proposed as agent for memory restoration, in treatment of dementia [36]. A correlation was also specifically observed with salvianolic acid B, whose role in AChE inhibition and possible development in drugs against neurological diseases has been recently documented [32].

The compounds putatively responsible of the anti-acetylcholinesterase activity observed in the extracts are not the most abundant in these populations. This result, in agreement with the general finding that flavonoids are AChE inhibitors, highlights that specific components, although minor, could be responsible of the observed bioactivity.

3. Material and Methods

3.1. General

Optical rotations were measured on a Jasco P2000 digital polarimeter. UV spectra were acquired on a Jasco V-650 Spectrophotometer, CD spectra were registered on a Jasco J-815 polarimeter. NMR spectra were recorded on a Bruker Avance DRX 600 operating at 600 MHz for proton, equipped with an inverse TCI CryoProbe fitted with a gradient along the Z-axis or on an Avance III HD operating at 400 MHz for proton, equipped with a CryoProbe Prodigy. Chemical shifts values are reported in ppm (δ) and referenced to internal signals of residual protons (for CD_3OD ^{1}H δ 3.34, ^{13}C 49.0 ppm). HPLC separations were performed on a Shimadzu high-performance liquid chromatography system using a Shimadzu liquid chromatograph (Shimadzu, Kyoto, Japan) LC-20ADXR equipped with a Diode Array Detector SPDM-20A and a Kromasil RP-18 column 250 mm × 10 mm, 5 μm (Phenomenex, Castel Maggiore (BO), Italy). Polyphenolic standards (purity > 98%), formic acid (LC–MS grade) Folin-Ciocalteu reagent, sodium carbonate (Na_2CO_3), aluminum chloride ($AlCl_3$), 2,2-diphenyl-1-picrylhydrazyl, 2,4,6-*tris*-(2-pyridyl)-*S*-triazine (TPTZ), lyophilized acetylcholinesterase (AChE, electric eel, type VI-S) and acetylthiocholine iodide (ATCI) were purchased from Sigma Aldrich. ACN LC–MS grade was purchased from Merck. β-carotene was obtained from Fluka. 5,5′-dithio-bis-[2-nitrobenzoic acid] (DTNB) was purchased from MP Biomedicals. Water for LC–MS was obtained by a MilliQ apparatus (Millipore, Milan, Italy).

3.2. Plant Collection and Extraction

Aerial parts of plants taxonomically identified as *M. rotundifolia* (L.) Huds. were collected in February 2014 from diverse geographical regions of Tunisia (Table S1). The samples were collected from rivers, lands and temporarily flooded areas. Ten individuals per population were harvested. Ten populations (MROT-1 to MROT-10) were considered. Voucher specimens (M.r.N°1-10, INSAT14) were deposited at the Herbarium of the Laboratory of National Institute of Applied Sciences (Tunis, Tunisia). The plants were dried at room temperature for two weeks. Methanolic extracts for chemical measurements and biological assays were prepared using 1 g of dry leaves. After maceration in 10 mL of methanol for 24 h at room temperature, the samples were filtered, dried under vacuum and the extracts stored at −20 °C until analysis.

3.3. Total Phenolic Content (TPC) and Total Flavonoid Content (TFC).

The total phenols for each individual were determined using a spectrophotometric method [37]. An aliquot of each diluted sample extract (0.5 mL) was mixed with 2 mL Folin-Ciocalteu reagent. After 5 min, 2.5 mL of sodium carbonate solution (7.5%) was added. After incubation (90 min) in dark, the absorbance of samples versus that of the blank was read at 760 nm. Total phenols were expressed as gallic acid equivalents (mg GAE/g DW).

The total flavonoid content was determined according to Chetoui et al. [38]. One mL of the sample was mixed with 1 mL of 2% $AlCl_3$. After incubation for 15 min, the absorbance was measured at 430 nm. The percentage content of flavonoids was expressed as mg rutin equivalent/g DW (mg ER/g DW), using the calibration curve of rutin (0–400 µg/mL range).

3.4. Identification and Quantification of Phenolic Compounds

3.4.1. Identification of Phenolic Compounds

For isolation and identification of metabolites, air-dried aerial parts (18 g) of *M. rotundifolia* from Tamra (MROT-1) were powdered and exhaustively extracted at room temperature with a hydro-alcoholic solution (MeOH/H_2O, 80:20, 200 mL × six times). The extracts obtained were combined and concentrated under vacuum to afford an aqueous solution (100 mL), which was sequentially partitioned with *n*-hexane (100 mL × six times), ethyl acetate (EtOAc; 100 mL × five times) and *n*-butanol (*n*-BuOH; 60 mL × four times). The corresponding three crude extracts from *n*-hexane (0.32 g), EtOAc (0.28 g) and *n*-BuOH (0.50 g) were obtained. The butanolic extract was loaded onto a LH-20 Sephadex column (150 cm length, 3 cm diameter) packed and eluted isocratically with MeOH. Homogeneous fractions were combined to give 12 sub-fractions (A-N). Isolation of pure metabolites was carried out by HPLC on a reversed phase Kromasil RP-18 column (250 mm × 10 mm) using a gradient elution of A (water/ACN 75:25, 0.25% FA) and B (ACN, 0.25% FA) as follows: 0–3 min, 100% A; 3–11 min, 65% A: 35% B; 11–18 min 65% A: 35% B; 18–22 min 100% B; 22–28 min, 100% B. Flow 2 mL/min. PDA detection 190–800 nm, extracted wavelength at 280 nm. Salvianolic acid W (**15**, 1.0 mg, t_R = 14.6 min) was purified from LH-20 fraction M (25 mg); luteolin-3'-glucuronide (**11**, 1.1 mg, t_R = 12.2 min) and salvianolic acid L (**8**, 2.5 mg, t_R = 11.8 min) were obtained from LH-20 fraction N (10 mg).

Salvianolic acid W (**15**). NMR data: see Table 2. HR-ESIMS [M − H]$^-$ 537.10386 *m/z*, $C_{27}H_{22}O_{12}$ (calc. 537.10385); $[\alpha]_D$ 10.0 (c = 0.1, MeOH); UV (MeOH) λ_{max} (ε): 219 (8740), 289 (5795), 324 (5560) nm; CD (c = 0.125 mg/mL) MeOH λ_{max} (θ): 222 (11400), 240 (−1562), 255 (285), 276 (−1591), 296 (5492), 330 (−898) and 351 (516) nm.

Identity of compounds in raw extracts was based on comparison of retention time, high resolution *m/z* measurements and fragmentation pattern with pure standards, when available. Identification of isomeric compounds was performed on the basis of HR-ESIMS and fragmentation data. Identification of salvianolic acid L, luteolin-3'-glucuronide and the new salvianolic acid W, were based on spectroscopic and spectrometric data of pure compounds obtained by HPLC.

3.4.2. Quantification of Phenolic Compounds

For the quantification of polyphenols, *Mentha* samples (0.05 g dried leaves) spiked with 125 µg of IS were extracted with methanol (3 mL × 1 mL). The extracts were dried under vacuum, resuspended in 1 mL of MeOH and diluted 1:25 for LCMS analysis to obtain a final IS concentration of 5 µg/mL. MS recovery was established on IS by comparing the response of a defined amount (5 µg/mL) in *Mentha* samples spiked before and after extraction. Structural isomers were quantified by using the calibration curve of the corresponding known standard compounds. UHPLC analysis was performed on Infinity 1290 UHPLC System (Agilent, Milan, Italy). Chromatographic separation was achieved on a Kinetex Core-Shell C18 column (75 mm × 2.1 mm, 100 A, 2.6 µm) (Phenomenex, Castel Maggiore (BO), Italy). Elution solvents: (A) water 0.1% FA, (B) ACN 0.1% FA. Gradient: 0–1 min, 10% to 20% B; 1–8 min to 50% B; 8–8.5 min to 100% B; 8.5–10 min 100% B; then in 1 min return to initial condition and equilibration for 2 min. Flow 0.6 mL/min. The injection volume was 5µL. The UHPLC system was coupled to Q Exactive Mass Spectrometer (Thermo Scientific, San Jose, CA, USA) equipped with a HESI source operating in negative ion mode. Spectra were acquired over the range 100–800 *m/z*. Optimum values were as follow: Spray voltage 3 kV; Capillary temperature 320 °C; S-lens RF level 60;

Aux gas heater temp 320 °C; Sheath gas flow rate 50; Aux gas flow rate 30. Resolution in Full Scan 70000, MS/MS experiments was performed with NCE at 20, 30 and 40. Resolution in MS/MS mode was set at 17500. MS data were processed by Xcalibur Software (vers. 3.0.63, Thermo Scientific, San Jose, CA, USA).

Validation of the UHPLC-MS Quantitative Method

The validation of the analytical procedure was performed in accordance with ICH guidelines considering as validation characteristics: Linearity, range, detection limit (LOD), quantitation limit (LOQ), precision and accuracy. Seven calibration solutions in the range 0.125–10 μg/mL containing a pool of polyphenolic standards with a spiked amount of 5 μg/mL of IS (sinapic acid), were prepared by serial dilution of a stock solution of 2 mg/mL in MeOH. Each analysis was performed in triplicate. Results were plotted considering as response the area ratio of each polyphenol standard/IS. Peak area was measured on the extracted ion chromatogram (XIC) of molecular ion $[M - H]^-$. A least-square linear regression weighting by the reciprocal of the concentration was used to best fit the linearity curve. LOD and LOQ were calculated by considering the standard deviation of the response (σ) and the slope (S) of the calibration curve: LOD was expressed as 3.3 σ/S and LOQ as 10 σ/S. Recovery was calculated as percent ± SD on sinapic acid, spiked in the organic matrix before and after methanol extraction, in duplicate for each sample. QC samples were prepared at three different concentrations for each phenolic standard (low, middle and high) of the working range and used for repeatability and intermediate precision and accuracy. Each solution was injected six times on the same day for intra-assay precision and three times for three consecutive days for intermediate precision. The instrumental precision was expressed as percentage of relative standard deviation (% RSD). Accuracy was evaluated with the above QC samples within day and inter-day as percentage of the ratio between measured mean concentration and nominal concentration.

3.5. Antioxidant Activity

The antioxidant activity was assessed by 1,1-diphenyl-2-picrylhydrazyl (DPPH), β-carotene bleaching method systems and ferric reducing ability (FRAP).

3.5.1. Free Radical-Scavenging Assay

The free radical-scavenging activity of methanolic extracts was evaluated with the DPPH assay [39]. Three mL of 4.10^{-5} M DPPH were added to 1 mL of the extract at different concentration. The mixture was shaken and allowed to stand at room temperature for 30 min. The decrease in absorbance at 517 nm was measured against a blank. The radical-scavenging activity of samples, expressed as percentage inhibition of DPPH, was calculated according to the formula:

$$\% \text{ inhibition} = [(AB - AA)/AB] \times 100 \tag{1}$$

where AB and AA are the absorbance values of the control and of the test sample, respectively.

3.5.2. β-Carotene Bleaching Assay

The β-carotene method was carried out according to Mata et al. [12]. Two mL of β-carotene solution (0.2 mg/mL in chloroform) were pipetted into around bottomed flask containing 20 μL linoleic acid and 200 μL Tween 20. The mixture was then evaporated at 40 °C for 10 min to remove the solvent, the addition of distilled water (100 mL) followed immediately. After agitating the mixture, 1.5 mL aliquot of the resulting emulsion was transferred into test tubes containing 150 μL of extract and the absorbance was measured at 470 nm against a blank. The tubes were placed in a water bath at 50 °C and the oxidation of the emulsion was monitored by measuring absorbance at 470 nm over a 60 min. The same procedure was repeated with the synthetic antioxidant, butylhydroxytoluene (BHT) as positive control. The antioxidant activity (%) was evaluated in terms of bleaching of β-carotene using the following formula:

$$\% \text{ Inhibition} = [(AA_{t120} - AB_{t120})/(BB_{t0} - AB_{t120})] \times 100 \tag{2}$$

where AA and AB are the absorbance values measured for the test sample and control, respectively, after incubation for 120 min (t120), and BB is the absorbance value for the control measured at time zero (t0). A concentration of extract providing 50% inhibition (IC50) was obtained plotting inhibition percentage versus extract solution concentrations.

3.5.3. Ferric Reducing Power Activity (FRAP Assay)

The ferric reducing ability was assessed following the method described by Benzie and Strain [40]. FRAP reagent containing 2.5 mL of 10 mM of 2,4,6-tris(2-pyridyl)-1,3,5-triazine (TPTZ) solution in 40 mM HCl plus 2.5 mL of 20 mM $FeCl_3$ and 25 mL of 0.3 M acetate buffer (pH 3.6) was warmed prior to the analysis. FRAP Reagent (900 μL) was mixed with 90 μL distilled water and 30 μL of diluted extracts (1:10 *v/v*) and then was warmed to 37 °C in a water bath for 30 min, and the absorbance was read at 593 nm. A standard curve was prepared using different concentrations of $FeSO_4$–$7H_2O$ (200–2000 μmol/L). Results were corrected for dilution and expressed in mmol Fe^{2+}/g of plant extract.

3.6. Acetylcholinesterase Inhibition Assay

The anti-acetylcholinesterase activity was measured using an adaptation of the methods described by Eldeen et al. [41] and Ferreira et al. [42]. Briefly, 355 μL of Tris-HCl buffer (50 mM, pH 8; containing 0.1% bovine serum albumine), 20 μL of methanolic extract (at different concentrations) and 25 μL of the enzyme solution (AChE, 0.28 U/mL) were incubated during 15 min. Subsequently, 100 μL of AChI solution (0.15 mM) and 500 μL of DTNB (0.3 mM) were added. The final mixture was incubated for another 30 min at 37 °C. Absorbance of the mixture was measured at 405 nm. A control mixture was performed without addition of the extract. The anti-acetylcholinesterase activity was calculated using the following formula:

$$\text{AChE inhibition } (\%) = [(A_c - A_s)/A_c] \times 100 \tag{3}$$

where, A_c and A_s are the absorbance of the control and the sample, respectively. All tests were performed in triplicate and results were expressed as IC50 (concentration providing 50% AChE inhibition) obtained by plotting the methanolic extract concentration versus inhibition percentage. Donepezil was used as positive control.

3.7. Statistical Analyses

Correlation analysis (CA) with PROC CORR procedure 9.3.1 (SAS, Cary, NC, USA) was used. All determinations were performed in triplicates and results were expressed as mean ± standard deviation.

4. Conclusions

The present study reports for the first time a comprehensive analysis of the phenolic composition in ten different Tunisian native populations of *M. rotundifolia* (L.) Huds. (MROT-1 to 10). The raw extracts, profiled and quantified by a novel and here validated UHPLC-HRESIMS analysis based on a Q Exactive platform, showed significant qualitative and quantitative phenolic variability among the analyzed populations. Few polyphenolic acids were recurrent in all the examined extracts, i.e., rosmarinic acid, caffeic acid, salvianolic acid L, isosalvianolic acid A and the novel salvianolic acid W. Common flavonoids were represented by luteolin and its glucuronides. Furthermore, for the first time salvianolic acid B, luteolin glicosides, apigenin-7-glucoside, hesperidin and isonaringin were reported from *M. rotundifolia*. This different chemical composition was correlated to the antioxidant and anticholinesterase potential of the extracts. The population MROT-1 from Tamra (Beja) displaying the highest contents in all investigated polyphenolic classes, exhibited the highest antioxidant activity of the polyphenolic extracts evaluated by the DPPH, β-carotene and FRAP tests. This population could be selected as starting material for crop improvement program exploiting antioxidant potential of mint extracts. On the other hand, the highest anticholinesterase inhibition activity observed for populations MROT-2, -3, -4 and -5 was correlated to the presence of salvianolic acid B, luteolin, hesperidin and/or diosmin.

The wide range and heterogeneity of the ecological factors characterizing the sites surveyed seem to influence the content of the compounds. The ongoing studies on the genetic diversity based on molecular markers of populations cultivated in the same ecological conditions should help in clarifying chemical variability, in selecting interesting genotypes and optimizing their use for human health.

Supplementary Materials: The following are available online at http://www.mdpi.com/1420-3049/24/13/2351/s1, Table S1. Geographical distribution of the investigated populations of *M. rotundifolia*. Figure S1. ^{1}H NMR spectrum (600 MHz, CD_3OD) of luteolin-3′-glucuronide. Figure S2. ^{1}H,^{1}H- COSY NMR spectrum (600 MHz, CD_3OD) of luteolin-3′-glucuronide. Figure S3. ^{1}H,^{1}H TOCSY- NMR spectrum (600 MHz, CD_3OD) of luteolin-3′-glucuronide. Figure S4. HSQC-edited NMR spectrum (600 MHz, CD_3OD) of luteolin-3′-glucuronide. Figure S5. HMBC NMR spectrum (600 MHz, CD_3OD) of luteolin-3′-glucuronide. Figure S6. ^{1}H NMR spectrum (400 MHz, CD_3OD) of salvianolic acid L. Figure S7. ^{1}H,^{1}H- COSY NMR spectrum (400 MHz, CD_3OD) of salvianolic acid L. Figure S8. HSQC-edited NMR spectrum (400 MHz, CD_3OD) of salvianolic acid L. Figure S9. HMBC NMR spectrum (400 MHz, CD_3OD) of salvianolic acid L. Figure S10. ^{1}H NMR spectrum (400 MHz, CD_3OD/D_2O) of salvianolic acid L. Figure S11. ^{1}H NMR spectrum (600 MHz, CD_3OD) of salvianolic acid W. Figure S12. ^{1}H,^{1}H- COSY NMR spectrum (400 MHz, CD_3OD) of salvianolic acid W. Figure S13. HSQC-edited NMR spectrum (600 MHz, CD_3OD) of salvianolic acid W. Figure S14. HMBC NMR spectrum (600 MHz, CD_3OD) of salvianolic acid W. Figure S15. ^{13}C NMR spectrum (100 MHz, CD_3OD) of salvianolic acid W. Figure S16. ^{1}H NMR spectrum (400 MHz, CD_3OD/D_2O) of salvianolic acid W. Figure S17. ^{1}H,^{1}H- COSY NMR spectrum (600 MHz, CD_3OD/D_2O) of salvianolic acid W Figure S18. HSQC-edited NMR spectrum (600 MHz, CD_3OD/D_2O) of salvianolic acid W. Figure S19. HMBC NMR spectrum (600 MHz, CD_3OD/D_2O) of salvianolic acid W. Figure S20. CD spectrum of salvianolic acid W. Figure S21. Calibration curve for Total Phenolic Content (TPC) analysis. Figure S22. Calibration curve for Total Flavonoid Content (TFC) analysis.

Author Contributions: I.B.H.Y. performed plant extraction and biological assays; Y.Z. performed statistical analysis; A.C. carried out LCMS analysis, compound purification and structural characterization; M.L.C. helped in chromatographic purification; R.J. collected plant material and M.B. did taxonomic classification. A.L. contributed analytical standards; I.B.H.Y. and A.C. analyzed data and wrote the paper. All authors read and approved the final version of the manuscript.

Funding: This research received no external funding.

Acknowledgments: Servizio NMR at ICB is gratefully acknowledged for recording spectra. IBHY was recipient of a grant from the Tunisian Ministry of Higher Education and Scientific Research.

Conflicts of Interest: The authors declare no conflicts of interest.

References

1. Attiya, J.; Bin, G.; Bilal, H.A.; Zabta, K.S.; Tariq, M. Phylogenetics of selected *Mentha* species on the basis of rps8, rps11 and rps14 chloroplast genes. *J. Med. Plants Res.* **2012**, *6*, 30–36.
2. Pereira, R.O.; Cardoso, M.S. Overview on *Mentha* and *Thymus* polyphenols. *Curr. Anal. Chem.* **2013**, *9*, 382–396. [CrossRef]
3. De Sousa Barros, A.; de Morais, S.M.; Ferreira, P.A.T.; Vieira, Í.G.P.; Craveiro, A.A.; dos Santos Fontenelle, R.O.; Silva Alencar de Menezes, J.E.; Ferreira da Silva, F.W.; de Sousa, H.A. Chemical composition and functional properties of essential oils from *Mentha* species. *Ind. Crops Prod.* **2015**, *76*, 557–564. [CrossRef]
4. Kanatt, S.R.; Chander, R.; Sharma, A. Antioxidant potential of mint (*Mentha spicata* L.) in radiation-processed lamb meat. *Food Chem.* **2007**, *100*, 451–458. [CrossRef]
5. Oroian, M.; Escriche, I. Antioxidants: characterization, natural sources, extraction and analysis. *Food Res. Int.* **2015**, *74*, 10–36. [CrossRef] [PubMed]
6. Joseph, S.V.; Edirisinghe, I.; Burton-Freeman, B.M. Fruit polyphenols: A review of anti-inflammatory effects in humans. *Crit. Rev. Food Sci. Nutr.* **2016**, *56*, 419–444. [CrossRef] [PubMed]

7. Boffetta, P.; Couto, E.; Wichmann, J.; Ferrari, P.; Trichopoulos, D.; Bueno-de-Mesquita, H.B. Fruit and vegetable intake and overall cancer risk in the European prospective investigation into cancer and nutrition (EPIC). *J. Nat. Canc. Inst.* **2010**, *102*, 529–537. [CrossRef]
8. Brahmi, F.; Hauchard, D.; Guendouze, N.; Madani, K.; Kiendrebeogo, M.; Kamagaju, L.; Duez, P. Phenolic composition, in vitro antioxidant effects and tyrosinase inhibitory activity of three Algerian *Mentha* species: *M. spicata* (L.), *M. pulegium* (L.) and *M. rotundifolia* (L.) Huds (Lamiaceae). *Ind. Crops Prod.* **2015**, *74*, 722–730.
9. Thilakarathna, S.H.; Rupasinghe, H.P.V. Anti-atherosclerotic effects of fruit bioactive compounds: A review of current scientific evidence. *Can. J. Plant Sci.* **2012**, *92*, 407–419. [CrossRef]
10. Cardona, F.; Andrés-Lacueva, C.; Tulipani, S.; Tinahones, F.J.; Queipo-Ortuño, M.I. Benefits of polyphenols on gut microbiota and implications in human health. *J. Nutr. Biochem.* **2013**, *24*, 1415–1422. [CrossRef]
11. Espín, J.C.; González-sarrías, A.; Tomás-barberán, F.A. The gut microbiota: a key factor in the therapeutic effects of (poly) phenols. *Biochem. Pharm.* **2017**, *139*, 82–93. [CrossRef]
12. Mata, A.T.; Proença, C.; Ferreira, A.R.; Serralheiro, M.L.M.; Nogueira, J.M.F.; Araújo, M.E.M. Antioxidant and antiacetylcholinesterase activities of five plants used as Portuguese food spices. *Food Chem.* **2007**, *103*, 778–786. [CrossRef]
13. Silva, L.; Rodrigues, A.M.; Ciriani, M.; Falé, P.L.V.; Teixeira, V.; Madeira, P.; Machuqueiro, M.; Pacheco, R.; Florencio, M.H.; Ascencao, L.; et al. Antiacetylcholinesterase activity and docking studies with chlorogenic acid, cynarin and arzanol from *Helichrysum stoechas* (Lamiaceae). *Med. Chem. Res.* **2017**, *26*, 2942–2950. [CrossRef]
14. Sutour, S.; Bradesi, P.; Casanova, J.; Tomi, F. Composition and chemical variability of *Mentha suaveolens* ssp. suaveolens and M. suaveolens ssp. insularis from Corsica'. *Chem. Biodiv.* **2010**, *7*, 1002–1008.
15. Riahi, L.; Elferchichi, M.; Ghazghazi, H.; Jebali, J.; Ziadi, S.; Aouadhi, C.; Chograni, H.; Zaouali, Y.; Zoghlami, N.; Mliki, A. Phytochemistry, antioxidant and antimicrobial activities of the essential oils of *Mentha rotundifolia* L. in Tunisia. *Ind. Crops Prod.* **2013**, *49*, 883–889. [CrossRef]
16. Liu, W.; Yin, D.; Li, N.; Hou, X.; Wang, D.; Li, D.; Liu, J. Influence of environmental factors on the active substance production and antioxidant activity in *Potentilla fruticosa* L. and its quality assessment. *Sci. Rep.* **2016**, *6*, 1–18.
17. Brahmi, F.; Hadj-Ahmed, S.; Zarrouk, A.; Bezine, M.; Nury, T.; Madani, K.; Lizard, G. Evidence of biological activity of *Mentha* species extracts on apoptotic and autophagic targets on murine RAW264.7 and human U937 monocytic cells. *Pharm. Biol.* **2017**, *55*, 286–293. [CrossRef]
18. Pares, M.E. A pharmacognostic study on *Mentha rotundifolia* (L.) Hudson. *Circle Farm.* **1983**, *41*, 133–152.
19. Benabdallah, A.; Rahmoune, C.; Boumendjel, M.; Aissi, O.; Messaoud, C. Total phenolic content and antioxidant activity of six wild *Mentha* species (Lamiaceae) from northeast of Algeria. Asian Pac. *J. Trop. Biomed.* **2016**, *6*, 760–766. [CrossRef]
20. Boussouf, L.; Boutennoune, H.; Kebieche, M.; Adjeroud, N.; Al-Qaoud, K.; Madani, K. Anti-inflammatory, analgesic and antioxidant effects of phenolic compound from Algerian *Mentha rotundifolia* L. leaves on experimental animals. *South Afr. J. Bot.* **2017**, *113*, 77–83. [CrossRef]
21. Ai, C.B.; Deng, Q.H.; Song, W.Z.; Li, L.N. Salvianolic acid J, a depside from Salvia flava. *Phytochemistry* **1994**, *37*, 907–908. [CrossRef]
22. Chatzopoulou, A.; Karioti, A.; Gousiadou, C.; Lax Vivancos, V.; Kyriazopoulos, P.; Golegou, S.; Skaltsa, H. Depsides and other polar constituents from *Origanum dictamnus* L. and their in vitro antimicrobial activity in clinical strains. *J. Agric. Food Chem.* **2010**, *58*, 6064–6068. [CrossRef]
23. Matsumoto, K.; Takahashi, H.; Miyake, Y.; Fukuyama, Y. Convenient syntheses of neurotrophic americanol A and isoamericanol A by HRP catalyzed oxidative coupling of caffeic acid. *Tetrahedron Lett.* **1999**, *40*, 3185–3186. [CrossRef]
24. Lin, Y.-L.; Wang, C.-N.; Shiao, Y.-J.; Liu, T.-Y.; Wang, W.-Y. Benzolignanoid and polyphenols from Origanum vulgare. *J. Chin. Chem. Soc.* **2003**, *50*, 1079–1083. [CrossRef]
25. Di Ferdinando, M.; Brunetti, C.; Agati, G.; Tattini, M. Multiple functions of polyphenols in plants inhabiting unfavorable Mediterranean areas. *Environ. Exp. Bot.* **2014**, *103*, 107–116. [CrossRef]
26. Moldovan, R.I.; Oprean, R.; Benedec, D.; Hanganu, D.; Duma, M.; Oniga, I.; Vlase, L. LC-MS analysis, antioxidant and antimicrobial activities for five species of *Mentha* cultivated in Romania. *Digest J. Nanomater. Biostr.* **2014**, *9*, 559–566.

27. Seladji, M.; Bekhechi, C.; Bendimerad, N. Antioxidant and antimicrobial activity of aqueous and methanolic extracts of *Mentha rotundifolia* L. from Algeria. *Int. J. Pharm. Sci. Rev. Res.* **2014**, *26*, 228–234.
28. Khaled-Khodja, N.; Boulekbache-Makhlouf, L.; Madani, K. Phytochemical screening of antioxidant and antibacterial activities of methanolic extracts of some Lamiaceae. *Ind. Crops Prod.* **2014**, *61*, 41–48. [CrossRef]
29. Dinis, P.C.; Falé, P.L.; Madeira, P.J.A.; Florêncio, M.H.; Serralheiro, M.L. Acetylcholinesterase inhibitory activity after in vitro gastrointestinal digestion of infusions of *Mentha* species. *Eur. J. Med. Plants* **2013**, *3*, 381–393. [CrossRef]
30. Uriarte-Pueyo, I.; Calvo, M.I. Flavonoids as acetylcholinesterase inhibitors. *Curr. Med. Chem.* **2011**, *18*, 5289–5302. [CrossRef]
31. Vladimir-Knezevic, S.; Blazekovic, B.; Kindl, M.; Vladic, J.; Lower-Nedza, A.D.; Brantner, A.H. Acetylcholinesterase inhibitory, antioxidant and phytochemical properties of selected medicinal plants of the Lamiaceae family. *Molecules* **2014**, *19*, 767–782. [CrossRef]
32. Habtemariam, S. Molecular pharmacology of rosmarinic and salvianolic acids: potential seeds for Alzeihmeir's and vascular dementia drugs. *Int. J. Mol. Sci.* **2018**, *19*, 458. [CrossRef]
33. Roseiro, L.B.; Rauter, A.P.; Serralheiro, M.L.M. Polyphenols as acetylcholinesterase inhibitors: Structural specificity and impact on human disease. *Nutr. Aging* **2012**, *1*, 99–111.
34. Katalinic, M.; Rusak, G.; Barovic, J.D.; Sinko, G.; Jelic, D.; Antolovic, R.; Kovarik, Z. Structural aspects of flavonoids as inhibitors of human butyrylcholinesterase. *Eur. J. Med. Chem.* **2010**, *45*, 186–192. [CrossRef]
35. Sezer Senol, F.; Ankli, A.; Reich, R.; Erdogan Orhan, I. HPTLC fingerprinting and cholinesterase inhibitory and metal-chelating capacity of various *Citrus* cultivars and *Olea europaea*. *Food Technol. Biotechnol.* **2016**, *54*, 275–281.
36. Shabani, S.; Mirshekar, M.A. Diosmin is neuroprotective in a rat model of scopolamine-induced cognitive impairment. *Biomed Pharm.* **2018**, *108*, 1376–1383. [CrossRef]
37. Singleton, V.L.; Orthofer, R.; Lamuela-Raventos, R.M. Analysis of total phenols and other oxidation substrates and antioxidants by means of Folin-Ciocalteu reagent. *Methods Enzym.* **1999**, *299*, 152–178.
38. Chetoui, I.; Messaoud, C.; Boussaid, M.; Zaouali, Y. Antioxidant activity, total phenolic and flavonoid content variation among Tunisian natural populations of *Rhus tripartita* (Ucria) Grande and *Rhus pentaphylla* Desf. *Ind. Crops Prod.* **2013**, *51*, 171–177.
39. Gulluce, M.; Sahin, F.; Sokmen, M.; Ozer, H.; Daferera, D.; Sokmen, A.; Polissiou, M.; Adiguzel, A.; Ozkan, H. Antimicrobial and antioxidant properties of the essential oils and methanol extract from *Mentha longifolia* L. ssp. *longifolia*. *Food Chem.* **2007**, *103*, 1449–1456. [CrossRef]
40. Benzie, F.F.; Strain, J.J. The ferric reducing ability of plasma (FRAP) as a measure of "Antioxidant Power": The FRAP Assay. *Anal. Biochem.* **1996**, *239*, 70–76. [CrossRef]
41. Eldeen, I.M.S.; Elgorashi, E.E.; Van Staden, J. Antibacterial, anti-inflammatory, anti-cholinesterase and mutagenic effects of extracts obtained from some trees used in South African traditional medicine. *J. Ethnopharm.* **2005**, *102*, 457–464. [CrossRef] [PubMed]
42. Ferreira, A.; Proenca, C.; Serralheiro, M.L.M.; Araujo, M.E.M. The in vitro screening for acetylcholinesterase inhibition and antioxidant activity of medicinal plants from Portugal. *J. Ethnopharm.* **2006**, *108*, 31–37. [CrossRef] [PubMed]

Sample Availability: Sample of novel pure compound salvianolic acid W is available from AC.

Article

Discerning between Two Tuscany (Italy) Ancient Apple cultivars, 'Rotella' and 'Casciana', through Polyphenolic Fingerprint and Molecular Markers

Ermes Lo Piccolo [1], Ambra Viviani [1], Lucia Guidi [1,2], Damiano Remorini [1,2,*], Rossano Massai [1,2], Rodolfo Bernardi [1,2] and Marco Landi [1]

[1] Department of Agriculture, Food & Environment, University of Pisa, Via del Borghetto, 80–56124 Pisa, Italy; ermes.lopiccolo@gmail.com (E.L.P.); vivianiambra@outlook.it (A.V.); lucia.guidi@unipi.it (L.G.); rossano.massai@unipi.it (R.M.); rodolfo.bernardi@unipi.it (R.B.); marco.landi@agr.unipi.it (M.L.)

[2] Interdepartmental Research Center Nutrafood "Nutraceuticals and Food for Health", University of Pisa, Via del Borghetto, 80–56124 Pisa, Italy

* Correspondence: damiano.remorini@unipi.it; Tel.: +050-2216155

Academic Editor: Maurizio Battino
Received: 16 April 2019; Accepted: 5 May 2019; Published: 7 May 2019

Abstract: Ancient apple cultivars usually have higher nutraceutical value than commercial ones, but in most cases their variability in pomological traits does not allow us to discriminate among them. Fruit of two Tuscany ancient apple cultivars, 'Casciana' and 'Rotella', picked from eight different orchards (four for each cultivar) were analyzed for their pomological traits, organoleptic qualities, polyphenolic profile and antiradical activity. The effectiveness of a polyphenol-based cluster analysis was compared to molecular markers (internal transcribed spacers, ITS1 and ITS2) to unequivocally discern the two apples. 'Casciana' and 'Rotella' fruit had a higher nutraceutical value than some commercial cultivars, in terms of phenolic abundance, profile and total antiradical activity. Although pedo-climatic conditions of different orchards influenced the phenolic profile of both apples, the polyphenolic discriminant analysis clearly separated the two cultivars, principally due to higher amounts of procyanidin B2, procyanidin B3 and *p*-coumaroylquinic acid in 'Casciana' than in 'Rotella' fruit. These three polyphenols can be used proficiently as biochemical markers for distinguishing the two apples when pomological traits cannot. Conversely, ITS1 and ITS2 polymorphism did not allow us to distinguish 'Casciana' from 'Rotella' fruit. Overall, the use of polyphenolic fingerprint might represent a valid tool to ensure the traceability of products with a high economic value.

Keywords: ancient cultivars; antiradical activity; apple; cluster analysis; molecular marker; organoleptic quality; pomology

---

1. Introduction

In Italy, the landscape complexity from ancient times has influenced the selection of numerous local apple cultivars. As long as the communities were isolated and the exchanges were limited, these cultivars had a strictly local value, but with the opening of the communities and the intensification of the exchanges, these local cultivars began to spread, mix and sometimes get confused. 'Rotella' and 'Casciana' are two ancient apple varieties typically found in Tuscany and both are listed in the germplasm bank of the Tuscany region (Regional Law N. 64, 16 November 2004; Figure 1). 'Rotella' is cultivated in Lunigiana and is characterized by medium-small fruits, with a slightly flattened round shape at the ends; the 'Rotella' apple is very tender, with a white pulp of a sweet-sour taste. The 'Casciana' apple is an ancient cultivar largely cultivated in Garfagnana, which is similar to the 'Rotella', but generally characterized by a larger and less flattened size [1]. The color of both cultivars varies with streaks from a light green to yellow, to bright red when the tree is most exposed to the sun. Even though the two varieties in most cases can be distinguished by a trained eye, in some other cases the pedo-climatic conditions influence their pomological features and/or the skin color making the visual recognition impossible, which, in turn, creates risks for local producers and paves the way for possible fraudulence.

Figure 1. Apple fruits: from 'Rotella' (**A**) and 'Casciana' (**B**).

It has been clearly demonstrated that ancient apple fruits have higher polyphenolic content if compared to some commercial cultivars, such as Golden Delicious, Fuji and Jonagold [2,3]. In addition, ancient apples are excellent sources of polyphenols, such flavonols (i.e., quercetin glycosides), flavanols (i.e., procyanidins, epicatechin and catechin), dihydrochalcones (i.e., phloridzin and phloretin 2'-O-xylosyl-glucoside) and phenolic acids (i.e., chlorogenic acid, *p*-coumaroylquinic acid, caffeic acid) [4–6]. It has been well established that foods rich in polyphenols have powerful cardioprotective properties and show anti-cancer activities [7,8]. In apples, these benefits are mostly related to a high flavonoid content, more specifically catechin and epicatechin in skin and pulp [7,8].

Besides the excellent nutritional value of ancient apples, the safeguard of ancient genetic material also contributes to the preservation of crop biodiversity [9,10] which is of extreme importance for fruit breeders, given that ecological systems with high biodiversity maintain a high resilience to abiotic and biotic stressors [11,12]. In a world where climate change is affecting food production, autochthonous genetic heritage can be a source of specific resistance genes in order to improve the resilience of commercial varieties to both biotic and abiotic stressors [13]. Conversely, a dramatic loss of biodiversity in apple cultivation has occurred in the last century due to the selection of a few high-yield and profitable cultivars, such Gala, Golden Delicious, Fuji, Red Delicious and Stayman [9]. In Italy, more than 70% of apples produced belong to Golden groups [14], although interest in the preservation of the autochthonous genetic heritage of fruit species is continuously growing and some neglected apple cultivars have been successfully safeguarded, rediscovered and valorized [5,6,15].

Molecular markers are an efficient way to estimate genetic diversity and determine the genetic relationships among the germplasm accessions [16–19]. Several studies using different molecular markers such as Restriction Fragment Length Polymorphisms (RFLPs) that is based on the variations in the length of DNA fragments produced by a digestion of genomic DNAs and hybridization, have been extensively used for genome comparisons [20]. Nevertheless, these types of molecular markers are highly expensive and require a high yield of DNA [21] with respect to molecular markers based on the polymerase chain reaction (PCR), including Internal Transcribed Spacers (ITSs).

Therefore, in this paper the 'Casciana' and 'Rotella' fruits were characterized in terms of pomological and organoleptic features, polyphenolic profile and total antiradical activity. In addition, we utilized polyphenol-based discriminant analyses as well as the polymorphism of ITS1 and ITS2 in the attempt to discriminate the two cultivars and to establish biochemical and/or molecular markers to unequivocally ensure the origin and traceability of both apples. ITS1 and ITS2 polymorphic sequences, which have been commonly used in phylogenetic studies [16–24], were also utilized to establish ITS1- and ITS2-based phylogenetic relations of 'Casciana' and 'Rotella' with other species belonging to the genus *Malus*.

2. Results and Discussion

2.1. Pomological and Organoleptic Characteristics

Apple fruit properties (weight, width max and min, solid soluble content (SSC) and titratable acidity) are listed in Table 1. Regarding pomological parameters, 'Rotella' fruits were generally higher in size than 'Casciana' ones, with larger values for width max, width min and weight. However, RBE fruits had the smallest size and the lowest values of width max and min among all accessions of both apples (46.0 g) and CPE and CBR fruits had values of width max similar to some 'Rotella' accession's fruits (namely RFM and RFR). The aforementioned results indicate the impossibility of discerning the two apple cultivars by pomological features.

Table 1. Pomological and organoleptic characteristics of fruit of 'Casciana' and 'Rotella' apple cultivars. The first letter of the code of each sample is indicative of the cultivar, namely 'Casciana' (C) or 'Rotella' (R). Each value is the mean of eight (for weight and width) or three (for SSC and TA) replicates ± standard deviation. For each parameter, means flanked by the same letter are not significantly different after a one-way ANOVA test with accession as source of variability following an LSD test ($P = 0.05$).

Parameter	Apple Code							
	RKI	RBE	RFM	RFR	CMA	CBR	CGR	CPE
Weight (g)	132.81 ± 15.79 [a]	46.00 ± 3.42 [f]	110.64 ± 6.54 [c]	124.43 ± 6.04 [b]	69.05 ± 10.58 [e]	94.73 ± 5.16 [d]	69.33 ± 5.24 [e]	89.36 ± 5.26 [d]
Width min (mm)	70.02 ± 3.48 [a]	47.48 ± 3.14 [d]	66.07 ± 2.14 [ab]	66.67 ± 2.10 [ab]	53.80 ± 2.47 [c]	60.86 ± 2.95 [b]	57.66 ± 3.16 [bc]	60.01 ± 4.12 [b]
Width max (mm)	74.93 ± 5.11 [a]	55.08 ± 3.33 [c]	71.26 ± 8.50 [a]	71.11 ± 3.31 [a]	57.50 ± 1.34 [c]	64.46 ± 2.85 [b]	55.31 ± 2.29 [c]	68.22 ± 2.82 [ab]
SSC (°Brix)	15.60 ± 0.68	17.20 ± 1.40	15.40 ± 0.58	17.70 ± 0.98	16.80 ± 1.56	16.60 ± 1.10	14.80 ± 0.67	17.70 ± 1.90
Titratable acidity (mg malic acid g^{-1} FW)	4.28 ± 0.27 [c]	2.71 ± 0.32 [d]	3.83 ±0.82 [c]	4.50 ± 0.23 [c]	6.86 ± 0.29 [a]	5.79 ± 0.59 [b]	3.85 ± 0.39 [c]	3.92 ± 0.39 [c]

In terms of SSC, no significant differences between cultivars were observed. Notably, both these ancient apples are characterized by values of SSC similar to some highly appreciated commercial apples, such as 'Fuji', 'Golden Delicious' and 'Jonagored' [25]. Samples of CMA and CBR had higher values of TA (6.86 and 5.79 mg malic acid g^{-1}, respectively), while RBE showed the lowest (2.71 mg malic acid g^{-1}) among all the accessions. However, two 'Casciana' accession, CPR and CBE, had similar values of TA to two 'Rotella' types, namely RKI and RFM, highlighting that this parameter also cannot be used to distinguish 'Casciana' from 'Rotella' fruits. Titratable acidity is an important attribute to evaluate apples and it is remarkable that two 'Casciana' apples reach values of TA similar to those measured in 'Granny Smith' fruits [26–28].

In view of above, nor pomological or organoleptic features can be considered as affordable parameters to discriminate unequivocally between fruits belonging to the two cultivars.

2.2. Polyphenolic Profile, Total Antiradical Activity and Nutraceutical Attributes

In the last decades, "nutrafood" has received increasing demand from final consumers because of the strong connection between the intake of phytochemicals and the increase of human health [29]. Polyphenols are the most widely abundant secondary metabolites in the Planta kingdom and, at the same time, they represent a key source of antioxidant power for human health [7,30]. Polyphenol analysis of ancient 'Casciana' and 'Rotella' fruit showed 22 phenolic compounds belonging to four main groups (flavonols, flavanols, dihydrochalcones and phenolic acids) (Table 2). In both the cultivars, flavonols consisted of five quercetin glycosides (Q-galactoside, Q-glucoside, Q-arabinopyranoside, Q-arabinofuranoside and Q-rhamnoside). In particular, RFR and CGR had the highest content of Q-glucoside (0.44 and 0.43 µg g^{-1}, respectively), whereas RBE had the highest total flavonol content (2.85 µg g^{-1} FW). Quercetin is an important dietary bioactive compound for human nutrition as it might prevent some type of cancers as well as cardiovascular diseases (CVD) [31,32]. However, its bioavailability strongly depends to the glycoside which is linked to the quercetin molecule [33]. A study conducted on rats demonstrated that Q-glucoside is rapidly absorbed by the small intestine, whereas Q-galactoside and Q-arabinopyranoside are conversely poorly absorbed [34].

Flavanols, also called flavan-3-ols, are derivates of flavans constituted by 2-phenyl-3,4-dihydro-2H-chromen-3-ol skeleton [34]. In 'Casciana' and 'Rotella' apples, they were represented by catechin, epicatechin and procyanidin B1–B4 (Table 2). Procyanidins in apple fruits belong to the B-type and are mostly constituted by epicatechin and catechin [35]. Regarding their bioavailability, only 8%–17% is absorbed by the small intestine, while the rest is metabolized by intestinal flora (especially procyanidins) of the large intestine, generating several simple phenolic compounds [34,36]. In the present work, epicatechin was the most representative compound amongst flavanols (Table 2). Epicatechin is principally absorbed by the colon (about 82%) [7], probably due to the association of epicatechin-associated fibers that can only be metabolized by the large intestinal microflora [7]. For flavanols, there is a strong inverse association between their intake, especially of catechin and epicatechin, and CVD incidence [7,8]. As these chemical compounds are relevant for human health, it is important to emphasize that 'Casciana' apples, independently of their accession, have higher flavanol contents than 'Rotella' ones, except for RBE fruit (Table 2).

The dihydrochalcones group included phlor-xyl-glucose and phloridzin that are thought to be unique in apples [37]. Evidence suggests that a large part of phloridzin and phlor-xyl-glucose are absorbed by the small intestine [37], whereas phloridzin is known to be a potent inhibitor of sodium glucose transport and, therefore, is able to modulate the postprandial blood glucose levels [38]. 'Casciana' and 'Rotella accessions did not show a significant statistical differences in phlor-xyl-glucose content, whereas all accessions have similar values of phloridzin (average 17.99 µg g^{-1} FW), except for RKI which shows the lowest value (7.35 µg g^{-1} FW), therefore making both the cultivars promising sources of these compounds.

Phenolic acids represent another major group of polyphenols in apples; in the accessions of the two cultivars tested in the present experiment, 9 phenolic acids were detected: chlorogenic acid, neochlorogenic acid, cryptochlorogenic acid, *p*-coumaroyl glucose, *p*-coumaroylquinic acid, gallic acid, caffeoyl glucoside, protocatechuic acid and feruloyl glucose (Table 2). Overall, chlorogenic acid and *p*-coumaroylquinic acid are the two most representative phenolic acids, though with remarkable differences in their content in the eight apple groups. 'Casciana' fruit always had higher concentrations of *p*-coumaroylquinic than 'Rotella', independently of the orchard of origin. Chlorogenic acid is a powerful antioxidant, even though its physiological action strongly depends on its availability due to intestinal microflora that hydrolyses the molecule, giving origin to caffeic acid [39]. The hydroxycinnamic acid *p*-coumaroylquinic, in its free unconjugated form, is rapidly absorbed by the small intestine, whereas the unconjugated form is transformed by the gut microbiota in the colon [7,40]. Clinical tests conducted on *p*-coumaric acid and its conjugated forms showed antimicrobial and antiviral activities connected to its high antioxidant potential [40].

Total phenol analysis shows that two 'Casciana' groups, CMA and CBR, had the highest values (1120.54 and 1091.01 µg g^{-1} FW, respectively), whereas one 'Rotella' accession, RKI, reached the lowest value (597.71 µg g^{-1} FW). RKI was also the accession with the lowest value of phloridzin, suggesting the low nutraceutical value of fruit belonging to this accession. In any case, it should be noted that cultivars belonging to both cultivars (excluded RKI) are very rich in flavonoids and phenolic acids (flavonols and flavanols) if compared to commercial cultivars, such as Golden Delicious, Fuji and Jonagold [2,3].

Table 2. Polyphenols profile (μg g^{-1} FW) of fruit of 'Casciana' and 'Rotella' apple cultivars. The first letter of the code of each sample is indicative of the cultivar, namely 'Casciana' (C) or 'Rotella' (R). Each value is the mean of three replicates ± standard deviation. For each phenol, means flanked by the same letter are not significantly different after a one-way ANOVA test with cultivars as source of variability following an LSD test ($P = 0.05$).

Polyphenols	Apple Code							
	RKI	RBE	RFM	RFR	CMA	CBR	CGR	CPE
Flavonols								
Q-galactoside	0.03 ± 0.005 [c]	0.07 ± 0.01 [b]	0.04 ± 0.005 [c]	0.05 ± 0.01 [c]	0.05 ± 0.005 [c]	0.08 ± 0.004 [b]	0.14 ± 0.02 [a]	0.04 ± 0.01 [c]
Q-glucoside	0.17 ± 0.01 [b]	0.35 ± 0.08 [ab]	0.27 ± 0.04 [b]	0.44 ± 0.04 [a]	0.29 ± 0.01 [b]	0.36 ± 0.15 [ab]	0.43 ± 0.07 [a]	0.26 ± 0.05 [b]
Q-arabinopyranoside	0.07 ±0.006 [c]	0.14 ± 0.03 [b]	0.14 ± 0.02 [b]	0.20 ± 0.04 [a]	0.11 ± 0.002 [bc]	0.12 ± 0.03 [bc]	0.19 ± 0.01 [ab]	0.14 ± 0.06 [b]
Q-arabinofuranoside	0.08 ± 0.008 [c]	0.35 ± 0.01 [a]	0.20 ± 0.02 [b]	0.29 ± 0.03 [ab]	0.16 ± 0.01 [bc]	0.24 ± 0.12 [b]	0.31 ± 0.02 [ab]	0.17 ± 0.07 [b]
Q-rhamnoside	0.49 ± 0.054 [d]	1.94 ± 0.12 [a]	0.87 ± 0.07 [c]	1.40 ± 0.35 [b]	0.85 ± 0.04 [c]	0.88 ± 0.20 [c]	1.10 ± 0.18 [c]	1.26 ± 0.11 [bc]
Total	0.84 ± 0.06 [e]	2.85 ± 0.22 [a]	1.52 ± 0.15 [cd]	2.38 ± 0.41 [b]	1.46 ± 0.05 [d]	1.68 ± 0.49 [cd]	2.17 ± 0.27 [bc]	1.87 ± 0.30 [c]
Flavanols								
Catechin	37.07 ± 2.62 [cd]	42.42 ± 0.89 [c]	31.92 ± 7.57 [d]	61.53 ± 6.93 [b]	44.63 ± 2.27 [c]	79.36 ± 6.42 [a]	55.82 ± 11.06 [bc]	46.09 ± 2.43 [c]
Epicatechin	191.08 ± 39.88 [c]	307.65 ± 13.09 [b]	272.16 ± 29.17 [b]	305.63 ± 10.26 [b]	354.08 ± 28.82 [ab]	336.45 ± 14.31 [ab]	345.65 ± 35.30 [ab]	361.40 ± 29.53 [a]
Procyanidin B1	5.47 ± 0.40 [e]	6.99 ± 0.34 [de]	6.01 ± 0.92 [e]	9.39 ± 0.95 [c]	7.55 ± 0.87 [d]	13.11 ± 0.79 [a]	10.72 ± 0.96 [b]	9.91 ± 0.46 [bc]
Procyanidin B2	4.62 ± 0.17 [e]	5.97 ± 0.21 [d]	5.68 ± 1.12 [de]	8.94 ± 0.40 [c]	10.51 ± 0.92 [b]	10.75 ± 0.64 [b]	11.43 ± 1.50 [b]	12.52 ± 0.18 [a]
Procyanidin B3	43.53 ± 3.53 [e]	55.06 ± 2.41 [d]	42.35 ± 3.70 [e]	78.08 ± 9.45 [c]	105.19 ± 6.36 [b]	111.87 ± 7.04 [b]	112.99 ± 14.93 [ab]	123.93 ± 10.51 [a]
Procyanidin B4	-	1.03 ± 0.11 [b]	-	-	0.72 ± 0.08 [c]	1.30 ± 0.37 [a]	1.11 ± 0.10 [ab]	0.97 ± 0.01 [b]
Total	281.77 ± 35.03 [d]	419.12 ± 13.96 [b]	358.12 ± 41.02 [c]	463.57 ± 9.26 [b]	522.68 ± 31.97 [a]	552,84 ± 26.42 [a]	537.72 ± 60.36 [a]	554.82 ± 39.53 [a]
Dihydrochalcones								
Phlor-xyl-glucose	29.78 ± 7.35	40.34 ± 3.05	63.05 ± 26.89	46.69 ± 6.86	61.50 ± 3.56	44.64 ± 4.80	52.30 ± 25.58	45.59 ± 9.06
Phloridzin	7.35 ± 2.67 [c]	21.35 ± 1.96 [a]	15.58 ± 1.42 [ab]	15.39 ± 8.11 [b]	18.64 ± 4.25 [ab]	21.31 ± 4.37 [a]	16.17 ± 4.47 [ab]	17.49 ± 3.94 [ab]
Total	37.13 ± 8.44	61.69 ± 4.98	78.63 ± 28.31	62.08 ± 14.91	80.14 ± 7.81	65.95 ± 9.17	68.47 ± 28.72	63.08 ± 12.67
Phenolic acids								
Chlorogenic acid	195.97 ± 13.84 [d]	403.31 ± 10.80 [a]	246.08 ± 20.94 [c]	324.66 ± 42.79 [b]	309.55 ± 51.13 [b]	293.39 ± 12.96 [b]	191.61 ± 20.97 [d]	166.47 ± 13.33 [d]
Neochlorogenic acid	7.20 ± 0.56 [e]	6.78 ± 0.17 [e]	17.96 ± 1.62 [d]	19.05 ± 1.49 [cd]	26.09 ± 1.21 [a]	20.42 ± 0.93 [c]	22.27 ± 0.38 [b]	25.63 ± 2.77 [a]
Cryptochlorogenic acid	-	-	-	-	0.71 ± 0.05 [a]	-	0.32 ± 0.03 [b]	0.34 ± 0.05 [b]
p-Coumaroyl glucose	1.05 ± 0.14 [e]	2.00 ± 0.24 [c]	2.51 ± 0.44 [b]	1.31 ± 0.10 [de]	2.23 ± 0.44 [bc]	2.58 ± 0.19 [b]	3.35 ± 0.23 [a]	1.60 ± 0.05 [d]
p-Coumaroylquinic acid	64.96 ± 5.10 [e]	27.52 ± 2.12 [f]	80.82 ± 8.98 [d]	86.21 ± 9.01 [d]	154.50 ± 5.51 [a]	138.28 ± 11.86 [b]	131.00 ± 9.13 [b]	118.79 ± 13.18 [c]
Gallic acid	0.03 ± 0.005	0.03 ± 0.01	-	-	-	-	-	-
Caffeoyl glucoside	0.10 ± 0.01 [c]	0.93 ± 0.10 [a]	0.25 ± 0.04 [b]	0.16 ± 0.01 [c]	0.22 ± 0.05 [bc]	0.17 ± 0.01 [c]	0.25 ± 0.02 [b]	0.20 ± 0.02 [bc]
Protocatechuic acid	0.04 ± 0.002 [d]	0.06 ± 0.01 [c]	-	0.06 ± 0.01 [c]	0.05 ± 0.005 [c]	0.10 ± 0.01 [a]	0.08 ± 0.01 [b]	0.07 ± 0.01 [bc]
Feruloyl glucose	8.64 ± 0.90 [d]	75.47 ± 6.61 [a]	15.56 ± 1.91 [c]	10.47 ± 2.18 [d]	22.91 ± 3.88 [b]	15.60 ± 1.11 [c]	23.68 ± 2.46 [b]	12.63 ± 0.21 [cd]
Total	277.97 ± 19.55 [e]	516.10 ± 6.55 [a]	363,18 ± 22.63 [cd]	441.92 ± 53.28 [b]	516.26 ± 48.26 [a]	470.54 ± 3.72 [b]	372.56 ± 28.43 [c]	325,73 ± 6,64 [d]
Total polyphenols	**597.71 ± 61.94 [d]**	**999.76 ± 11.81 [b]**	**801.45 ± 59.47 [c]**	**969.95 ± 60.13 [b]**	**1120.54 ± 63.53 [a]**	**1091.01 ± 37.78 [a]**	**980.92 ± 23.74 [b]**	**945.50 ± 53.04 [b]**

A high total polyphenolic content is often associated with a high antiradical activity. Values of Total Antiradical activity (TAA), measured by the DPPH radical scavenge ability (Figure 2), are found to be higher in flesh of both the ancient cultivars when compared to commercial apples [2], and RBE had the highest value (790.6 mM TE eq 100 g^{-1} FW). Furthermore, Table 3 summarizes the data obtained from the correlation analysis between the content of singular phenols versus the values of TAA. Table 3 only reports the phenols for which a significant correlation was found with TAA. According to the correlation analysis, a strong positive correlation between Q-arabinofuranoside, feruloyl glucose, caffeoyl glucoside, chlorogenic acid, Q-rhamnoside and epicatechin content and TAA was found (Table 3).

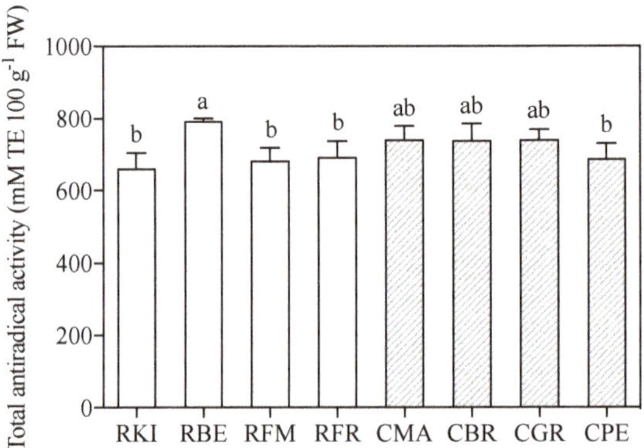

Figure 2. Antiradical activity determined by DPPH assay in the flesh of eight groups belonging to 'Casciana' and 'Rotella' ancient apple cultivars. Each value is the mean of three replicates ± standard deviation. The first letter of the code of each sample is indicative of the cultivar, namely 'Casciana' (C) or 'Rotella' (R). Bars with the same letter are not significantly different after a one-way ANOVA test with accession as source of variability following an LSD test ($P = 0.05$). TE: Trolox equivalent.

Table 3. Correlation coefficients (r) between selected phenols and total antiradical activity of fruit of 'Casciana' and 'Rotella' apple accessions. Table only reports the phenols for which a significant correlation was found with the total antiradical activity (*: $P < 0.05$, **: $P < 0.01$; ***: $P < 0.001$).

Phenol	Correlation
Caffeoyl glucoside	0.61 **
Chlorogenic acid	0.58 **
Epicatechin	0.51 *
Feruloyl glucose	0.66 ***
Q-arabinofuranoside	0.68 ***
Q-rhamnoside	0.58 **

The obtained results are in agreement with other previous works which investigated the antioxidant properties of several apple cultivars [5,41,42]. To note, Tsao et al. [42] reported that flavan-3-ols, and especially procyanidins and epicatechin were the major contributors to the TAA. Although a good correlation between TAA and epicatechin was found, in our work the coefficient of correlation between procyanidins and TAA was not significant (*data not showed*). It seems therefore conceivable that values of TAA of a fruit are dependent on the whole polyphenol profile rather than on a single or a few compounds [43,44].

2.3. Discriminatory Analysis polyphenolic fingerprint

Hierarchical cluster analysis (based on flavanol, dihydrochalcone and phenolic acid data matrix) and a polyphenol heatmap are shown in Figure 3. For cluster analyses, the samples (reported with all the three replicates) were separated into two homogenous groups by choosing a relatively and large safe cutting value at the linkage distance of 15. The obtained two major cluster groups fully corresponded to 'Casciana' and 'Rotella' cultivars. Through the heatmap, 'Casciana' and 'Rotella' cultivars are visually distinct. All the 'Casciana' accessions show higher levels of procyanidin B2, procyanidin B3 and *p*-coumaroylquinic acid than 'Rotella', which might be related to a constitutive preference of 'Casciana' apples to produce these compounds. Polyphenolic profile in apple is influenced by a plethora of environmental factors such as light, pedo-climatic conditions, agronomical practices and biotic stresses especially in the skin given that it represents the first fruit defense line [45,46]. However, the fruit responses to these external factors are strictly dependent on the interaction between the genetic background and environment and these interactions are yet to be explored in depth. Different polyphenol compositions are related to distinct apple genotypes and also the tissue-specificity of polyphenol fingerprint (i.e., in skin and flesh) is under genetic control [47]. Indeed, some authors found that apple flesh phenols and antiradical activity were linked to the different apple genotype [48,49]. Volz & McGhie [47] also showed that the variation in apple genotypes depends on polyphenol groups, in particular, chlorogenic acid, flavan-3-ols, procyanidins, dihydrochalcones and anthocyanins, whereas flavonol variation was more independent from the genotype. In our experiment, although the eight apple accessions (four for each cultivars) come from different orchards, and therefore plants were grown under different pedo-climatic and agronomical conditions, the polyphenol matrix allow us to clearly discern the two cultivars. Therefore, procyanidin B2, procyanidin B3 and *p*-coumaroylquinic seem not to be strictly influenced by environmental or agronomical conditions, and can be utilized as valid biochemical markers for a cheap and rapid methodology to discriminate and trace 'Casciana' and 'Rotella' fruits.

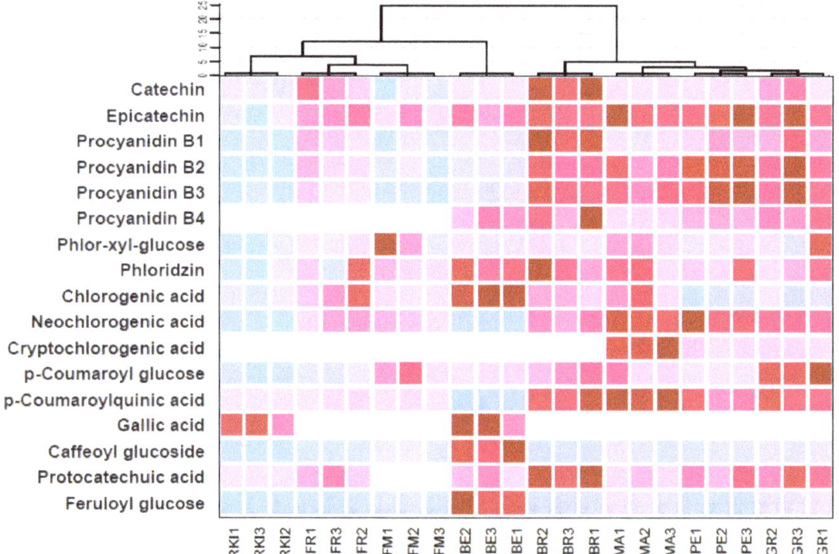

Figure 3. Heatmap visualization of the twenty-two phenolic compounds detected in the flesh of eight groups belonging to 'Casciana' and 'Rotella' ancient apple cultivars. The first letter of the code of each sample is indicative of the cultivar, namely 'Casciana' (C) or 'Rotella' (R). The intensity of different colors represents the phenol content (cyan = low content, red = high content). On the top side of the figure the hierarchical cluster is reported according to the phenolic profile of each group, by excluding flavonols.

2.4. Polymorfism of ITS1 and ITS2

The DNA extracted from leaves of apple trees was amplified using as primers M15/M17. The amplification products after sequencing were not readable. Therefore, the PCR products were cloned into pGEM-T Easy Vector, then 12 clones were pick up from each cloning and, after PCR colony screening, the clones that showed different molecular weights, of about 900 bp, (Figure 4) were sequenced. The sequences obtained have shown the reason for the impossibility of directly sequencing the PCR product, since the primers also amplified agents used in biological control and fungi presents in the biological materials used (data not shown). The sequences were deposited in GenBank with the accession numbers MH633843–MH633854.

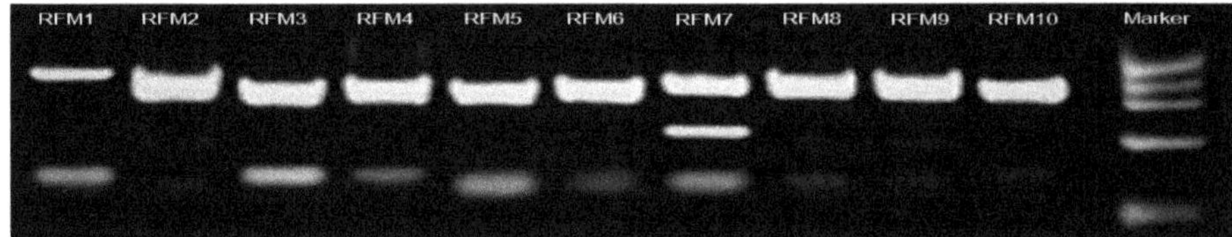

Figure 4. Electrophoresis of PCR colony screening of 10 colonies obtained after cloning the RFM amplified. In the right line the Marker ΦX174 DNA-HaeIII digest (Thermo Fisher Scientific, Waltham, MA, USA).

2.5. Phylogenetic Analysis Based on ITS1 and ITS2 Polymorfisms

The alignment of our sequences with sequences of *Malus* found in GeneBank allowed the construction of a phylogenetic tree using the MEGA7 program (Figure 5); as an outgroup we used the sequence of *Platanus acerifolia* found in GeneBank. The evolutionary relationships among the accessions were estimated by the statistical model Neighbor-Joining and the bootstrap was estimated with 1000 replications. The dendrogram shows that single accession of both 'Casciana' and 'Rotella' had multiple forms, such as two forms for CPE and CMA ('Casciana') and three forms in RFR ('Rotella'). In addition, different forms belonging to the same accession did not form a distinct cluster but are interspersed among other accessions and also among other *Malus* species and some forms of 'Casciana' and 'Rotella' cluster together with higher level of confidence than different forms belonging to the same accession. The evolutionary relationships evaluated with these molecular markers did not enable us to distinguish 'Casciana' from 'Rotella' fruits.

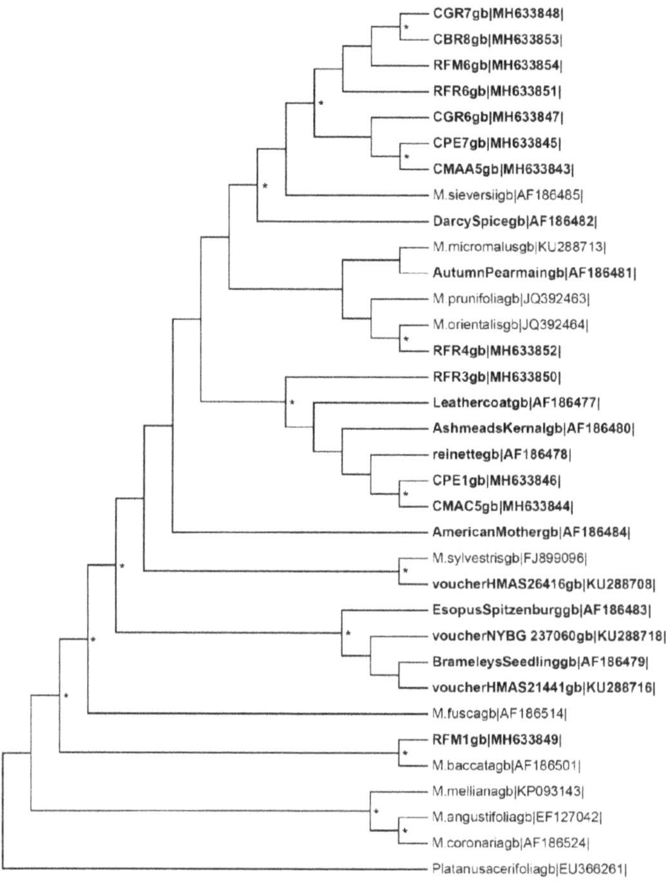

Figure 5. Molecular phylogenetic relationship between sequences of 'Rotella' and 'Casciana' nuclear ribosomal ITS1-5.8S-ITS2 and other sequences belonging to the genus *Malus*. The first letter of the code of each sample is indicative of the cultivar, namely 'Casciana' (C) or 'Rotella' (R). The taxa highlighted in bold are cultivar of *Malus domestica*. The evolutionary relationships were estimated by the statistical model Neighbor-Joining and the bootstrap was estimated with 1000 replications with the MEGA7 program. The sequence of *Platanus acerifolia* was used as outgroup. Asterisks represent a bootstrap of more than 70%.

3. Materials and Methods

3.1. Chemicals

Ultrapure standards used for determination of polyphenol profiles, as well as 2,2-diphenyl-1-picrydazil (DPPH), ethylenediaminetetraacetic acid (EDTA), Tris(hydroxymethyl)aminomethane hydrochloride (Tris-HCl), cetyltrimethylammonium bromide (CTAB), 2-mercaptoetanol, NaCl, were purchased from Sigma-Aldrich (Milan, Italy). Methanol, formic acid and acetronile were purchased from Carlo Erba Reagents (Cornaredo, Milan, Italy). The pGEM-T Easy Vector System were purchased by Promega (Madison, WI, USA) whereas DreamTaq and DreamTaq buffer were purchased from Thermo Fisher Scientific (Waltham, MA, USA).

3.2. Plant Material, Pomological and Organoleptic Properties

Fruit of 'Casciana' and 'Rotella' accessions were picked from different orchards (four for each cultivar) localized in eight different geographical places in Garfagnana and Lunigiana, (Tuscany, Italy) for 'Casciana' and 'Rotella', respectively. These were identified by a sitee code: KI (44°10′51″), BE (44°15′22″), FM (44°17′17″), FR (44°14′07″), MA (44°09′30″), BR (44°09′56″), GR (44°08′02″), PE (44°06′36″). Before the site code, every accession code was completed with the letters 'C' and 'R', which represent 'Casciana' and 'Rotella', respectively. Therefore, every code contains the apple cultivar followed by the accession name.

About 3 kg of fruits were randomly selected from different plants in each orchard at commercial maturity. Then apples were stored, one month before the analysis, in a cold chamber (4 °C and 95% of relative humidity). Three different fruits per accession were peeled and sliced with a sharping knife by removing the core portion. Little slices were cut in small portions (about 1 cm × 1 cm × 0.5 cm), mixed together and randomized in falcon tubes, frozen in liquid nitrogen and stored at −80 °C until biochemical analysis. This represented a sample replicate. Three replicates were produced and stored for biochemical analyses. Fresh weight (FW) (g) and width (mm) were calculated on eight randomly selected fruits. Solid soluble content (SSC, Brix) was analyzed on flesh juice of three randomly selected samples using a digital refractometer (refractometer Mod. 53 011, Turoni, Forli', Italy). Titratable acidity (TA) was measured following the method reported in Landi et al. [50] on three randomly selected fruits. Fruit juice samples were diluted with deionized water (1:10) and microtritated to pH 8.0 with 0.1 NaOH and expressed as mg acid malic g^{-1}.

3.3. Polyphenol Extraction

Flesh apple samples (about 1 g FW per sample obtained as described in Section 3.2) were homogenized with 10 mL of 70% aqueous methanol (*v/v*; 99.5% HPLC grade) by sonication for 30 min, keeping the temperature from 0 to 4 °C. After centrifugation (6000 *g* for 10 min at 4 °C), the supernatant was collected and filtered with PTFE filters (0.20 μm pore size; Sarstedt, Verona, Italy). Extracts were stored at −80 °C until analysis.

3.4. UPLC–MS Analysis

Phenolic profile was determined according to Assumpção et al. [51] method with few modifications. The UPLC–MS analysis was performed using an Agilent 1290 Infinity II LC system (Agilent Technologies Italia S.p.A., Cernusco sul Naviglio, Italy) consisting of a degasser, a binary pump, an autosampler, a column oven and equipped with an Agilent 6495A triple quadrupole. A C18 column, 2.1 × 50 mm, 1.8 μm (Agilent Zorbax Eclipse Plus, Santa Clara, CA, USA) was used for separation of phenolic compounds. Solvent A consisted of 0.2% formic acid in water, whereas solvent B was 0.2% formic acid in acetonitrile. The elution gradient was: 6% B (3 min), from 6 to 30% B in 11 min, from 30 to 100% B in 2 min, 100% B (2 min). The column temperature was 35 °C, the flow rate was 0.3 mL min^{-1}, and the injection volume was 2×10^{-6} L. MS parameters employed were as follows in ESI(+): gas temp: 150 °C; gas flow: 13 L min^{-1}; nebulizer: 50 psi; sheath gas heater: 350 °C; sheath gas flow: 12 L min^{-1}; capillary: 3500 V, HPRF funnel: 120; LPRF funnel: 40; in ESI(−): gas temp: 150 °C; gas flow: 13 L min^{-1}; nebulizer: 50 psi; sheath gas heater: 350 °C; sheath gas flow: 12 L min^{-1}; capillary: 1500 V; HPRF funnel: 120; LPRF funnel: 80. For quantification, an external standard method was used. A calibration curve in at least five different concentrations from 1 to 500 μg L^{-1} was constructed for each compound analyzed and utilized to quantify each compound. Data are expressed as μg g^{-1} FW.

3.5. Total Antiradical Activity

Total antiradical activity (TAA), was measured using the method of Brand-Williams, Cuvelier and Berset [52]. Briefly, 10 μL of phenolic extract were added to 990 μL of a solution containing 3.12×10^{-5} M DPPH in methanol. The decrease in absorbance at 515 nm was measured against a blank (without extract) after reaction time of 30 min (that was preliminary optimized to observe the highest antiradical effect of the extract) using a spectrophotometer (Ultrospec 2100 pro, GE Healthcare Ltd., Chalfont St. Giles, Buckinghamshire, UK). Results (n=3) were expressed as percentage of reduction of the initial DPPH absorption by the extracts and expressed as mM Trolox Equivalents (TE) 100 g^{-1} FW.

3.6. DNA Extraction

DNA was extracted from leave of apple trees using a modified CTAB extraction method of Gawel & Jarret [53]. One hundred milligrams of leaf tissue were finely crushed using mortar and pestle and homogenized with 1 mL of CTAB extraction buffer [NaCl 1.4 M, EDTA 20 mM, Tris-HCl 100 mM, (pH 8.0), CTAB 3% (*w/v*) and 2-Mercaptoetanol 0.2% (*v/v*) in a 6:1 ratio (*v/w*)]. The mixture was recovered and transferred to 14 mL tubes and incubated for 20 min at 60 °C then extracted twice with isoamyl alcohol chloroform. After adding isopropyl alcohol to the upper phase, the DNA was precipitated at 4 °C for 1 h. The pellet obtained after centrifugation was washed with ethanol at 70% (*v/v*) and dissolved in DNase free water. The concentration of each DNA sample was measured using a WPA biowave DNA spectrophotometer (Biochrom Ltd.,Cambridge, UK), and their integrity was evaluated by agarose gel electrophoresis. The DNA was stored at −20 °C until further analysis.

3.7. PCR Amplification

The nuclear rDNA region, comprising the first internal transcribed spacer (ITS1), the 5.8S rRNA gene and the second internal transcribed spacer (ITS2), was amplified by the polymerase chain reaction (PCR) by using two primers, respectively complementary to the 18S and 25S rDNA near the ITS1 and ITS2 borders, M15 (5′-AAGTCGTAACAAGGTTTCCGTAGG-3′) and M17 (5′-CTTTTCCTCCGCTTATTGATATG-3′) [27].

Amplification was carried out with conventional PCR in 20 μL reactions containing 1× 10X DreamTaq Buffer and 0.5 μM of each primer, 1U of DreamTaq and 20 ng of template DNA. PCR was run in a PCR system 2700 (Applied Biosystem, Waltham, MA, USA): Thermocycling consisted of an initial denaturation step at 95 °C (5 min), which was followed of cycles for: M15/M17 (95 °C for 45 s, 60 °C for 45 s and 72 °C for 80 s), with final extension step at 72 °C (10 min).

All reactions were checked for amplification by gel electrophoresis. Amplified DNA sequences were directly cloned in pGEM-T Easy Vector System (Promega, Madison, WI, USA). Colony PCR screening was performed on individual white colonies using as primers M13 Forward and M13 Reverse. The clones that shown inserts with different weight were sequenced by automated sequencing (MWG Biotech, Ebersberg, Germany). The sequences were analyzed using BLASTN, for their identification in GeneBank.

3.8. Phylogenetic Analyses

The sequences are multi-aligned using CLUSTALW program [54]. The multi-alignment of the sequences of 'Rotella' and 'Casciana' accessions with the sequences of the different *Malus* species already present in the database allowed us to construct a phylogenetic tree using the MEGA7 program [55]. The evolutionary relationships were estimated by the statistical model Neighbor-Joining and the bootstrap was estimated with 1000 replications with the MEGA7 program. The sequence of *Platanus acerifolia* was used as outgroup. Asterisks represent a bootstrap higher than 70%.

3.9. Statistical Analysis

Data are expressed as mean ± standard deviation and are subjected to one-way ANOVA test and statistical differences among the eight groups of two apple cultivars were calculated by least significant difference (LSD) test at 95% confidence with GraphPad Software (GraphPad, La Jolla, CA, USA). Linear correlation between phenolic compounds and total antiradical activity was carried out with the same software. Hierarchical clusters were carried out using Ward's method on normalized data matrix (using all biological replicates), in order to see similarities between cultivars by using their phenolic content. Cluster analysis (CA) was conducted using IBM SPSS Statistics 24 (IBM, New York, NY, USA). Heatmap was elaborated using GraphPad Software on the normalized polyphenol data matrix (showing all the biological replicates).

4. Conclusions

The current study reports new information about the nutraceutical properties of two ancient apple cultivars, 'Rotella' and 'Casciana'. The polyphenolic content and the total antiradical activity found in these two ancient apple cultivars were higher than values reported for some commercial cultivars, drawing attention to the need to rediscover and re-evaluate old varieties in the attempt to find new "nutrafood" sources. When polyphenolic fingerprint was used for a cluster analyses, it allowed us to clearly separate the two cultivars and individuated three polyphenols (procyanidin B2, procyanidin B3 and *p*-coumaroylquinic acid) that were higher in 'Casciana' than in 'Rotella' accessions, independently of the orchard of origin and, therefore, of different pedo-climatic and agronomic factors. The three polyphenols mentioned above can be used proficiently as biochemical markers and their simultaneous presence can be considered as a sensitive, rapid, cheap and reliable methodology for discrimination and traceability of 'Casciana' and 'Rotella' fruit. Conversely, the molecular marker used in the present experiment, ITS1 and ITS2, did not enable us to distinguish 'Casciana' from 'Rotella' fruits.

Although the importance of these results might seem to be circumscribed at the local level, the use of chemometric parameters based on polyphenol fingerprint and the identification of valid biochemical markers is certainly of broader interest and can allow us to ensure the traceability of products with high economic value and to contrast the fraudulence phenomena.

Author Contributions: Conceptualization, D.R., L.G., M.L. and R.M.; methodology, A.V., E.L.P., M.L. and R.B.; inspiration and discussions, A.V., D.R., E.L.P., L.G., M.L., R.B., R.M.; writing—original draft preparation, E.L.P. and M.L.; writing—review and editing, A.V., D.R., L.G., R.B. and R.M.

Funding: This research did not receive any specific grant from funding agencies in the public, commercial, or not-for-profit sectors.

Acknowledgments: The authors are thankful to Gaetano Bernardi, Enrico Bersanelli, Michele Bruzzi, Francesca Chinca, Salvatore Farnese, Guido Grisanti, Raffaello Grisanti, Fernando Pedreschi, Michele Pieretti and Ivo Poli for providing the apple samples.

Conflicts of Interest: The authors declare no conflict of interest.

References

1. Bartolini, S.; Viti, R.; Ducci, E. Local fruit varieties for sustainable cultivations: Pomological, nutraceutical and sensory characterization. *Agrochimica* **2015**, *59*, 281–284. [CrossRef]
2. Wojdylo, A.; Oszmianski, J.; Laskowski, P. Polyphenolic compounds and antioxidant activity of new and old apple varieties. *J. Agric. Food Chem.* **2008**, *56*, 6520–6530. [CrossRef] [PubMed]
3. Masi, E.; Taiti, C.; Vignolini, P.; Petrucci, A.W.; Giordani, E.; Heimler, D.; Romani, A.; Mancuso, S. Polyphenols and aromatic volatile compounds in biodynamic and conventional "Golden Delicious" apples (*Malus domestica* Bork.). *Eur. Food Res. Technol.* **2017**, *243*, 1519–1531. [CrossRef]
4. Escarpa, A.; González, M.C. High-performance liquid chromatography with diode-array detection for the determination of phenolic compounds in peel and pulp from different apple varieties. *J. Chromatogr. A* **1998**, *823*, 331–337. [CrossRef]
5. Iacopini, P.; Camangi, F.; Stefani, A.; Sebastiani, L. Antiradical potential of ancient Italian apple varieties of *Malus×domestica* Borkh. in a peroxynitrite-induced oxidative process. *J. Food Compos. Anal.* **2010**, *23*, 518–524. [CrossRef]
6. Jakobek, L.; Barron, A.R. Ancient apple varieties from Croatia as a source of bioactive polyphenolic compounds. *J. Food Compos. Anal.* **2016**, *45*, 9–15. [CrossRef]
7. Hyson, D.A. A Comprehensive Review of Apples and Apple Components and Their Relationship to Human Health. *Adv. Nutr.* **2012**, *2*, 408–420. [CrossRef] [PubMed]
8. Bondonno, N.P.; Bondonno, C.P.; Ward, N.C.; Hogdson, J.M.; Croft, K.D. The cardiovascular health benefits of apples: Whole fruit vs. isolated compounds. *Trends Food Sci. Technol.* **2017**, *69*, 243–256. [CrossRef]

9. Alcázar, E.J. Protecting crop genetic diversity for food security: Political, ethical and technical challenges. *Nat. Rev. Genet.* **2005**, *6*, 946–953. [CrossRef]

10. Horrigan, L.; Lawrence, R.S.; Walker, P. How sustainable agriculture can address the environmental and human health harms of industrial agriculture. *Environ. Health Perspect.* **2002**, *110*, 445–456. [CrossRef] [PubMed]

11. Fischer, J.; Lindenmayer, D.B.; Manning, A.D. Biodiversity, ecosystem function, and resilience: Ten guiding principles for commodity production landscapes. *Front. Ecol. Environ.* **2006**, *4*, 80–86. [CrossRef]

12. Di Falco, S.; Chavas, J.P. Rainfall shocks, resilience, and the effects of crop biodiversity on agroecosystem productivity. *Land Econ.* **2008**, *84*, 83–96. [CrossRef]

13. Tartarini, S.; Gennari, F.; Pratesi, D.; Palazzetti, C.; Sansavini, S.; Parisi, L.; Fouillet, A.; Durel, C.E. Characterisation and genetic mapping of a major scab resistance gene from the old Italian apple cultivar "Durello di Forlì". *Acta Hortic.* **2004**, *663*, 129–134. [CrossRef]

14. Donno, D.; Beccaro, G.L.; Mellano, M.G.; Torello Marinoni, D.; Cerutti, A.K.; Canterino, S.; Bounous, G. Application of sensory, nutraceutical and genetic techniques to create a quality profile of ancient apple cultivars. *J. Food Qual.* **2012**, *35*, 169–181. [CrossRef]

15. Maragò, E.; Michelozzi, M.; Calamai, L.; Camangi, F.; Sebastiani, L. Antioxidant properties, sensory characteristics and volatile compounds profile of apple juices from ancient Tuscany (Italy) apple varieties. *Eur. J. Hortic. Sci.* **2016**, *81*, 255–263. [CrossRef]

16. Tuna, M.; Khadka, D.K.; Shrestha, M.K.; Arumuganathan, K.; Golan-Goldhirsh, A. Characterization of natural orchardgrass (*Dactylis glomerata* L.) populations of Thrace region of Turkey base on ploidy and DNA polymorphisms. *Euphytica* **2004**, *135*, 39–46. [CrossRef]

17. Meudt, H.M.; Clarke, A.C. Almost forgotten or latest practice? AFLP applications, analyses and advances. *Trends Plant Sci.* **2007**, *12*, 106–117. [CrossRef]

18. Mondini, L.; Noorani, A.; Pagnotta, M.A. Assessing plant genetic diversity by molecular tools. *Diversity* **2009**, *1*, 19–35. [CrossRef]

19. Poczai, P.; Hyvönen, J. Nuclear ribosomal spacer regions in plant phylogenetics: Problems and prospects. *Mol. Biol. Rep.* **2010**, *37*, 1897–1912. [CrossRef]

20. Thormann, C.E.; Fereira, M.E.; Camargo, L.E.A.; Tivang, J.G.; Osborn, T.C. Comparison of RFLP and RAPD markers to estimating genetic relationships within and among cruciferous species. *Theor. Appl. Genet.* **1994**, *88*, 973–980. [CrossRef]

21. Collard, B.C.Y.; Jahufer, M.Z.Z.; Brouwer, J.B.; Pang, E.C.K. An introduction to markers, quantitative trait loci (QTL) mapping and marker-assisted selection for crop improvement: The basic concepts. *Euphytica* **2005**, *142*, 169–196. [CrossRef]

22. Ahmad Haji, R.F.; Tiwari, S.; Gandhi, S.G.; Kumar, A.; Brindavanam, N.B.; Verma, V. Genetic diversity analysis among accessions of *Desmodium gangeticum* (L) DL with Simple Sequence Repeat (SSR) and Internal Transcribed Spacer (ITS) Regions for species conservation. *J. Biodivers., Bioprospect. Dev.* **2016**, *3*, 2–5.

23. Le Cam, B.; Devaux, M.; Parisi, L. Specific polymerase chain reaction identification of *Venturia nashicola* using internally transcribed spacer region in the ribosomal DNA. *Phytopathology* **2001**, *91*, 900–904. [CrossRef]

24. Bernardi, R.; Manzo, M.; Durante, M.; Petrucelli, R.; Bartolini, G. Molecular markers for cultivar characterisation in olea *Olea europaea* L. *Acta Hortic.* **2002**, *586*, 97–100. [CrossRef]

25. Drogoudi, P.D.; Michailidis, Z.; Pantelidis, G. Peel and flesh antioxidant content and harvest quality characteristics of seven apple cultivars. *Sci. Hortic.* **2008**, *115*, 149–153. [CrossRef]

26. Aguilar-Rosas, S.F.; Ballinas-Casarrubias, M.L.; Nevarez-Moorillon, G.V.; Martin-Belloso, O.; Ortega-Rivas, E. Thermal and pulsed electric fields pasteurization of apple juice: Effects on physicochemical properties and flavour compounds. *J. Food Eng.* **2007**, *83*, 41–46. [CrossRef]

27. Vangdal, E. Quality criteria for fresh consumption. *Acta Agric. Scand.* **1985**, *35*, 41–47. [CrossRef]

28. Skendrović, B.M.; Kresimir, I.; Druzic, J.; Kovac, A.; Voca, S. Chemical and sensory characteristics of three apple cultivars (*Malus x domestica* Borkh.). *Agric. Conspec. Sci.* **2007**, *72*, 317–322.

29. Rodriguez-Casado, A. The health potential of fruits and vegetables phytochemicals: Notable examples. *Crit. Rev. Food Sci. Nutr.* **2016**, *56*, 1097–1107. [CrossRef]

30. Rasouli, H.; Farzaei, M.H.; Khodarahmi, R. Polyphenols and their benefits: A review. *Int. J. Food Prop.* **2017**, *20*, 2647–2659. [CrossRef]

31. Verhoeyen, M.E.; Bovy, A.; Collins, G.; Muir, S.; Robinson, S.; de Vos, C.H.R.; Colliver, S. Increasing antioxidant levels in tomatoes through modification of the flavonoid biosynthetic pathway. *J. Exp. Bot.* **2002**, *53*, 2099–2106. [CrossRef]
32. Toh, J.Y.; Tan, V.M.; Lim, P.C.; Lim, S.T.; Chong, M.F. Flavonoids from fruit and vegetables: A focus on cardiovascular risk factors. *Curr. Atheroscler. Rep.* **2013**, *15*, 368–375. [CrossRef]
33. Arts, I.C.W.; Sesink, A.L.A.; Faassen-Peters, M.; Hollman, P.C.H. The type of sugar moiety is a major determinant of the small intestinal uptake and subsequent biliary excretion of dietary quercetin glycosides. *Br. J. Nutr.* **2004**, *91*, 841–847. [CrossRef]
34. Santhakumar, A.B.; Battino, M.; Alvarez-Suarez, J.M. Dietary polyphenols: Structures, bioavailability and protective effects against atherosclerosis. *Food Chem. Toxicol.* **2018**, *113*, 49–65. [CrossRef]
35. Renard, C.M.G.C.; Dupont, N.; Guillermin, P. Concentrations and characteristics of procyanidins and other phenolics in apples during fruit growth. *Phytochemistry* **2007**, *68*, 1128–1138. [CrossRef]
36. Zanotti, I.; Dall'Asta, M.; Mena, P.; Mele, L.; Bruni, R.; Ray, S.; Del Rio, D. Atheroprotective effects of (poly)phenols: A focus on cell cholesterol metabolism. *Food Funct.* **2015**, *6*, 13–31. [CrossRef]
37. Marks, S.C.; Mullen, W.; Borges, G.; Crozier, A. Absorption, metabolism, and excretion of cider dihydrochalcones in healthy humans and subjects with an ileostomy. *J. Agric. Food Chem.* **2009**, *57*, 2009–2015. [CrossRef]
38. Walle, T.; Walle, U.K. The ß-D-glucoside and sodium-dependent glucose transporter 1 (SGLT1)-inhibitor phloridzin is transported by both SGLT1 and multidrug resistance-associated proteins 1/2. *Drug Metab. Dispos.* **2003**, *31*, 1288–1291. [CrossRef]
39. Sato, Y.; Itagaki, S.; Kurokawa, T.; Ogura, J.; Kobayashi, M.; Hirano, T.; Sugawara, M.; Iseki, K. In vitro and in vivo antioxidant properties of chlorogenic acid and caffeic acid. *Int. J. Pharm.* **2011**, *403*, 136–138. [CrossRef]
40. Pei, K.; Ou, J.; Huang, J.; Ou, S. p-Coumaric acid and its conjugates: Dietary sources, pharmacokinetic properties and biological activities. *J. Sci. Food Agric.* **2016**, *96*, 2952–2962. [CrossRef]
41. Chinnici, F.; Bendini, A.; Gaiani, A.; Riponi, C. Radical scavenging activities of peels and pulps from cv. Golden Delicious apples as related to their phenolic composition. *J. Agric. Food Chem.* **2004**, *52*, 4684–4689. [CrossRef]
42. Tsao, R.; Yang, R.; Xie, S.; Sockovie, E.; Khanizadeh, S. Which polyphenolic compounds contribute to the total antioxidant activities of apple? *J. Agric. Food Chem.* **2005**, *53*, 4989–4995. [CrossRef]
43. Reber, J.D.; Eggett, D.L.; Parker, T.L. Antioxidant capacity interactions and a chemical/structural model of phenolic compounds found in strawberries. *Int. J. Food Sci. Nutr.* **2011**, *65*, 445–452. [CrossRef]
44. Bolling, B.W.; Chen, Y.Y.; Chen, C.Y.O. Contributions of phenolics and added vitamin C to the antioxidant capacity of pomegranate and grape juices: Synergism and antagonism among constituents. *Int. J. Food Sci. Technol.* **2013**, *48*, 2650–2658. [CrossRef]
45. Treutter, D. Biosynthesis of phenolic compounds and its regulation in apple. *Plant Growth Regul.* **2001**, *34*, 71–89. [CrossRef]
46. Awad, M.A.; Wagenmakers, P.S.; De Jager, A. Effects of light on flavonoid and chlorogenic acid levels in the skin of "Jonagold" apples. *Sci. Hortic.* **2001**, *88*, 289–298. [CrossRef]
47. Volz, R.K.; McGhie, T.K. Genetic variability in apple fruit polyphenol composition in *Malus* × domestica and *Malus sieversii* germplasm grown in New Zealand. *J. Agric. Food Chem.* **2011**, *59*, 11509–11521. [CrossRef] [PubMed]
48. Łata, B.; Przeradzka, M.; Bínkowska, M. Great differences in antioxidant properties exist between 56 apple cultivars and vegetation seasons. *J. Agric. Food Chem.* **2005**, *53*, 8970–8978. [CrossRef]
49. Łata, B.; Trampczynska, A.; Paczesna, J. Cultivar variation in apple peel and whole fruit phenolic composition. *Sci. Hortic.* **2009**, *121*, 176–181. [CrossRef]
50. Landi, M.; Massai, R.; Remorini. Effect of rootstock and manual floral bud thinning on organoleptical and nutraceutical properties of sweet cherry (*Prunus avium* L.) cv 'Lapins'. *Agrochimica* **2014**, *58*, 335–351. [CrossRef]
51. Assumpção, C.F.; Hermes, V.S.; Pagno, C.; Castagna, A.; Mannucci, A.; Sgherri, C.; Pinzino, C.; Ranieri, A.; Flôres, S.H.; Rios, A.O. Phenolic enrichment in apple skin following post-harvest fruit UV-B treatment. *Postharvest Biol. Tech.* **2018**, *138*, 37–45. [CrossRef]

52. Brand-Williams, W.; Cuvelier, M.E.; Berset, C. Use of free radical method to evaluate antioxidant activity. *Lebenson Wiss. Technol.* **1995**, *28*, 25–30. [CrossRef]

53. Gawel, N.J.; Jarret, R.L. A modified CTAB DNA extraction procedure for *Musa* and *Ipomoea*. *Plant Mol. Bio. Rep.* **1991**, *9*, 262–266. [CrossRef]

54. Thompson, J.D.; Higgins, D.G.; Gibson, T.J. CLUSTAL W: Improving the sensitivity of progressive multiple sequence alignment through sequence weighting, position-specific gap penalties and weight matrix choice. *Nucleic Acids Res.* **1994**, *22*, 4673–4680. [CrossRef]

55. Kumar, S.; Stecher, G.; Tamura, K. MEGA7: Molecular Evolutionary Genetics Analysis version 7.0 for bigger datasets. *Mol. Biol. Evol.* **2016**, *33*, 1870–1874. [CrossRef]

Sample Availability: Not available.

 *molecules*

Article

Variability in Catechin and Rutin Contents and Their Antioxidant Potential in Diverse Apple Genotypes

Wajida Shafi [1], Sheikh Mansoor [2], Sumira Jan [1], Desh Beer Singh [1], Mohsin Kazi [3], Mohammad Raish [3], Majed Alwadei [3], Javid Iqbal Mir [1,*] and Parvaiz Ahmad [4,5,*]

[1] Indian Council of Agricultural and Research Central Institute of Temperate Horticulture, Old Airport Road, Rangreth, Srinagar 190007, J&K, India; wajida.shafi@gmail.com (W.S.); sumira.sam@gmail.com (S.J.); deshbsingh@yahoo.co.in (D.B.S.)

[2] Department of Biochemistry, Sher-e-Kashmir University of Agricultural Sciences and Technology, Jammu 180009, J&K, India; mansoorshafi21@gmail.com

[3] Department of Pharmaceutics, College of Pharmacy, King Saud University, Riyadh 11451, Saudi Arabia; mkazi@ksu.edu.sa (M.K.); mraish@ksu.edu.sa (M.R.); ph.alwadei@gmail.com (M.A.)

[4] Botany and Microbiology Department, College of Science, King Saud University, P.O. Box. 2460, Riyadh 11451, Saudi Arabia

[5] Department of Botany, S.P. College Srinagar, Srinagar 190001, Jammu and Kashmir, India

* Correspondence: javidiqbal1234@gmail.com (J.I.M.); parvaizbot@yahoo.com (P.A.); Tel.: +91-985-837-6669 (P.A.)

Received: 22 January 2019; Accepted: 1 March 2019; Published: 7 March 2019

Abstract: Catechins and rutin are among the main metabolites found in apple fruit. Sixty apple genotypes, harvested in 2016 and 2017, were analyzed for their phenolic content and antioxidant activity. The HPLC analysis showed that the catechin concentration ranged from 109.98 to 5290.47 µg/g, and the rutin concentration ranged from 12.136 to 483.89 µg/g of apple fruit. The level of DPPH activity ranged from 9.04% to 77.57%, and almost half of the 15 genotypes showed below 30–40% DPPH activity. The apple genotypes 'Lal Ambri', 'Green Sleeves', and *'Mallus floribunda'* showed the highest DPPH activity of between 70% and 80%, while 'Schlomit', 'Luxtons Fortune', 'Mayaan', 'Ananas Retrine', and 'Chaubatia ambrose' showed the lowest ferric reducing antioxidant power (FRAP) activity (0.02–0.09%). Statistical analysis showed a correlation between DPPH activity and catechin content ($r = 0.7348$) and rutin content ($r = 0.1442$). Regarding antioxidant activity, fractionated samples of apple genotypes revealed significant activity comparable to that of ascorbic acid. There was also a consistent trend for FRAP activity among all apple genotypes and a significant positive correlation between FRAP activity and rutin content ($r = 0.244$). Thus, this study reveals a significant variation in antioxidant potential among apple genotypes. This data could be useful for the development of new apple varieties with added phytochemicals by conventional and modern breeders.

Keywords: apple; DPPH; FRAP; polyphenolics; catechins; rutin

---

1. Introduction

Apples are cultivated in temperate countries and are one of the most important fruits [1]. Worldwide, apples are consumed throughout the year because of their organoleptic qualities as well as due to technological advancements in the area of conservation [2]. Significant concentrations of phenolic compounds are present in apples and their products, and these play critical roles in maintaining human health due to their preventive effect against various diseases, such as cardiovascular diseases, neuropathies and diabetes [3]. The main phenolic acids found in apples are chlorogenic acid and *p*-coumaroylquinic acid, and the major flavonoids are epicatechins, catechins, procyanidins (B1 and B2), quercetin glycosides, anthocyanins, and phloridzin [4]. In recent years, there has been a rising inclination towards the use of bio-active compounds and in this context, extraction of such compounds from tissues rich in their content is desired [5,6]. Different plant materials have different extraction conditions, as they are affected by several parameters, such as the chemical nature of the sample, the type of solvent used, agitation, the time of extraction, the solute/solvent ratio and the presence of an optimum temperature [7,8]. Furthermore, validation of the extraction method for phenolic compounds is needed to avoid enzymatic oxidation during the process, as this leads to loss of phenol function and antioxidant potential [9]. Accordingly, in order to counteract oxidation, frozen or lyophilised samples are taken to prevent enzymatic oxidation [10].

Polyphenolic compounds are responsible for the aroma and organoleptic properties of apples. Phenolic acids in apples are subdivided into benzoic acids and hydroxycinnamic acids [11,12]. Flavonoids possess a nucleus comprising two phenolic rings and oxygenated heterocycle compounds, and they can be categorized into different types, e.g., anthocyanins, flavonols, flavanols (e.g., catechins), flavones, and chalcones [13]. Catechins and rutin are the predominant phytochemicals in apples; they not only confer color but also aroma to different genotypes. These compounds vary greatly in diverse apple genotypes depending on the place, season, light, and altitude [14,15]. The rationale of this study was to determine the variation in the concentrations of catechins and rutin in apple genotypes cultivated in the same location but harvested in two different years. Thus, the variation in catechin and rutin content and the antioxidant activity of apples were determined. The contribution of single phenolic compounds to the antioxidant capacity was estimated with special respect to standards. We intended to compare the antioxidant properties of catechins with those of other pure standards and synthetic antioxidants in all apple genotypes thriving in the same location under similar climatic and geographical conditions. The antioxidant activities were determined by commonly used methods of radical scavenging: DPPH (2,2-diphenyl-2-picrylhydrazyl) and ferric reducing antioxidant power (FRAP) assays.

2. Results

2.1. Phytochemical Determinations

Total Phenols, Total Flavanols, and Flavonoids

The total phenolic content of apples in this study ranged from 31.5 to 980.8 GAE/g, which is comparatively higher than the concentration in grape extract, a beverage known for its polyphenolic content. The flavanol content varied from 0.004 to 0.185 QEA (mg/g), while that of flavonoids ranged from 0.36 to 0.3584 QEA (mg/g). The maximum phenolic content, 980.8 GAE/g, was observed in the wild apple genotype *Mallus floribunda*, followed by 722.0 GAE/g in *Tydemans Early Worcestor*, and the minimum phenolic content of 31.5 mg L^{-1} was observed in Starking Delicious. The rest of the genotypes had moderate ranges. The maximum flavonoid content of 0.3884 QEA (mg/g) was observed in the wild apple genotype *Mallus floribunda* followed by *Ambri* (0.367 QEA (mg/g), and the minimum flavonoid content of 0.024 QEA (mg/g) was observed in Star Summer Gold, followed by 0.027 QEA (mg/g) in wealthy apple. The rest of the genotypes had moderate concentrations. Similarly, the maximum (0.351 QEA (mg/g)) and minimum (0.002 QEA (mg/g)) flavanol contents were observed in Orange Val and Red Fuji, respectively (Table 1).

Table 1. Variability in rutin and catchin concentrations in different apple genotypes. The data is represented in mean ± SD (*n* = 10) and letter in the superscript symbolize the letters of significance with respect to each other using Tukey' test.

S.NO.	VARIETIES	RUTIN (μg/g)	CATECHINS (μg/g)
1	VISTA BELLA	57.727 [UT] ± 5.87	1228.61 [KLM] ± 15.26
2	TYDEMANS EARLY WORCESTOR	87.023 [R] ± 5.34	665.10 [XY] ± 11.15

Table 1. *Cont.*

S.NO.	VARIETIES	RUTIN (µg/g)	CATECHINS (µg/g)
3	BENONI	28.70 ABC ± 2.34	1745.92 ED ± 18.79
4	MICHAL	140.17 J ± 9.87	1805.47 D ± 19.04
5	SUMMER RED	131.29 LJK ± 7.87	109.98 C ± 4.56
6	LEMON GUARD	46.86 VWX ± 4.32	1157.21 MNOP ± 11.26
7	LAXTONS FURTUNE	136.13 JK ± 7.65	832.52 TU ± 10.86
8	MAYAAN	20.31C D ± 1.23	1438.02 HI ± 12.34
9	JUNE EATING	12.136 D ± 2.32	851.43 ST ± 11.08
10	MOLLIES DELICIOUS	88.603 R ± 7.32	1250.30 KLM ± 16.04
11	PRIMA	53.251 UV ± 4.87	1498.37 GH ± 12.45
12	STAR SUMMER GOLD	63.08 T ± 6.12	824.47 TU ± 9.32
13	BLACK BEN DAVIS	28.459 ABC ± 2.45	1940.44 C ± 20.12
14	GALA MAST	65.598 ST ± 6.78	1199.00 LMNO ± 12.04
15	AKBER	72.423 S ± 7.02	1309.31 JK ± 14.05
16	RED BARON	100.588 P ± 8.12	373.43 A ± 10.45
17	FANNY	40.529 XY ± 3.45	1209.10 LMN ± 14.98
18	MALLUS BACCATA	212.087 E ± 9.12	1771.83 ED ± 18.98
19	FUJI	30.211 ABZ ± 3.12	357.43 A ± 11.08
20	VANCE DELICIOUS	124.454 LM ± 6.89	1409.72 I ± 12.01
21	COE RED FUJI	24.453 BC ± 2.13	982.03 Q ± 10.20
22	COOPER IV	59.31 UT ± 5.01	1253.16 KL ± 15.89
23	GRANNY SMITH	245.318 D ± 9.89	1608.65 F ± 17.78
24	AMBRI	332.405 B ± 10.98	2028.94 B ± 22.34
25	LAL AMBRI	483.888 A ± 11.23	1717.66 E ± 17.89
26	RED DELICIOUS	115.691 MN ± 7.08	1810.28 D ± 18.09
27	MALLUS FLORIBUNDA	85.961 R ± 7.23	5290.47 A ± 34.43
28	AMARTARAPRIDE	148.792 I ± 8.07	975.51 Q ± 12.03
29	ANANAS RETRINE	113.684 N ± 8.98	940.90 QR ± 11.87
30	ANTINOVIKA	36.763 AYZ ± 3.45	1180.84 LMNO ± 21.08
31	BELLE DE BESCOPE	101.316 OP ± 9.06	1239.24 KLM ± 15.78
32	BISSBEE SPUR	160.584 H ± 9.87	1117.38 OP ± 10.98
33	CHAUBATIA AMBROSE	124.388 LM ± 6.98	381.21 A ± 11.56
34	CHECK AMBRI	29.883 ABZ ± 3.01	1499.82 GH ± 12.01
35	FIRDOUS	200.225 F ± 8.98	366.60 A ± 9.89
36	GREEN SLEEVES	43.533 WXY ± 3.89	1085.48 P ± 11.05
37	HARDIMAN	119.226 MN ± 7.19	723.42 WXY ± 7.98
38	JONICA	50.889 UVW ± 4.56	1533.91 FG ± 15.32
39	MAHARAJI	240.65 D ± 9.06	863.66 RST ± 10.09
40	NEEMA DELICIOUS	86.995 R ± 8.12	238.57 B ± 10.04
41	ORANGE VAL	166.55 H ± 9.89	691.49 XY ± 6.78
42	OREGON SPUR	94.541 PQR ± 8.98	675.30 XY ± 5.98
43	PRINCE NOBLE	58.415 UT ± 5.10	416.41 A ± 12.09
44	RED CHIEF	112.096 N ± 8.08	640.32 Y ± 5.23
45	RED FUJI	330.08 B ± 10.09	1169.94 LMNOP ± 20.98
46	RED GOLD	47.608 WXV ± 4.56	328.76 A ± 10.09
47	RED SPUR	65.101 ST ± 6.28	1088.67 P ± 10.98
48	ROME BEAUTY	96.025 PQR ± 9.03	730.73 VWX ± 8.12
49	ROYAL DELICIOUS	87.085 R ± 7.79	547.12 Z ± 10.98
50	SCHLOMIT	98.579 PQ ± 8.76	1252.15 KL ± 17.86
51	SHIREEN	87.69 R ± 6.98	1371.73 IJ ± 18.78
52	SILVER SPUR	129.172 LK ± 7.46	1000.94 Q ± 18.78
53	SPARTAN	90.711 QR ± 8.90	782.16 TUVW ± 8.09
54	STARKRIMSON	38.951 XYZ ± 3.02	921.98 QRS ± 10.57
55	STARK EARLIEST	302.247 C ± 8.98	967.42 Q ± 10.45
56	STARKING DELICIOUS	109.75 NO ± 9.05	745.32 UVWX ± 7.16
57	TOP RED	167.432 H ± 9.14	361.08 A ± 9.08
58	TROPICAL BEAUTY	58.411 TU ± 4.78	1130.16 NOP ± 15.98
59	WEALTHY APPLE	179.283 G ± 6.98	1210.38 LMN ± 14.23
60	WELL SPUR	309.153 C ± 9.12	816.61 TUV ± 9.76

2.2. Phytochemical Determinations

Quantification of Catechins and Rutin in Apple Genotypes by Reverse Phase-High Performance Liquid Chromatography (RP-HPLC)

Three types of polyphenol were detected in apple samples representing sixty apple genotypes (Table 2). Catechins and rutin were the predominant bioactive compounds in all apple genotypes. The maximum catechin content (5290.47 (μg/g)) was observed in the apple genotype *Malus floribunda* followed by Ambri (2028.94), and the minimum catechin content (238.57 (μg/g)) was observed in Neema Delicious, while the maximum (483.89 (μg/g)) and minimum (12.13 (μg/g)) rutin contents were found in Lal Ambri and June Eating, respectively.

Table 2. Variability in antioxidant efficacy of diverse apple genotypes. The data is represented in mean $\pm$ SD (n = 10) and letter in the superscript symbolize the letters of significance with respect to each other using Tukey' test.

S.No.	VARIETIES	DPPH (%)	FRAP (μmol Fe^{+2} g^{-1} FW)	FLAVONOIDS (QEA)	FLAVANOLS QEA (mg/g)	PHENOLS GAE
1	VISTA BELLA	30.47 $\pm$ 7.23	1.48 $\pm$ 0.17	0.178 $\pm$ 0.09	0.185 $\pm$ 0.06	545 $\pm$ 14.56
2	TYDEMANS EARLY WORCESTOR	29.14 $\pm$ 3.28	0.504 $\pm$ 0.16	0.163 $\pm$ 0.04	0.122 $\pm$ 0.03	722 $\pm$ 23.89
3	BENONI	19.08 $\pm$ 1.86	0.062 $\pm$ 0.006	0.147 $\pm$ 0.07	0.141 $\pm$ 0.06	523 $\pm$ 14.68
4	MICHAL	36.12 $\pm$ 8.34	1.75 $\pm$ 0.15	0.063 $\pm$ 0.005	0.021 $\pm$ 0.003	174.8 $\pm$ 9.89
5	SUMMER RED	50.2 $\pm$ 9.32	1.005 $\pm$ 0.90	0.201 $\pm$ 0.10	0.114 $\pm$ 0.04	395.1 $\pm$ 10.98
6	LEMON GUARD	24.35 $\pm$ 3.78	0.039 $\pm$ 0.002	0.078 $\pm$ 0.0021	0.064 $\pm$ 0.01	356.1 $\pm$ 9.87
7	LAXTONS FURTUNE	34.46 $\pm$ 8.20	0.03 $\pm$ 0.002	0.0627 $\pm$ 0.0018	0.032 $\pm$ 0.006	272.7 $\pm$ 7.67
8	MAYAAN	29.54 $\pm$ 3.19	0.049 $\pm$ 0.005	0.105 $\pm$ 0.09	0.108 $\pm$ 0.02	618.9 $\pm$ 20.90
9	JUNE EATING	36.63 $\pm$ 3.56	0.297 $\pm$ 0.080	0.112 $\pm$ 0.07	0.13 $\pm$ 0.01	583.9 $\pm$ 15.10
10	MOLLIES DELICIOUS	47.87 $\pm$ 8.10	0.437 $\pm$ 0.097	0.045 $\pm$ 0.009	0.012 $\pm$ 0.04	311.2 $\pm$ 9.12
11	PRIMA	38.83 $\pm$ 3.78	1.348 $\pm$ 0.13	0.057 $\pm$ 0.007	0.039 $\pm$ 0.007	160.8 $\pm$ 7.89
12	STAR SUMMER GOLD	39.1 $\pm$ 9.01	0.485 $\pm$ 0.078	0.024 $\pm$ 0.0014	0.006 $\pm$ 0.0005	85.7 $\pm$ 5.16
13	BLACK BEN DAVIS	49.94 $\pm$ 8.98	0.398 $\pm$ 0.056	0.03 $\pm$ 0.0011	0.009 $\pm$ 0.0011	106.6 $\pm$ 4.56
14	GALA MAST	40.85 $\pm$ 8.08	0.311 $\pm$ 0.049	0.164 $\pm$ 0.0012	0.008 $\pm$ 0.00013	201 $\pm$ 8.08
15	AKBER	32.75 $\pm$ 2.98	1.41 $\pm$ 0.12	0.1364 $\pm$ 0.005	0.151 $\pm$ 0.00012	150.3 $\pm$ 7.67
16	RED BARON	17.72 $\pm$ 1.17	0.494 $\pm$ 0.054	0.087 $\pm$ 0.006	0.054 $\pm$ 0.0065	335.7 $\pm$ 9.09
17	FANNY	66.21 $\pm$ 8.76	1.082 $\pm$ 0.70	0.1225 $\pm$ 0.0010	0.075 $\pm$ 0.0045	115.4 $\pm$ 5.46
18	MALLUS BACCATA	37.05 $\pm$ 3.15	1.551 $\pm$ 0.18	0.3584 $\pm$ 0.0054	0.13 $\pm$ 0.0052	108.4 $\pm$ 6.12
19	FUJI	51.06 $\pm$ 6.45	1.563 $\pm$ 0.16	0.108 $\pm$ 0.0045	0.092 $\pm$ 0.0034	174.8 $\pm$ 8.10
20	VANCE DELICIOUS	44.66 $\pm$ 7.67	2.132 $\pm$ 0.18	0.0836 $\pm$ 0.0054	0.068 $\pm$ 0.0059	562.9 $\pm$ 14.34
21	COE RED FUJI	9.04 $\pm$ 1.23	0.462 $\pm$ 0.013	0.147 $\pm$ 0.009	0.057 $\pm$ 0.0008	281.5 $\pm$ 6.78
22	COOPER IV	35.69 $\pm$ 2.17	2.995 $\pm$ 0.034	0.196 $\pm$ 0.007	0.078 $\pm$ 0.0004	169.6 $\pm$ 5.78
23	GRANNY SMITH	18.34 $\pm$ 1.09	0.51 $\pm$ 0.010	0.0455 $\pm$ 0.0078	0.021 $\pm$ 0.00015	648.6 $\pm$ 19.87
24	AMBRI	75.73 $\pm$ 9.04	2.241 $\pm$ 0.134	0.360 $\pm$ 0.030	0.114 $\pm$ 0.0005	244.8 $\pm$ 5.76
25	LAL AMBRI	77.57 $\pm$ 9.87	2.639 $\pm$ 0.167	0.192 $\pm$ 0.0076	0.171 $\pm$ 0.0009	232.5 $\pm$ 5.10
26	RED DELICIOUS	14.69 $\pm$ 1.02	0.498 $\pm$ 0.010	0.0195 $\pm$ 0.006	0.005 $\pm$ 0.00012	75.2 $\pm$ 8.98
27	MALLUS FLORIBUNDA	70.32 $\pm$ 6.78	1.377 $\pm$ 0.13	0.387 $\pm$ 0.045	0.144 $\pm$ 0.0020	980.8 $\pm$ 24.56
28	AMARTARAPRIDE	38.65 $\pm$ 3.10	2.313 $\pm$ 0.178	0.201 $\pm$ 0.098	0.131 $\pm$ 0.007	212.6 $\pm$ 7.65
29	ANANAS RETRINE	10.52 $\pm$ 2.05	0.081 $\pm$ 0.009	0.159 $\pm$ 0.005	0.162 $\pm$ 0.0065	455.2 $\pm$ 13.20
30	ANTINOVIKA	40.78 $\pm$ 3.19	1.434 $\pm$ 0.19	0.049 $\pm$ 0.0012	0.018 $\pm$ 0.0015	383.6 $\pm$ 8.79
31	BELLE DE BESCOPE	49.43 $\pm$ 4.14	1.185 $\pm$ 0.16	0.1505 $\pm$ 0.004	0.121 $\pm$ 0.006	248.3 $\pm$ 5.87
32	BISSBEE SPUR	63.83 $\pm$ 6.76	2.433 $\pm$ 0.198	0.1188 $\pm$ 0.007	0.096 $\pm$ 0.0006	166.1 $\pm$ 4.98
33	CHAUBATIA AMBROSE	35.92 $\pm$ 2.98	0.092 $\pm$ 0.187	0.1406 $\pm$ 0.006	0.127 $\pm$ 0.0004	493.7 $\pm$ 14.1
34	CHECK AMBRI	49.29 $\pm$ 4.09	1.617 $\pm$ 0.20	0.1938 $\pm$ 0.009	0.103 $\pm$ 0.0006	577.6 $\pm$ 15.78
35	FIRDOUS	46.51 $\pm$ 7.89	2.186 $\pm$ 0.12	0.1792 $\pm$ 0.007	0.138 $\pm$ 0.0005	209.8 $\pm$ 6.78
36	GREEN SLEEVES	40.27 $\pm$ 3.10	0.793 $\pm$ 0.070	0.0899 $\pm$ 0.006	0.027 $\pm$ 0.0008	418.5 $\pm$ 10.89
37	HARDIMAN	30.83 $\pm$ 2.56	2.232 $\pm$ 1.23	0.078 $\pm$ 0.0009	0.036 $\pm$ 0.0007	541.6 $\pm$ 14.23
38	JONICA	13.07 $\pm$ 3.04	2.331 $\pm$ 1.40	0.0875 $\pm$ 0.0008	0.048 $\pm$ 0.0008	199.3 $\pm$ 6.89
39	MAHARAJI	19.5 $\pm$ 3.98	1.142 $\pm$ 0.25	0.135 $\pm$ 0.009	0.06 $\pm$ 0.0009	653.8 $\pm$ 20.78
40	NEEMA DELICIOUS	14.16 $\pm$ 2.98	0.446 $\pm$ 0.065	0.07 $\pm$ 0.006	0.045 $\pm$ 0.0007	618.9 $\pm$ 18.89
41	ORANGE VAL	30.63 $\pm$ 2.98	0.102 $\pm$ 0.004	0.178 $\pm$ 0.0069	0.351 $\pm$ 0.054	171.7 $\pm$ 5.23
42	OREGON SPUR	47.27 $\pm$ 8.05	3.750 $\pm$ 0.198	0.213 $\pm$ 0.099	0.129 $\pm$ 0.006	342.7 $\pm$ 6.78
43	PRINCE NOBLE	28.68 $\pm$ 3.98	0.535 $\pm$ 0.078	0.036 $\pm$ 0.0078	0.021 $\pm$ 0.0008	103.1 $\pm$ 7.12
44	RED CHIEF	40.02 $\pm$ 2.98	2.022 $\pm$ 0.098	0.12 $\pm$ 0.007	0.087 $\pm$ 0.0007	218.5 $\pm$ 4.98
45	RED FUJI	24.63 $\pm$ 3.43	1.096 $\pm$ 0.007	0.099 $\pm$ 0.0006	0.002 $\pm$ 0.0001	540.6 $\pm$ 13.98
46	RED GOLD	38.64 $\pm$ 2.99	1.163 $\pm$ 0.19	0.0924 $\pm$ 0.0006	0.092 $\pm$ 0.0003	225.9 $\pm$ 8.10
47	RED SPUR	38.07 $\pm$ 3.19	1.396 $\pm$ 0.17	0.117 $\pm$ 0.0004	0.072 $\pm$ 0.0002	141.6 $\pm$ 3.52
48	ROME BEAUTY	30.02 $\pm$ 1.98	1.384 $\pm$ 0.14	0.06 $\pm$ 0.0004	0.024 $\pm$ 0.0001	290.2 $\pm$ 5.23
49	ROYAL DELICIOUS	10.45 $\pm$ 2.34	1.610 $\pm$ 0.20	0.101 $\pm$ 0.0005	0.061 $\pm$ 0.0005	342.5 $\pm$ 6.89
50	SCHLOMIT	14.06 $\pm$ 3.78	0.028 $\pm$ 0.056	0.084 $\pm$ 0.0005	0.315 $\pm$ 0.0064	356.8 $\pm$ 7.23
51	SHIREEN	35.27 $\pm$ 2.54	0.511 $\pm$ 0.070	0.1845 $\pm$ 0.0009	0.126 $\pm$ 0.006	664.0 $\pm$ 21.09
52	SILVER SPUR	22.22 $\pm$ 3.19	0.469 $\pm$ 0.067	0.0952 $\pm$ 0.00067	0.063 $\pm$ 0.00056	536.7 $\pm$ 12.87
53	SPARTAN	37.98 $\pm$ 2.98	0.318 $\pm$ 0.045	0.174 $\pm$ 0.0006	0.122 $\pm$ 0.004	218.5 $\pm$ 7.23

Table 2. *Cont.*

S.No.	VARIETIES	DPPH (%)	FRAP (μmol Fe^{+2} g^{-1} FW)	FLAVONOIDS (QEA)	FLAVANOLS QEA (mg/g)	PHENOLS GAE
54	STARKRIMSON	45.95 ± 7.67	0.491 ± 0.07	0.054 ± 0.0004	0.021 ± 0.0015	162.6 ± 4.10
55	STARK EARLIEST	36.56 ± 2.20	2.568 ± 1.87	0.153 ± 0.0007	0.105 ± 0.002	347.9 ± 6.10
56	STARKING DELICIOUS	26.76 ± 4.12	0.931 ± 0.078	0.072 ± 0.0008	0.051 ± 0.0001	31.5 ± 3.04
57	TOP RED	22.78 ± 3.10	2.427 ± 1.87	0.091 ± 0.00067	0.141 ± 0.0006	533.9 ± 14.56
58	TROPICAL BEAUTY	50.06 ± 9.67	2.334 ± 1.23	0.063 ± 0.00023	0.039 ± 0.00013	564.3 ± 16.10
59	WEALTHY APPLE	48.51 ± 8.17	0.331 ± 0.043	0.027 ± 0.0067	0.009 ± 0.0001	181.8 ± 4.13
60	WELL SPUR	38.40 ± 3.19	2.010 ± 0.005	0.165 ± 0.0014	0.096 ± 0.00013	129.4 ± 1.43
CD		0.911	0.053	0.005	0.004	11.915
SE(d)		0.459	0.027	0.002	0.002	6.009
SE(m)		0.325	0.019	0.002	0.001	4.249
CV		1.550	2.752	2.245	2.881	2.162

2.3. Determination of the Antioxidant Potential of Apple Genotypes

With reference to the antioxidant potential of apple samples, the DPPH assay exhibited a minimum of 10.45% scavenging activity in Royal Delicious and a maximum of 77.57% in Michal, with Star Summer Gold presenting 39.10% scavenging activity on average (Table 1). Neema Delicious, Maharaji, and Ananas Retrine exhibited the lowest DPPH activity, while Benoni, Luxtons Fortune, Mayan and Chaubatia Ambrose exhibited the lowest FRAP activity. The antioxidant potential in terms of DPPH activity was found to be consistent among the apple genotypes, while the FRAP assay showed large variation among apple genotypes.

2.4. Correlation between Polyphenol Content and Antioxidant Assay

The antioxidant potential estimated by the DPPH assay, which involves an electron transfer mechanism from polyphenols, such as catechins and rutin, to DPPH, showed significant relations with rutin ($r = 0.14424$) and catechins ($r = 0.7348$). The FRAP assay, which measures the reducing potential of apple samples, also showed significant correlations with rutin ($r = 0.244$) and catechins ($r = 0.9067$). The most significant correlation was observed between FRAP and catechins ($r = 0.9067$) (Table 3).

Table 3. Correlation matrix between total phenolics, diphenyl-2-picrylhydrazyl (DPPH), and the ferric reducing antioxidant power (FRAP) with catechin and rutin in diverse apple genotypes.

	DPPH	PHENOLS	FLAVANOLS	FLAVONOIDS	FRAP	RUTIN	CATECHIN
DPPH		0.3789	0.0782	0.3528	0.3238	0.14424	0.7348
PHENOL			0.3023	0.2616	0.4241	0.8851	0.8614
FLAVANOLS				0.79049	0.3082	0.7655	0.6642
FLAVONOIDS					0.1899	0.802	0.6534
FRAP						0.24479	0.9067

2.5. Principal Component and Hierarchical Cluster Analysis of Bioactive Molecules and Antioxidant Assays

All observations recorded from the sixty apple genotypes were subjected to principal component analysis (PCA). The first three components explained 72.75% of the total variation (PC1 = 32.05, PC2 = 21.71%, and PC3 = 19.63%, respectively, Figure 1). The first principal component (PC1) was mainly contributed to by rutin and catechins, and PC3 was linked with the antioxidant assays DPPH (40.84%) and FRAP (67.33%). The PCA scatter plot revealed the distribution between 60 apple genotypes of diverse origins. The results obtained showed a comparatively discrete distribution of data points, thereby ascertaining that native apple genotypes, e.g., Ambri, Check Ambri, Lal Ambri, and Maharaji, together with wild ones such as *Malus baccata* and *Malus floribunda*, exhibit an extensive range of total antioxidant potential. The results for the native apple genotypes Ambri, Lal Ambri, and Red Delicious significantly deviated from those of other genotypes and also displayed the highest catechin and rutin contents and FRAP activity subsequent to the wild genotypes *Malus baccata* and *Malus floribunda*. The presence of the longest diagonal interception between catechins and rutin demonstrates that the higher disparity among apple genotypes and phenolic compounds varied significantly as a result of the apples' diverse genetic backgrounds. Catechins and DPPH were closely associated, whereas rutin, flavanoids, flavanols, phenols, and FRAP were dispersed over the whole scatter plot, displaying a highly intricate association with regard to genotype.

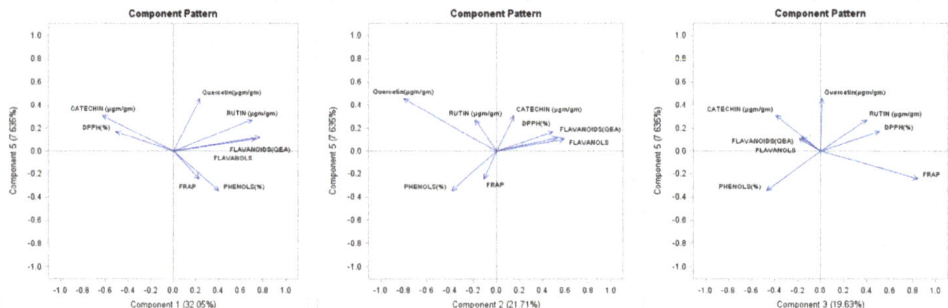

Figure 1. Principal component analysis showing the variability in phenolics and varied antioxidant assays among different apple varieties.

Hierarchical Cluster analysis (HCA) analysis was carried out to evaluate the similarities between 60 apple genotypes, which were categorized into two main clusters (Figure 2): Cluster 1, which was characterized by relatively high FRAP activity, and Cluster 2, which was characterized by high levels of catechins and total phenols and had comparatively high DPPH activity. Cluster 2 was further divided into two clusters, with the wild genotype *Malusbaccata* exhibiting a close association with Red Delicious. Ambri and Lal Ambri shared the same sub-cluster under Cluster 1. Cluster 2 was divided into a sub-cluster bearing only the wild genotype *Malus Floribunda* and another sub-cluster comprising two clusters with genotypes such as Granny Smith, Chaubatia Ambrose, Check Ambri, and hybrid Shireen, which exhibited high FRAP activity.

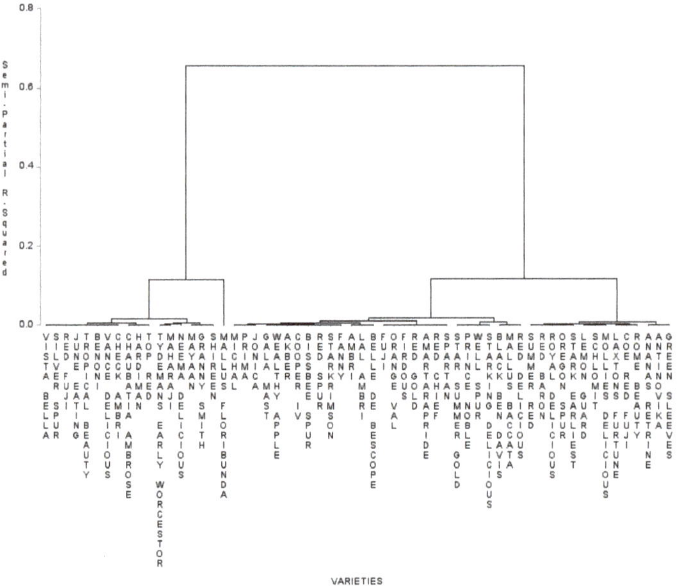

Figure 2. Schematic representation of 60 apple genotypes displaying the metabolically rich Cluster 2 (phenol rich/high DPPH), Cluster 5 (flavonoid rich/high FRAP), and Cluster 6 (quercetin rich) with their respective antioxidant potential values.

3. Discussion

The association between the prospective health effects of apples and their flavor and aroma still remains very controversial. It would be exciting if the most flavorsome apple extract offered the most abundant bioactive metabolites and the highest antioxidant potential [16]. In the present research, in terms of the general quality of the extracts of diverse botanical origin, Cluster 2 presented the highest antioxidant potential and highest values for catechins and rutin, as estimated by both HPLC and UV-Visible spectrophotometry. Rutin is known to increase in some plants as the main strategy to protect against microclimatic variations [17]. The pronounced accumulation of rutin and catechins in our indigenous apple genotype, Ambri, puts it among the most desirable apple genotypes. Although the market demand for Ambri is still low, our data corroborate the possible health effects of extract. Even though the wild genotypes *Malus floribunda* and *Malusbaccata* accumulate catechins and rutin at maximum levels, due to their unpalatability as a result of high bitterness and astringency, they have low market acceptability [18]. However, these genotypes can function as parents to transfer desirable traits to diverse genotypes. Breeding programs with the aim of introgressing such fruits to commercial genotypes from wild species will play an important role in the development of new genetic material with higher contents of catechins and rutin. Additionally, the identification of molecular markers that are tightly linked to traits, such as high catechin and rutin content, could also pave the way for marker-assisted breeding programs in apples to transfer traits such as high catechin and rutin content. Hence, in this regard, it can be ascertained that Ambri containing a high content of polyphenols can be perceived as a nutritionally valuable extract.

To better understand the data obtained and the functionality of the apple samples selected in this study, PCA and WCA were employed (Figures 1 and 2). Using two-dimensional projections of apple genotypes procured by PCA, the clustering of extracts from the apples based on chemical composition and antioxidant power was likely; this projection was able to elucidate up to 75.93% of data variability. Apple extract exhibited the highest contents of total phenolic compounds, rutin, and catechins as well as the greatest antioxidant capacity in comparison with other extracts [19,20]. When WCA was applied to the whole data set and a Euclidean distance of 10.5 was considered, four different clusters were identified (Figure 2). Cluster 1 exhibited high levels of flavonoids and flavanols and low levels of phenolics as well as relatively high FRAP activity. Cluster 2 was characterized by high levels of catechins, total phenols and comparatively high DPPH activity. When variables were used for WCA, inferences about the associations among all responses could be made. The DPPH seemed to be highly associated with catechins, and rutin was strongly correlated with FRAP activity.

Although we did not conduct a clinical study on apple extracts, research studies have determined the medicinal value of apples. The clinical significance of quercetin and catechins has been correlated with the antioxidant potential and radical scavenging activity of the apple samples. There is a significant disparity between the composition of phytochemicals such as catechins and quercetin in the different varieties of apple depending on the maturation and ripening stage of apple fruit [18]. Different research experiments have determined the relationship between the incidences of various diseases, such as cancer and coronary mortality, with daily intake of apples. There is a significantly inverse relationship between flavonoid intake and lung cancer development. The relationship between dietary catechins and epithelial cancer has been associated with 87% of the total catechin intake while apples contribute to 8.0% of catechin consumption [21].

Granato et al. [20] evaluated the chemical composition and antioxidant activity of Brazilian red wines by DPPH and FRAP assays and confirmed that flavonoids are the main phenolic class driving the antioxidant capacity. Further, HCA was applied to the variables and a quantitative measure of the degree of association between variables was performed using Pearson correlation coefficients; the results are presented in Table 3. Principal component analysis (PCA) is used to emphasize the variation and demonstrate strong patterns in a dataset [21]. PCA, using two antioxidant assays (DPPH and FRAP) and three antioxidants (catechins and rutin), showed that the first two components explained 78.31% of the total variation. Principal component (PC1) accounted for 83.04% of the total variation and the variables responsible for separation along the PC1 included DPPH (0.31), TP (0.25), rutin (0.43), and FRAP (0.40), and catechins (0.39).

The hierarchical cluster analysis grouped 60 apple genotypes in clusters based on the antioxidant activity and anti-oxidative bioactive metabolites (Figure 2). The number of genotypes in the cluster varied from 2 (Cluster 4) to 24 (Cluster 2). The contribution of individual genotypes to the antioxidant grouping and the relationships between the clusters were assessed by plotting PC1 and PC2. PC1 and PC2 contributed average catechin values of 104.567 and 107.756, respectively, while PC3 was found to be rich in catechins, exhibiting a value of 111.859. On the contrary, PC2 was rich in rutin, displaying a value of 12.08, and PC3 was rich in catechins and antioxidants, 111.859 and 43.213, respectively. PC2 and PC4 were equal in terms of the average value of rutin accumulation in apple genotypes. Cluster 2, which was divided into sub-clusters, formed a separate group on the bi-plot and appeared to be distant from both Clusters 1 and 4. Like hierarchical clustering, accessions from cluster 4 grouped separately and formed a distant group on the PCA plot. Both PCA and cluster analysis were found to be equally effective for grouping the apple genotypes based on their antioxidant contents. The wild genotype *Malus floribunda* from Cluster 2 should be promoted for apple breeding programs and consumption because of its high antioxidant activities. Similar groupings of crop plants according to their antioxidant potential by means of cluster analysis and PCA have been done by other researchers worldwide [20,22].

4. Materials and Methods

4.1. Materials

A total of 60 apple genotypes, originating from different geographical areas but cultivated under similar conditions, were obtained from the research farm of the Central Institute of Temperate Horticulture, Srinagar, India. The reagents used were Folin–Ciocalteau and DPPH (2,2-diphenyl-2-picrylhydrazyl). Methanol, acetone, and acetic acid were purchased from Hi media (Phillipsburg, NJ, USA). The aqueous solutions were prepared using ultra-pure water (Milli-Q, Millipore, São Paulo, SP, Brazil).

4.2. Methods

4.2.1. Preparation of Extracts

Extraction of Phenolic Compounds (Catechin and Rutin)

The fragmentation of apples 10 fruits for each variety was carried out in a mortar and pestle, and fragments were immediately frozen with liquid nitrogen (1:2, w/v) in order to avoid oxidation of the phenolic compounds [23]. Homogenization of freeze-dried material without seeds was carried out by crushing in a mortar. An amount of 10 g of the powdered apple was transferred in Oakridge tubes and mixed with 10 mL of methanol or acetone in different concentrations, which was followed by incubation at −10 °C for 10 min. The mixture was centrifuged (8160× g, 20 min at 4 °C) (Sigma 3–30 K, Munich, Germany), concentrated by evaporation under vacuum (40 °C) in a rotary evaporator (IKA, HB-10, Germany), and freeze dried. The samples were reconstituted with 2 mL of 2.5% acetic acid and methanol (3:1, v/v) and filtered through a 0.22 µm (Nylon, Mumbai, India) syringe filter (Moxcare, Haryana, India) prior to analysis.

4.2.2. HPLC Analysis

The HPLC analysis of samples was carried out in a Shimadzu HPLC (Kyoto, Japan) equipped with quaternary pumps, a degasser coupled to a photo-diode-array detector, and an injection valve with a 20 µL loop. An injection volume of 20 µL and a flow rate of 1.0 mL min^{-1} with 1 h of run time were used for the separation process. The analysis was carried out in triplicate for each sample. Chromatographic separations were performed on C18 (250 × 4.6 mm) with a 5 µm column using a solvent system in gradient mode followed by isocratic run, as represented in Table 4. The filtration of the mobile phase was done through a 0.45 μm membrane filter (Millipore, Bedford, MA, USA) and was subjected to 40 min ultrasonication. Instrument control, data acquisition, and data processing were done by using Class WP software (version 6.1) from Shimadzu (Columbia, SC, USA). Quantitative determinations were made by taking into account the respective peak areas of standards at a particular retention time versus the concentration and are expressed in mg/g of apple fruit.

Table 4. The table represents mobile phase and their gradient mode.

Compound	Mobile Phase	Gradient	λ_{max}
Rutin Catechins	Solvent A—2.5% acetic acid Solvent B—aceto-nitrile	3–9% B (0–5 min) 9–16% B (5–15 min) 16–36.4% B (15–33 min) 100% B (5 min)	280 320 350

4.2.3. Determination of Total Polyphenolic Content (TPC)

The total phenol contents of extracts from different apple genotypes were determined by the modified Folin-Ciocalteau method [24]. Absorbance was then measured at 765 nm using a spectrophotometer. The results are expressed as mg of gallic acid equivalents μg/g FW.

4.2.4. Determination of Total Flavonoid and Flavonol Contents

The total flavonoid content in apples (fruit) was determined using the method of Chang, et al. [25]. The absorbance was then measured at 415 nm using a spectrophotometer (Shimadzu, Columbia, SC, USA). Results are expressed in terms of the quercetin equivalent (mg/g). The same method was employed for flavonol determination, but the incubation period was 150 min instead of 40 min, and the absorbance was measured at 440 nm. The total flavonol content was also expressed in terms of the quercetin equivalent (mg/g).

4.3. Antioxidant Activity

4.3.1. Ferric Reducing Antioxidant Potential (FRAP) Assay

The FRAP assay was done using the Benzie and Strain method with minor modifications [26]. The absorbance was then measured at 593 nm after 40 min. $FeSO_4$ solution (0, 40, 80,160, 320, 640 μmol/L) was used for calibration of the standard curve. The results are expressed as μmol Fe^{+2} g^{-1} FW.

4.3.2. DPPH (2,2-diphenyl-1-picrylhydrazyl) Scavenging Activity

DPPH free radical scavenging assay was measured using the procedure with slight modifications (Xu et al., 2012). Percentage inhibition was calculated by the formula

$$(\%IP) = [(A_{t=0} - A_{t=15})]/(A_{t=0}) \times 100 \tag{1}$$

where $A_{t=15}$ is the absorbance of the test sample after 15 min, and $A_{t=0}$ is the absorbance of the control after 15 min.

Furthermore, the scavenging activity percentage (AA%) was determined.

$$AA\% = 100 - [(Abs_{sample} - Abs_{control})/Abs_{blank} \times 100] \tag{2}$$

where a mixture of methanol and DPPH in the ratio of 1:1 served as a blank, and a mixture of the standard (ascorbic acid) and DPPH in the ratio of 1:1 was used as the control. Here, the concentration of the test sample as well as that of the standard used was 15 μg/mL.

4.4. Statistical Analysis

In this study, the data was subjected to various statistical tests, such as cluster analysis, and correlations were determined to ascertain the superlative genotypes exhibiting high antioxidant potential estimated via DPPH and FRAP assays. All experiments were carried out in triplicate. The results are shown as mean values and standard error of the mean. The existence of significant differences among the results for total catechin and rutin contents was determined. The results obtained were subjected to one-way analysis of variance (ANOVA) and Duncan's test. All statistical tests were done using SAS Enterprise Guide 4.2, SPSS 13 (SPSS Inc., New Orchard Road, NY, USA) and OP-STAT software (2.0, IBM, New Orchard Road, NY, USA) at a 5% significance level. Correlation analysis was carried out using Pearson's test.

5. Conclusions

Overall, by using correlation analysis and multivariate statistical techniques (WCA and PCA), we verified that the phenolic compounds catechins and rutin are involved in the antioxidant activity of the commercial extract under study, although the contribution of other phytochemicals cannot be excluded. Apple samples from Ambri, Lal Ambri, and Red Delicious genotypes presented the highest antioxidant activity, as measured by antioxidant assays. On the other hand, apple extracts from Orange Val, Royal Delicious, and AnanasRetrine had the lowest DPPH and FRAP values. In this sense, the utilization of unsubstantiated statistical techniques, coupled to the ANOVA procedure, was demonstrated to be a suitable approach for evaluating the quality of commercial fruit extract based on various analytical measurements.

Author Contributions: Conceptualization, W.S. and J.I.M.; methodology, S.M.; software, M.K., M.R., M.A.; validation, M.R., M.A. and P.A.; formal analysis, S.J., D.B.S.; investigation, W.S.; data curation, J.I.M.; writing—original draft preparation, W.S., S.M., S.J.; writing—review and editing, S.J., P.A.; supervision, J.I.M.; project administration, J.I.M.; funding acquisition, M.K., M.R., M.A.

Funding: Research group number (RG-1435-017).

Acknowledgments: The authors are thankful to ICAR for supporting this research project. The authors would also like to extend their sincere appreciation to the Deanship of Scientific Research at King Saud University for its funding through research group number (RG-1435-017).

Conflicts of Interest: The authors declare no conflict of interest.

References

1. Alberti, A.; Zielinski, A.A.F.; Zardo, D.M.; Demiate, I.M.; Nogueira, A.; Mafra, L.I. Optimisation of the extraction of phenolic compounds from apples using response surface methodology. *Food Chem.* **2014**, *149*, 151–158. [CrossRef] [PubMed]
2. Braga, C.M.; Zielinski, A.A.F.; Silva, K.M.d.; de Souza, F.K.F.; Pietrowski, G.d.A.M.; Couto, M.; Granato, D.; Wosiacki, G.; Nogueira, A. Classification of extracts and fermented beverages made from unripe, ripe and senescent apples based on the aromatic profile using chemometrics. *Food Chem.* **2013**, *141*, 967–974. [CrossRef] [PubMed]
3. Shahidi, F. Nutraceuticals, functional foods and dietary supplements in health and disease. *J. Food Drug Anal.* **2012**, *20*, 226–230.
4. Khanizadeh, S.; Tsao, R.; Rekika, D.; Yang, R.; Charles, M.T.; Rupasinghe, H.V. Polyphenol composition and total antioxidant capacity of selected apple genotypes for processing. *J. Food Comp. Anal.* **2008**, *21*, 396–401. [CrossRef]
5. Wijekoon, M.M.J.O.; Bhat, R.; Karim, A.A. Effect of extraction solvents on the phenolic compounds and antioxidant activities of bunga kantan (*Etlingera elatior* Jack.) inflorescence. *J. Food Comp. Anal.* **2011**, *24*, 615–619. [CrossRef]
6. Kchaou, W.; Abbès, F.; Blecker, C.; Attia, H.; Besbes, S. Effects of extraction solvents on phenolic contents and antioxidant activities of Tunisian date varieties (*Phoenix dactylifera* L.). *Ind. Crops Prod.* **2013**, *45*, 262–269. [CrossRef]
7. Haminiuk, C.W.I.; Maciel, G.M.; Plata-Oviedo, M.S.V.; Peralta, R.M. Phenolic compounds in fruits—An overview. *Int. J. Food Sci. Tech.* **2012**, *47*, 2023–2044. [CrossRef]

8. Luthria, D.L. Influence of experimental conditions on the extraction of phenolic compounds from parsley (*Petroselinum crispum*) flakes using a pressurized liquid extractor. *Food Chem.* **2008**, *107*, 745–752. [CrossRef]

9. Kroll, J.; Rawel, H.M.; Rohn, S. Reactions of Plant Phenolics with Food Proteins and Enzymes under Special Consideration of Covalent Bonds. *Food Sci. Technol. Res.* **2003**, *9*, 205–218. [CrossRef]

10. Escribano-Bailon, M. Polyphenol Extraction from Foods. In *Methods in Polyphenol Analysis*; Saltmarsh, M., Santos-Buelga, C., Williamson, G., Eds.; Royal Society of Chemistry: Cambridge, UK, 2003; pp. 1–12.

11. Soares, M.C.; Ribeiro, É.T.; Kuskoski, E.M.; Gonzaga, L.V.; Lima, A.; Mancini Filho, J.; Fett, R. Composição do conteúdo de ácidos fenólicos no bagaço de maçã (*Malus sp*). *Semina: Ciências Agrárias* **2008**, *29*, 339. [CrossRef]

12. Masuda, I. Apple polyphenols protect cartilage degeneration through modulating mitochondrial function in mice. *Osteoarthritis Cartilage* **2016**, *24*, S354–S355. [CrossRef]

13. Chandrasekar, V.; Martín-González, M.F.S.; Hirst, P.; Ballard, T.S. Optimizing Microwave-Assisted Extraction of Phenolic Antioxidants from Red Delicious and Jonathan Apple Pomace. *J. Food Process. Eng.* **2015**, *38*, 571–582. [CrossRef]

14. Łata, B.; Trampczynska, A.; Paczesna, J. Cultivar variation in apple peel and whole fruit phenolic composition. *Sci. Hort.* **2009**, *121*, 176–181. [CrossRef]

15. Wang, X.; Li, C.; Liang, D.; Zou, Y.; Li, P.; Ma, F. Phenolic compounds and antioxidant activity in red-fleshed apples. *J. Func. Foods* **2015**, *18*, 1086–1094. [CrossRef]

16. Francini, A.; Sebastiani, L. Phenolic Compounds in Apple (*Malus x domestica* Borkh.): Compounds Characterization and Stability during Postharvest and after Processing. *Antioxidants* **2013**, *2*, 181–193. [CrossRef] [PubMed]

17. Thomas, A.L.; Byers, P.L.; Finn, C.E.; Chen, Y.-C.; Rottinghaus, G.E.; Malone, A.M.; Applequist, W.L. In Occurrence of rutin and chlorogenic acid in elderberry leaf, flower, and stem in response to genotype, environment, and season. In Proceedings of the XXVII International Horticultural Congress-IHC2006: International Symposium on Plants as Food and Medicine: The Utilization and Development of Horticultural Plants for Human Health, Seoul, Korea, 13–19 August 2006; pp. 197–206.

18. Gygax, M.; Gianfranceschi, L.; Liebhard, R.; Kellerhals, M.; Gessler, C.; Patocchi, A. Molecular markers linked to the apple scab resistance gene Vbj derived from *Malus baccata* jackii. *Theor. Appl. Genet.* **2004**, *109*, 1702–1709. [CrossRef] [PubMed]

19. Mishra, K.; Ojha, H.; Chaudhury, N.K. Estimation of antiradical properties of antioxidants using DPPH assay: A critical review and results. *Food Chem.* **2012**, *130*, 1036–1043. [CrossRef]

20. Granato, D.; Karnopp, A.R.; van Ruth, S.M. Characterization and comparison of phenolic composition, antioxidant capacity and instrumental taste profile of extracts from different botanical origins. *J. Sci. Food Agric.* **2015**, *95*, 1997–2006. [CrossRef] [PubMed]

21. Iezzoni, A.F.; Pritts, M.P. Applications of principal component analysis to horticultural research. *HortScience* **1991**, *26*, 334–338. [CrossRef]

22. Patras, A.; Brunton, N.P.; Downey, G.; Rawson, A.; Warriner, K.; Gernigon, G. Application of principal component and hierarcahical cluster analysis to classify fruits and vegetables commonly consumed in Ireland based on in vitro antioxidant activity. *J. Food Comp. Anal.* **2011**, *24*, 250–256. [CrossRef]

23. Guyot, S.; Le Bourvellec, C.; Marnet, N.; Drilleau, J.F. Procyanidins are the most Abundant Polyphenols in Dessert Apples at Maturity. *LWT Food Sci. Technol.* **2002**, *35*, 289–291. [CrossRef]

24. Omoruyi, B.E.; Bradley, G.; Afolayan, A.J. Antioxidant and phytochemical properties of *Carpobrotus edulis* (L.) bolus leaf used for the management of common infections in HIV/AIDS patients in Eastern Cape Province. *BMC Comp. Alt. Med.* **2012**, *12*. [CrossRef] [PubMed]

25. Chang, C.-C.; Yang, M.-H.; Wen, H.-M.; Chern, J.-C. Estimation of total flavonoid content in propolis by two complementary colorimetric methods. *J. Food Drug Anal.* **2002**, *10*, 178–182.

26. Benzie, I.F.F.; Strain, J.J. The Ferric Reducing Ability of Plasma (FRAP) as a Measure of "Antioxidant Power": The FRAP Assay. *Anal. Biochem.* **1996**, *239*, 70–76. [CrossRef] [PubMed]

Sample Availability: Not available.

Article

Molecular Docking Studies of Coumarins Isolated from Extracts and Essential Oils of *Zosima absinthifolia* Link as Potential Inhibitors for Alzheimer's Disease

Songul Karakaya [1], Mehmet Koca [2], Serdar Volkan Yılmaz [1], Kadir Yıldırım [1], Nur Münevver Pınar [3], Betül Demirci [4], Marian Brestic [5] and Oksana Sytar [5,6,*]

[1] Department of Pharmacognosy, Faculty of Pharmacy, Ataturk University, 25240 Erzurum, Turkey; ecz-songul@hotmail.com (S.K.); svyilmaz@yandex.com (S.V.Y.); kdryldrm25@gmail.com (K.Y.)

[2] Department of Pharmaceutical Chemistry, Faculty of Pharmacy, Ataturk University, 25240 Erzurum, Turkey; kocamehmet@atauni.edu.tr

[3] Department of Biology, Faculty of Science, Ankara University, 06560 Ankara, Turkey; Nur.M.Pinar@science.ankara.edu.tr

[4] Department of Pharmacognosy, Faculty of Pharmacy, Anadolu University, 26210 Eskisehir, Turkey; betuldemirci@gmail.com

[5] Department of Plant Physiology, Slovak Agricultural University in Nitra, 94976 Nitra, Slovak; marian.brestic@uniag.sk

[6] Department of Plant Biology, Educational and Scientific Center "Institute of Biology and Medicine", Kiev National University of Taras Shevchenko, Hlushkova Avenue, 2, 03127 Kyiv, Ukraine

* Correspondence: oksana.sytar@gmail.com; Tel.: +82-63-570-3167

Academic Editor: Diego Muñoz-Torrero
Received: 18 January 2019; Accepted: 14 February 2019; Published: 17 February 2019

Abstract: Coumarins and essential oils are the major components of the Apiaceae family and the *Zosima* genus. The present study reports anticholinesterase and antioxidant activities of extracts and essential oils from aerial parts, roots, flowers, fruits and coumarins—bergapten (**1**); imperatorin (**2**), pimpinellin (**3**) and umbelliferone (**4**)—isolated of the roots from *Zosima absinthifolia*. The investigation by light and scanning electron microscopy of the structures of secretory canals found different chemical compositions in the various types of secretory canals which present in the aerial parts, fruits and flowers. The canals, present in the aerial parts, are characterized by terpene hydrocarbons, while the secretory canals of roots, flowers and fruits include esters. Novel data of a comparative study on essential oils constituents of aerial parts, roots, flowers and fruits of *Z. absinthfolia* has been presented. The roots and fruits extract showed a high content of total phenolics and antioxidant activity. The GC-FID and GC-MS analysis revealed that the main components of the aerial parts, roots, flowers and fruits extracts were octanol (8.8%), octyl octanoate (7.6%), octyl acetate (7.3%); *trans*-pinocarvyl acetate (26.7%), β-pinene (8.9%); octyl acetate (19.9%), *trans*-*p*-menth-2-en-1-ol (4.6%); octyl acetate (81.6%), and (*Z*)-4-octenyl acetate (5.1%). The dichloromethane fraction of fruit and flower essential oil was characterized by the highest phenolics level and antioxidant activity. The dichloromethane fraction of fruit had the best inhibition against butyrylcholinesterase enzyme (82.27 ± 1.97%) which was higher then acetylcholinesterase inhibition (61.09 ± 4.46%) of umbelliferone. This study shows that the flowers and fruit of *Z. absinthifolia* can be a new potential resource of natural antioxidant and anticholinesterase compounds.

Keywords: Apiaceae; antioxidant; anticholinesterase; essential oil; secretory canals; *Zosima absinthifolia*

1. Introduction

Alzheimer's disease (AD) is a degenerative brain disease and the most widespread reason for dementia. The characteristical symptoms of dementia are troubles with memory, language, problem-solution and other cognitive abilities that influence a person's ability to make daily activities. These troubles happen because nerve cells in parts of the brain involved in cognitive function have been ruined. In AD, neurons in other parts of the brain are finally damaged or destroyed as well, including those that permit a person to perform basic bodily functions such as walking and swallowing. People in the final stages of the disease are bed-bound and require around-the-clock care. AD is eventually fatal [1]. Many factors such as age are risk factors in AD. Due to the ageing population, it is expected that AD will become a serious socio-economic challenge globally in the coming years [2]. With reference to the World Health Organization data, AD, which affects about 47 million people worldwide, is the most pervasive form of dementia (60–80% of all cases) [3] with a proximate worldwide cost of US$818 billion [4]. Oxidative stress, occurring through a damage to neurons or metal accumulation has been related to the pathogenesis of AD. Drugs endowed with anticholinesterase and antioxidant capacity could thus be useful for the prevention/treatment of AD [5].

To overcome the limitations of current therapeutics for AD, extensive research is under way to identify drugs that are both effective and free of undesired side effects. In this context naturally occurring dietary polyphenolic phytochemicals have received remarkable attention as alternative options for AD treatment.

In particular, curcumin, resveratrol, and green tea catechins have been identified to have the potential to prevent AD owing to their anti-amyloidogenic, anti-oxidative, and anti-inflammatory characteristics. These polyphenolic phytochemicals also activate adaptive cellular stress responses, called 'neurohormesis', and supress illness processes [6]. Antioxidants may trap reactive oxygen species (ROS) and break inflammatory pathways. The utilization of antioxidants is useful to delay AD progress [7]. A particular and important preventive action against AD with hop iso-α-acids, which are reponsible for the bitterness in beer, was discussed lately. Besides, proof has appeared for anti-carcinogenic action from hops'prenylflavonoids, as well as from phenolic components extracted from both malt and hops [8]. Numerous investigations were performed on the biological activities of plants which are utilised traditionally as memory enhancers and acetylcholinesterase inhibitors [9,10]. Representatives of the family Apiaceae demonstrate acetylcholinesterase inhibitory activity [10,11]. Natural compounds of phenolic nature have shown a substantial role in the inhibition of acetylcholinesterase enzyme (AChE) [12,13]. Phenolic compounds of medicinal plants and dietary plants are present as coumarins, curcuminoids, flavonoids, lignans, phenolic acids, tannins, stilbenes, quinones, and others. The varied bioactivities of phenolic compounds are responsibe for their AChE inhibition capacities [12,14].

At the same time, essential oils are verified to provide varied pharmacological effects, like antiflatulent, antiviral, antispasmodic, anticarcinogenic, and hepatoprotective effects, etc. Essential oils have been reported to be natural antioxidants and proffered principally as potential substitutes of synthetic antioxidants used the in food conservation sectors. Further, biologically active natural compounds can be used in the pharmaceutical industry to check human sicknesses of microbial origin and treat lipid peroxidative damage, which is observed in certain pathological disorders, such as AD, carcinogenesis, ischemia-reperfusion injury, coronary atherosclerosis, and ageing processes [14]. Antioxidants comprise most of the active ingredients of the $80 billion anti-ageing product market, which is growing at >10% yearly growth rate. Olive polyphenols—whose hydroxytyrosol and verbascoside compounds share the highest grade of antioxidant activity ever reported for any natural compound—can be effectively utilized in for health, appearance enhancement, and fitness purposes as well as in the anti-ageing products market [15].

Representatives of the families Apiaceae and Lamiaceae are characterized by high phenolics content [16] and were demonstrated to have positive effects on the central nervous system [4].

Z. absinthifolia Link is the only member of the *Zosima* genus that grows in Turkey, where it is commonly known as 'ayı eli' or 'peynir otu'. The aerial parts of the plant are used up as a vegetable and added to a traditional cheese in East Anatolia. In folk medicine, fruits of the plant have digestive and sedative effects with anti-inflammatory properties. Moreover, the aerial parts cure dyspepsia, stomach gas, cough and intestinal disorders [17]. Coumarins, such as deltoin and columbianadin, have also been isolated from *Z. absinthifolia* [18]. It has been reported that *Z. absinthifolia* has biological activities such as cytotoxic, antioxidant, antibacterial, anti-inflammatory [19,20] and antimycobacterial effects [21]. Previous phytochemical studies have demonstrated that *Z. absinthifolia* contains alkaloids and coumarins such as deltoin, imperatorin, zosimine, pimpinellin, bergapten, isobergapten, sphondin isopimpinellin, and umbelliferone [17].

The presented research studied the cholinesterase inhibitory, antioxidant activity, and phenolics content of the methanol, hexane, dichloromethane, ethyl acetate, butanol and aqueous extracts and essential oils of aerial parts, roots, flowers and fruits of *Z. absinthifolia*. The AChE and BuChE inhibitory activities of the coumarins bergapten (**1**), imperatorin (**2**), pimpinellin (**3**) and umbelliferone (**4**) isolated from the roots of *Z. absinthifolia* were also assessed through molecular docking studies with parallel investigation of the structures of the plant's secretory canals.

2. Results

The CH$_3$OH extracts of aerial parts, roots, flowers and fruits of *Zosima absinthifolia* were fractionated with the use of different solvents (*n*-hexane, dichloromethane, ethyl acetate and butanol), to give the respective fractions and the individual coumarins bergapten (**1**), imperatorin (**2**), pimpinellin (**3**) and umbelliferone (**4**) isolated from roots which were assayed for antioxidant, acetylcholinesterase and butyrylcholinesterase inhibitory activities. Also, the AChE and BuChE inhibitory activities of the compounds were determined via molecular docking.

The active dichloromethane fraction of root was subjected to column chromatography over silica gel and Sephadex LH-20. As the result, four known coumarins namely, bergapten (**1**) [17], imperatorin (**2**) [17], pimpinellin (**3**) [22] and umbelliferone (**4**) [17] (Figure 1) were isolated and identified in several places before it says these were isolated from the roots—explain.

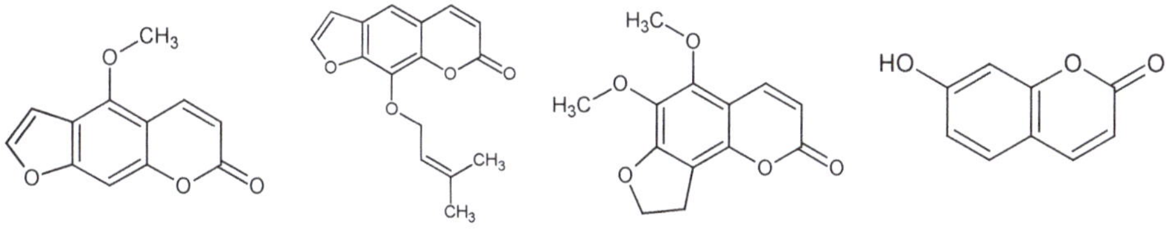

Bergapten (**1**)   Imperatorin (**2**)   Pimpinellin (**3**)   Umbelliferone (**4**)

Figure 1. Chemical structures of compounds **1–4**.

The extracts, fractions and essential oils of aerial parts, roots, flowers and fruits were studied regarding their antioxidant capacity potential. The findings of content of total phenolics from the samples are presented in Figure 2B. The highest level of total phenolics was seen in root and fruit (59.81 and 52.34 mg GAE g^{-1} DW, respectively) while the least content of phenolics was seen in the aerial parts of the plant (34.07 mg GAE g^{-1} DW). DPPH analysis results showed the presence of antioxidant activity in the range from 61.92–69.2% with the highest seen in the fruits compared to the extracts of the aerial parts of the plant (Figure 2A). The antioxidant activity results of extract from different *Z. absinthifolia* parts were quite high compared with the standards propyl gallate, chlorogenic acid, and rutin (Table 1).

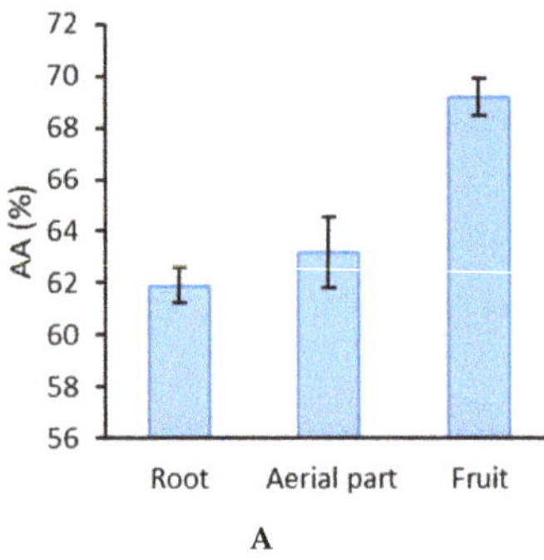
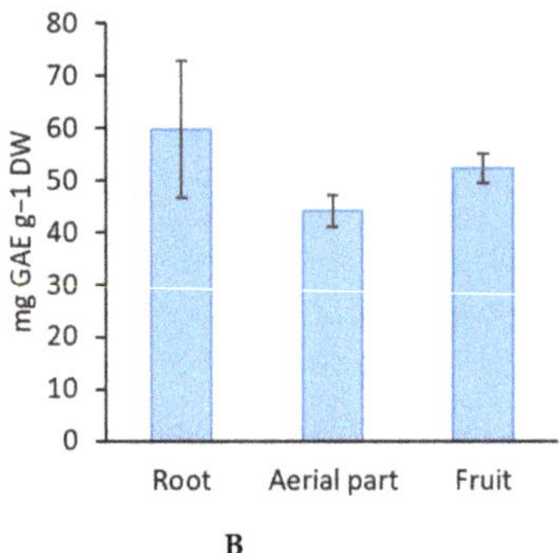

Figure 2. DPPH radical scavenging activity (**A**), total phenolic contents (**B**) of samples.

Table 1. Antioxidant activities of the samples from *Zosima absinthifolia* in thiobarbituric acid (TBA) test.

Tested Samples	IC$_{50}$ Values (µg/mL) ± SD *			
	Aerial Part	**Root**	**Flower**	**Fruit**
MeOH	204.16 ± 2.16	234.21 ± 4.26	199.43 ± 2.46	116.25 ± 3.51
Hexane	266.67 ± 2.97	198.15 ± 1.78	159.54 ± 1.48	500>
CH$_2$Cl$_2$	169.21 ± 4.22	500>	301.53 ± 3.01	48.98 ± 2.45
EtOAc	255.21 ± 3.43	217.99 ± 4.35	478.90 ± 1.78	196.25 ± 2.66
BuOH	500>	367.60 ± 3.21	298.33 ± 3.55	487.45 ± 3.12
Aqueous residue	500>	500>	500>	500>
Essential oils	225.17 ± 3.29	390.36 ± 1.56	97.11 ± 2.25	389.67 ± 2.86
Bergapten	56.99 ± 3.87			
Imperatorin	79.23 ± 3.48			
Pimpinellin	49.23 ± 2.19			
Umbelliferone	79.53 ± 3.98			
Chlorogenic acid	12.98 ± 4.89			
Propyl gallate	3.44 ± 2.05			
Rutin	9.65 ± 3.09			

* Standard deviation.

Table 1 shows the TBA assay results of the specimens reported as IC$_{50}$ values (µg mL^{-1}). The highest antioxidant potential in the TBA assay was seen in the fruit CH$_2$Cl$_2$ fraction and flower essential oil (IC$_{50}$ = 48.98 and 97.11 µg/mL, respectively). Among the isolated compounds pimpinellin and bergapten had strong antioxidant effects, with IC$_{50}$ values of 49.23 and 56.99 µg/mL. Many of samples indicated considerable antioxidant activity on liposomes but not comparable to rutin or chlorogenic acid. The correlation coefficient between antioxidant capacity and content of total phenolics is remarkable (0.96).

The anticholinesterase activity of the samples was assessed by means of the Ellman colourimetric method [23], with a few changes and using commercially available donepezil as reference [24]. The in vitro anti-acetylcholinesterase activities of the specimens at 20 µg/mL are presented in Table 2. The MeOH, hexane, CH$_2$Cl$_2$, EtOAc and BuOH extracts and fractions of essential oils from all plant parts demonstrated significative inhibitory activities towards butyrylcholinesterase. The fruit CH$_2$Cl$_2$ fraction and flower essential oil indicated considerable inhibition against BuChE (82.27 ± 1.97 and 78.65 ± 2.66%, respectively). The CH$_2$Cl$_2$ fractions of root and fruit also showed considerable inhibition against AChE (29.15 ± 2.45 and 31.46 ± 2.78%, respectively). Among the isolated compounds umbelliferone indicated strong inhibition against AChE (61.09 ± 4.46%) and pimpinellin had strong inhibition against BuChE with a 66.55 ± 2.61% value. On the other hand, none of the aqueous residues had activity against AChE, while only aerial part essential oil had no activity against this enzyme. The BuOH fraction of fruit and aqueous residue fraction of aerial parts and flowers displayed no butyrylcholinesterase inhibition activity. Amongst the essential oils the fruit (83.01%) and root (81.32%) ones indicated considerable inhibition towards BuChE.

Table 2. In vitro AChE and BuChE inhibitory activities of samples from *Zosima absinthifolia* at 20 μg/mL.

Samples	Enzymes	Percentile of inhibition ± S.E.M [a] against AChE and BuChE			
		Aerial Part	Root	Flower	Fruit
MeOH	AChE	6.45 ± 2.33	9.58 ± 2.55	[c]	[b]
	BuChE	14.44 ± 1.56	38.12 ± 4.05	27.33 ± 2.65	67.35 ± 1.56
Hexane	AChE	[b]	3.25 ± 1.57	[c]	[c]
	BuChE	17.35 ± 3.08	45.09 ± 2.66	24.97 ± 4.09	34.31 ± 2.76
CH$_2$Cl$_2$	AChE	[b]	29.15 ± 2.45	[c]	31.46 ± 2.78
	BuChE	64.66 ± 2.56	71.32 ± 3.09	69.25 ± 4.10	82.27 ± 1.97
EtOAc	AChE	3.34 ± 1.49	4.58 ± 4.66	[b]	9.03 ± 2.78
	BuChE	29.09 ± 2.66	28.05 ± 2.13	36.21 ± 2.35	43.44 ± 3.15
BuOH	AChE	3.55 ± 3.70	[c]	4.33 ± 1.65	[c]
	BuChE	17.56 ± 2.54	14.54 ± 3.44	28.23 ± 2.54	[b]
Aqueous residue	AChE	[c]	[b]	[c]	[b]
	BuChE	[b]	9.42 ± 1.97	[b]	17.21 ± 2.45
Essential oils	AChE	[b]	16.66 ± 3.21	6.45 ± 2.09	26.11 ± 2.13
	BuChE	34.56 ± 2.47	56. 30 ± 3.51	78.65 ± 2.66	72.24 ± 2.44
Bergapten	AChE	18.98 ± 2.98			
	BuChE	31.00 ± 3.02			
Imperatorin	AChE	20.44 ± 2.24			
	BuChE	44.23 ± 2.09			
Pimpinellin	AChE	23.54 ± 1.29			
	BuChE	66.55 ± 2.61			
Umbelliferone	AChE	61.09 ± 4.46			
	BuChE	40.99 ± 5.61			
Donepezil	AChE	82.45 ± 2.64			
	BuChE	90.33 ± 4.16			

[a] Standard error mean, [b] No activity, [c] Not detected because of turbidity in the wells of microplates.

The most active compounds (umbelliferone against AChE and pimpinellin against BuChE) were docked at the binding sites of 1-EVE and 1-P0I. The molecular interactions of the compounds possibly accounting for the inhibition are shown in Figures 3 and 4. Umbelliferone exhibits a good docking score for 1-EVE (−7.46 kcal/mol) when compared to the standard donepezil. Umbelliferone has three π-π stacking interactions (4.25 Å, 3.89 Å and 4.70 Å) with PHE330 and TRP84. In addition, hydrophobic interactions were formed between the molecule and TRP84, PHE330, PHE331, TYR121, TYR334. The polar interaction was realized by HIS 440. On the other hand the docking score of pimpinellin was −5.78 kcal/mol compared to the standard donepezil. Pimpinellin has two π-π stacking interactions (5.09 Å and 5.10 Å) with the phenyl ring of PHE 329. In addition, hydrophobic interactions were formed between the molecule and the PHE329, PRO285, LEU286, VAL288, TRP231, PHE398, ALA199 residues. The polar interactions were realized by SER287, GLN119, SER198.

Essential oil % yields of the various parts and the colours of these essential oils are presented in Table 3. The colours of essential oils from different parts of *Z. absinthfolia* varied. The flowers and fruits essential oils of *Z. absinthfolia* were yellow while the aerial part and roots gave light yellow and white coloured oils, respectively.

In general, the yield of the root oil was low compared to the aerial part, fruit and flower ones. The best yield results were obtained for fruit (Table 3). A total of thirty-three compounds totaling 94.7% of the oil were identified in the essential oil of aerial parts of *Z. absinthfolia*. Octanol, octyl octanoate and octyl acetate were the primary components, amounting to 8.8%, 7.6% and 7.3%, respectively. The analysis of the roots of *Z. absinthfolia* resulted in the identification of forty-four compounds totaling 81.6% of the oil. *trans*-Pinocarvyl acetate at 26.7% was the most abundant compound in the essential oil, followed by β-pinene (8.9%). Eighty-three compounds were characterized in the oil of the flowers of *Z. absinthfolia*, accounting for 82.5% of the oil. The primary constituents were identified as octyl acetate (19.9%), and *trans-p*-menth-2-en-1-ol (4.6%). The analysis on the fruits of *Z. absinthfolia* resulted in the determination of fifty-two essential compounds totaling 99.2%. Octyl acetate at 81.6% was the most abundant compound in the essential oil followed by (Z)-4-octenyl acetate (5.1%). The compositions of essential oils are presented in Table 4.

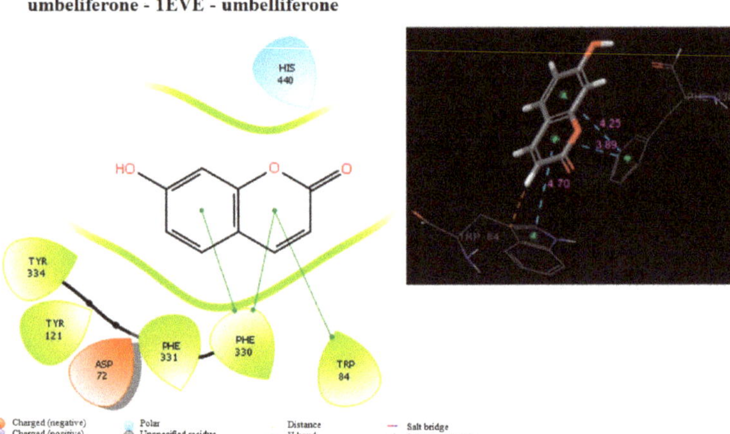

Figure 3. Schematic representation of main interaction of umbelliferone with AChE (1-EVE). Green color represents hydrophobic interactions, light blue represents polar interactions, blue represents positively charged residues, red represents negatively charged residues.

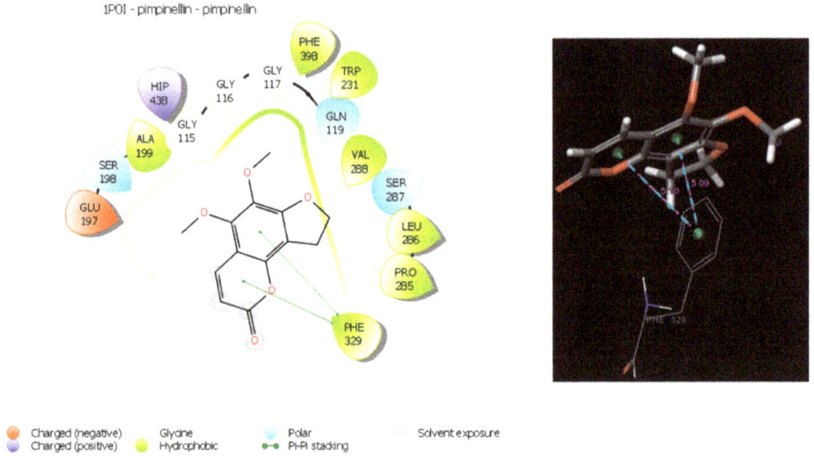

Figure 4. Schematic representation of the main interaction of pimpinellin with BuChE(1-P0I) Green colour represents hydrophobic interactions, light blue represents polar interactions, blue represents positively charged residues, red represents negatively charged residues.

Table 3. *Zosima absinthifolia* Essential oil yields (*w/v*, %).

Used Parts	Crushed (g)	Yields	Colour	Collection Time
Aerial	152	0.329	Light yellow	2017
Root	132	0.008	White	2017
Flower	35	0.057	Yellow	2018
Fruit	80	1.250	Yellow	2017

Table 4. The composition of the essential oils of *Zosima absinthifolia*.

RRI	Compound	Ap %	R %	Fl %	Fr %
1032	α-Pinene	4.4	1.3	2.2	0.1
1048	2-Methyl-3-buten-2-ol	-	-	tr	tr
1076	Camphene	0.2	-	0.1	tr
1093	Hexanal	0.3	-	tr	-
1118	β-Pinene	2.0	8.9	0.2	0.1
1132	Sabinene	0.3	0.1	0.1	tr
1151	δ-4-Carene	-	-	0.1	-
1174	Myrcene	1.0	3.0	1.3	tr
1176	α-Phellandrene	0.2	0.1	-	-
1194	Heptanal	-	0.4	-	-
1203	Limonene	1.8	2.7	1.5	0.1
1218	β-Phellandrene	1.0	0.4	0.7	0.1
1225	(Z)-3-Hexenal	-	-	tr	-
1244	2-Pentyl furan	-	0.2	0.1	tr
1246	(Z)-β-Ocimene	-	0.3	-	tr
1255	γ-Terpinene	-	0.2	tr	-
1266	(E)-β-Ocimene	-	-	0.4	-
1280	p-Cymene	0.5	2.2	0.1	-
1290	Terpinolene	0.4	1.1	0.1	tr
1296	Octanal	0.3	2.5	tr	0.2
1348	6-Methyl-5-hepten-2-one	-	-	tr	-
1360	Hexanol	-	-	tr	-
1398	2-Nonanone	-	2.6		-
1399	Methyl octanoate	-	-	tr	-
1400	Nonanal	-	0.3	tr	-
1444	Ethyl octanoate	-	-	0.2	-
1452	α,p-Dimethylstyrene	-	0.3	-	-
1483	Octyl acetate	7.3	1.0	19.9	81.6
1497	α-Copaene	-	-	0.1	tr
1506	Decanal	-	-	-	0.1
1516	(Z)-4-Octenyl acetate	0.3	-	0.5	5.1
1535	β-Bourbonene	1.3	-	0.1	0.3
1538	*trans*-Chrysanthenyl acetate	-	-	1.6	-
1553	Linalool	-	-	0.4	0.2
1562	Octanol	8.8	2.8	4.6	3.2
1571	*trans-p*-Menth-2-en-1-ol	-	-	1.5	0.1
1586	Pinocarvone	-	0.5	-	-
1589	β-Ylangene	-	-	-	tr
1591	Bornyl acetate	0.9	0.3	1.3	0.2
1597	β-Copaene	-	-	-	0.1
1600	β-Elemene	-	-	-	tr
1610	Calarene (=β-gurjunene)	-	0.2	-	-
1612	β-Caryophyllene	1.8	0.2	1.0	0.2
1614	Carvacrol methyl ether (= methyl carvacrol)	-	0.5	-	-
1623	Octyl butyrate	0.4	-	0.2	0.2
1634	Octyl 2-methyl butyrate	0.5	-	0.4	0.1
1638	*cis-p*-Menth-2-en-1-ol	-	-	0.7	0.1
1648	Myrtenal	-	0.4	-	-
1655	(E)-2-Decenal	-	1.5	-	-
1660	(Z)-4-Octenyl butyrate	-	-	0.2	-
1661	*trans*-Pinocarvyl acetate	-	26.7	-	0.1
1668	Citronellyl acetate	-	-	1.4	0.1
1670	*trans*-Pinocarveol	-	1.4	-	-
1687	Decyl acetate	-	-	-	0.1
1687	α-Humulene	-	-	0.1	-
1689	*trans*-Piperitol	-	-	0.4	-
1690	Cryptone	-	-	0.2	-
1704	Myrtenyl acetate	-	0.9	-	-
1706	α-Terpineol	-	0.3	-	-

Table 4. *Cont.*

RRI	Compound	Ap %	R %	Fl %	Fr %
1719	Borneol	-	-	0.1	-
1726	Germacrene D	2.3	-	2.0	0.5
1733	Neryl acetate	-	-	0.1	-
1747	*trans*-Carvyl acetate	-	0.2	-	-
1755	Bicyclogermacrene	0.7	-	0.7	0.1
1758	*cis*-Piperitol	-	-	0.5	-
1758	(*E,E*)-α-Farnesene	-	-	0.2	-
1772	Citronellol	-	-	0.4	0.1
1773	δ-Cadinene	-	-	0.1	-
1779	(*E,Z*)-2,4-Decadienal	-	0.3	-	-
1786	*ar*-Curcumene	0.2	0.3	0.2	0.1
1689	*trans*-Piperitol	-	-	-	tr
1804	Myrtenol	-	0.6	-	-
1827	(*E,E*)-2,4-Decadienal	-	0.9	-	-
1829	Octyl hexanoate	0.7	-	0.6	0.2
1849	Cuparene	-	0.6	0.1	-
1856	(*Z*)-4-octenyl hexanoate	0.7	-	0.7	-
1857	Geraniol	-	-	0.2	0.1
1868	(*E*)-Geranyl acetone	-	1.4	0.1	-
1878	2,5-Dimethoxy-*p*-cymene	-	3.6	tr	-
1945	1,5-Epoxysalvial(4)14-ene	-	-	tr	-
1958	(*E*)-β-Ionone	-	-	0.3	-
1981	Heptanoic acid	-	0.2	-	-
2000	Citronellyl hexanoate	-	-	0.3	-
2008	Caryophyllene oxide	1.9	0.8	0.4	0.1
2020	Octyl octanoate	7.6	0.8	0.3	0.9
2050	(*E*)-Nerolidol	0.8	-	0.1	-
2069	Germacrene D-4β-ol	-	-	0.3	-
2084	Octanoic acid	-	-	-	0.1
2100	Heneicosane	-	-	0.1	-
2127	10-*epi*-γ-Eudesmol	-	-	0.1	-
2131	Hexahydrofarnesyl acetone	0.4	-	0.1	tr
2144	Spathulenol	2.7	-	0.5	0.1
2170	β-Bisabolol	-	-	0.1	-
2183	γ-Decalactone	-	-	-	0.1
2187	T-Cadinol	-	-	0.1	-
2192	Nonanoic acid	-	-	-	0.1
2200	Docosane	-	-	0.1	-
2209	T-Muurolol	-	-	0.2	-
2214	(2E,6Z)-Farnesal	-	-	0.1	-
2219	δ-Cadinol (= torreyol)	-	-	0.1	-
2247	*trans*-α-Bergamotol	-	-	0.1	-
2255	α-Cadinol	-	-	0.6	-
2271	(2E,6E)-Farnesyl acetate	-	-	2.3	-
2278	(2E,6E)-Farnesal	-	-	0.4	-
2300	Tricosane	-	-	0.7	-
2373	Unknown I	12.5	5.0	15.4	1.0
2369	(2E,6E)-Farnesol	-	-	1.7	-
2450	Unknown II	26.4	2.3	8.9	1.4
2500	Pentacosane	-	-	-	0.1
2503	Dodecanoic acid	-	-	-	0.2
2622	Phytol	-	-	0.5	-
2670	Tetradecanoic acid	-	-	-	1.0
2700	Heptacosane	-	-	0.5	-
2900	Nonacosane	-	-	-	0.2
2931	Hexadecanoic acid	4.1	1.3	0.5	0.4
	Total Identified	55.8	74.3	58.2	96.8
	Total	*94.7*	*81.6*	*82.5*	*99.2*

RRI: Relative retention indices calculated against n-alkanes; % calculated from FID data; tr Trace (< 0.1 %); Ap: Aerial part; R: Root; Fl: Flower; Fr: Fruit; Unknown I: EIMS, 70 eV, m/z (rel. .int.): 270[M]$^+$(0.4), 227(41), 159(6), 141(37), 115(97), 98(100), 81(33), 69(23), 57(19), 43(66); Unknown II: EIMS, 70 eV, m/z (rel. .int.): 228[M]$^+$(0.5), 210(0.5), 116(25), 98(100), 87(19), 71(23), 57(18), 41(26).

The determined compounds were categorized into two main classes on the basis of their chemical structures: isoprenoids (oxygenated monoterpenes, terpene hydrocarbons) and nonisoprenoids variously functionalized (alkanes, aldehydes, lactones, ketones, alcohols, furans, fatty acids and esters). Terpene hydrocarbons, esters, fatty acids-esters, and alcohols were the dominant groups of compounds in the essential oils (Table 5).

Table 5. Chemical class distribution of the samples.

Compound Class	Ap %	R %	Fl %	Fr %
Esters	9.4	29.1	27.9	87.5
Alcohols	8.8	5.1	8.8	3.8
Aldehydes	0.6	6.3	0.5	0.3
Ketones	0.4	4.5	0.7	tr
Fatty acids+ esters	13.1	2.3	2.6	2.9
Terpene hydrocarbons	18.1	21.9	11.4	1.7
Oxygenated terpenes	5.4	4.9	4.8	0.2
Furans	-	0.2	0.1	tr
Alkanes	-	-	1.4	0.3
Lactones	-	-	-	0.1
Total Identified	55.8	74.3	58.2	96.8

The micrographs of the peduncles, rays, pedicels, and fruits of *Z. absinthifolia* were obtained from alcohol samples utilizing light microcopy (Figures 5–8) and from the dried samples through Scanning Electron Microscopy (SEM, Jeol JSM 6490LV) (Figure 9a–k). The number of secretory canals in the centre was less than in the cortex at the peduncle. At the ray and pedicel secretory canals were only found in the cortex and the number of canals are higher. The secretory canals in fruit were very large and wide.

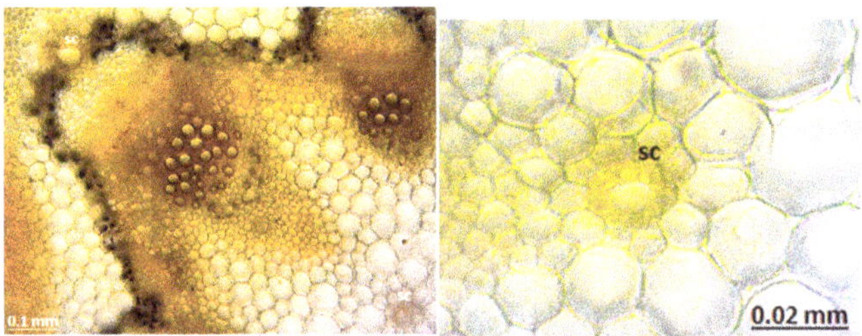

Figure 5. Secretory canals at the peduncle of *Zosima absinthifolia* by light microscopy.

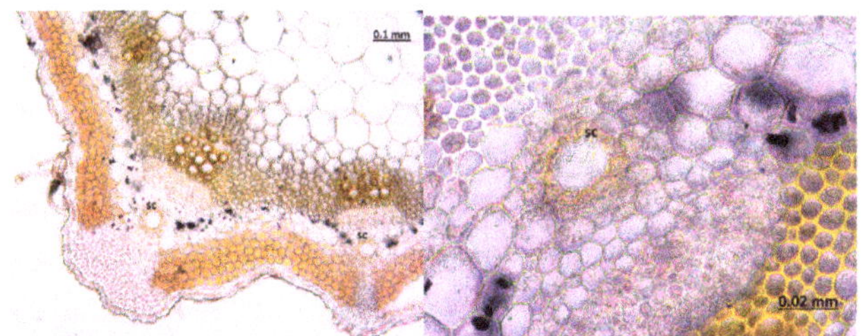

Figure 6. Secretory canals at the ray of *Zosima absinthifolia* by light microscopy.

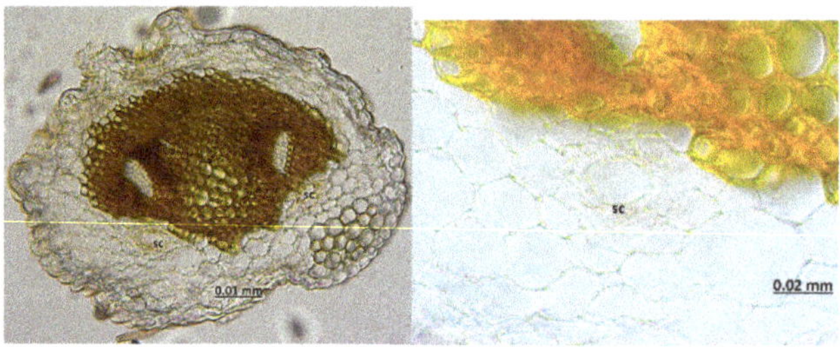

Figure 7. Secretory canals at the pedicel of *Zosima absinthifolia* by light microscopy.

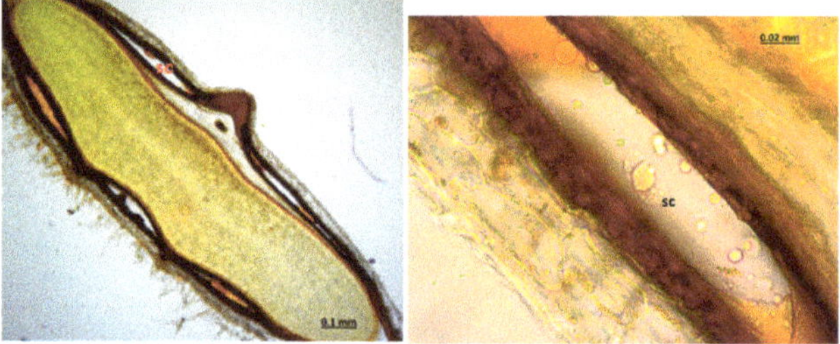

Figure 8. Secretory canals at the fruit of *Zosima absinthifolia* by light microscopy.

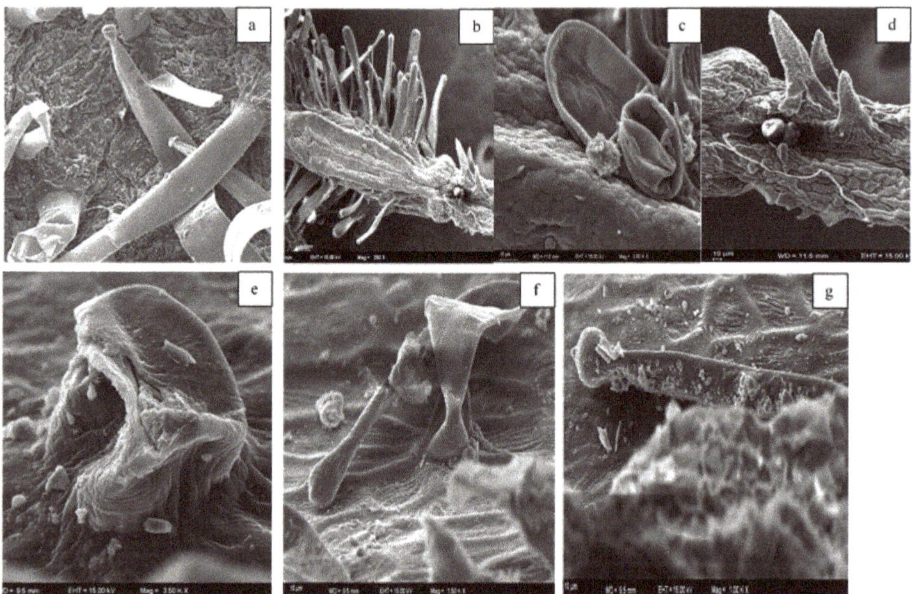

Figure 9. *Cont.*

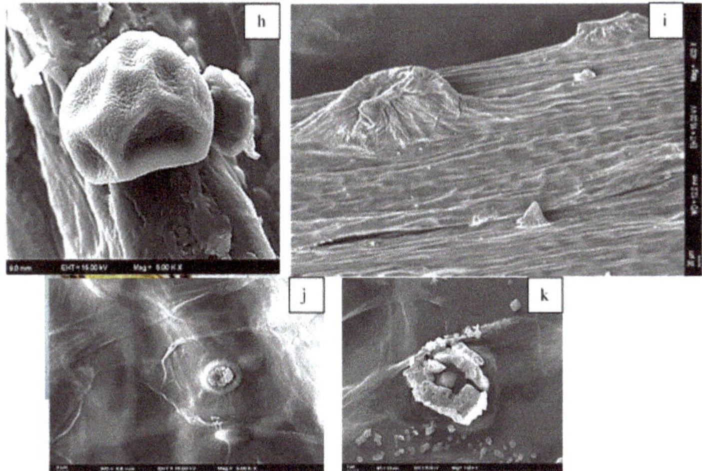

Figure 9. (**a**) Capitate trichomes on the leaf by SEM, (**b–d**) capitate trichomes on the pedicel by SEM, (**e–g**) capitate trichomes on the stem by SEM, (**h–k**) extrafloral nectaries, the secretory ducts and excretion secretory system on the fruit by SEM.

Secretory structures of stem, leaf, flower and fruit samples of *Z. absinthifolia* were studied in detail using light and scanning electron microscopy. The plant has secretory trichomes in the leaf, stem, pedicel and fruit. There are two types of glandular trichomes; capitate trichomes and sessile peltate trichomes. The capitate trichomes were identified on the leaf, pedicel and stem, peltate trichomes on pedicel and fruit. The capitate trichomes are composed of multi basal cells, a long stalk cell with the unicellular secretory head. Peltate trichomes exhibit a flattened head in the pedicel or a granular head in fruit formed by several cells arranged in a circle (Figure 9). Extrafloral nectaries are found on the pedicel. The secretory ducts show a lumen surrounded by a layer of specialized cells in fruit. Excretion secretory system organs including crystals are observed in the fruit.

3. Discussion

Coumarins are compounds naturally present in a great number of plants. Coumarin and its derivatives are prevalent in Nature. Coumarins are benzopyrones, which are compounds comprised of benzene rings linked to a pyrone moiety. Human dietary exposure to benzopyrones is quite considerable, as these compounds occur in fruits, vegetables, seeds, nuts, and higher plants. It has been determined that the mean Western diet includes ~1 g/day of mixed benzopyrones [25]. Coumarins have various biological activities such as anticancer, anticoagulant, anti-inflammatory, antitubercular, antihyperglycemic, antiadipogenic, antifungal, antibacterial, anticonvulsant, antihypertensive, antiviral, antioxidant, neuroprotective and antidiabetic effects [26].

In the current research, umbelliferone and pimpinellin were isolated from *Zosima absinthfolia*, and indicated activity against AChE and BuChE. We assumed that the inhibitory activity of umbelliferone was primary due to the hydroxyl group at the C-7 position. Also, we assumed that the inhibitory activity of pimpinellin was primary due to the methoxy groups at the C-5 and C-6 positions. Umbelliferone and pimpinellin presented the best activity at 20 µg/mL, while bergapten indicated weak inhibitory activity.

The roots fractions have been characterized by the significant higher content of total phenolics than aerial part fractions. Previously, it was observed that methanol fruit extract of *Z. absinthfolia* showed high antioxidant activities and a greater content of total phenols in comparison to the hexane and dichloromethane extracts of the plant [27]. A significant correlation was observed between antioxidant capacity and the total phenolics content in previous investigations [28,29] as well. The antioxidant capacity of *Z. absinthfolia* plant extracts was studied. It was found that peroxidation inhibition of MeOH extract of *Z. absinthfolia* was 143.5 RC_{50} [27]. However, we found that, fruit CH_2Cl_2 fraction showed higher antioxidant activity in comparison with the MeOH extract.

In a previous study, the major components of essential oils from different parts of *Z. absinthfolia* were studied and it was shown that the main volatile compounds were lavandulyl acetate (23.9%, an ester of the irregular monoterpenol), bornyl acetate (12.0%), octyl octanoate (11.7%), lavandulol (5.0%), octyl hexanoate (4.2%) and lavandulyl octanoate (3.1%) [30]. Another study reported that the major components of oil from the aerial part of *Z. absinthfolia* were octyl acetate (32.50%), octanol (20.60%) and α-pinene (10.90%) [31]. Octyl acetate (87.48%), octyl octanoate (5.03%) and 1-octanol (2.37%) were found the major components of fruit essential oil [19]. Octyl acetate (38.4%) and octyl hexanoate (31.9%) were detected as major components of fruit essential oil of *Z. absinthfolia* [32]. Our study is the first comparative study on essential oils constituents of aerial parts, roots, flowers and fruits of *Z. absinthfolia* and their anticholinesterase and antioxidant activities.

So far, no data are available about the presence of phenolic compounds in the essential oil from *Zosima* genus, especially such as α-pinene and octyl acetate which in previous studies showed high antioxidant capacity [33,34]. The high abundance of octyl acetate and possible antioxidative and anticancer effects were estimated in essential oils from the leaves of species of *Pittosporum* (Pittosporaceae) [35] and in essential oils from Ethiopian herbs *Boswellia carterii* and *Commiphora pyracanthoides* Engler [36].

The chemical composition of the different types of secretory canals present in the aerial parts, fruits and flowers were different. The canals present in the aerial parts are characterized by terpene hydrocarbons, while the secretory canals of roots, flowers and fruits include esters. Only canals of fruits contain lactones. Besides, canals of flowers and fruits contain alkanes. We can comment that the number of secretory channels, their location in the organs and their different shapes (broad, big, little, oval etc) could be related to the different chemical compositions of aerial part, root, flower and fruits. The chemical class distribution of the samples is presented in Table 5.

Members of the Apiaceae family are characterized by a specific type of essential oil secretory structure known as secretory canals. Their shapes and numbers can vary between species, within species or even in individual plants. They have large amounts of metabolic products in the area between their secretory canals. Particularly they generate and store essential oils in plants [37,38].

AD is a neurodegenerative disease induced by oxidative stress with a further cholinergic deprivation in the brain. Expressly, a decrease in the amount of acetylcholine delivered from cholinergic synapses has been identified as a cause. One cure methodology involves augmenting or protecting the ratio of acetylcholine via inhibiting acetylcholinesterase [39]. This research indicated that the CH_2Cl_2 fraction of fruit from *Z. absinthifolia* has AChE and BuChE inhibitory activity along with high antioxidant capacity. The use of antioxidants may be useful to treat AD. To the knowledge of the authors, this is the first study on the anticholinesterase activity of extracts from *Z. absinthifolia*.

Essential oil components represent a diverse family of low molecular weight organic compounds with remarkable biological activity. In accordance with their chemical structure, these active compounds can be divided into four major groups: terpenoids, phenylpropenes, terpenes, and "others". Besides, they might include diversified functional groups in accordance with which they can be categorised as hydrocarbons (monoterpenes, sesquiterpenes, and aliphatic hydrocarbons); oxygenated compounds (monoterpene and sesquiterpene alcohols, esters, ketones, aldehydes, and other oxygenated compounds); and sulfur and/or nitrogen sulfur including compounds (sulfides, nitriles, thioesters, isothiocyanates, and others). Components that act as cholinesterase inhibitors still represent the only pharmacological treatment of AD. Many in vitro investigations have demonstrated that some compounds present in essential oils such as α-pinene, α- and β-asarone, δ-3-carene, carvacrol, 1,8-cineole, thymohydroquinone, anethole, etc have certain cholinesterase inhibitory activity [40].

4. Materials and Methods

4.1. Plant Specimens

Zosima absinthifolia samples were gathered at the flowering and fruity period from Erzurum in the Palandöken Mountains in 2017 and 2018, and the verified by Prof. Dr Hayri Duman. Voucher specimens were stored at the Herbarium of the Atatürk University Faculty of Pharmacy (AUEF 1275 and AUEF 1283).

4.2. Extraction and Isolation

Aerial parts (150 g), roots (150 g), flowers (150 g) and fruits (500 g) were comminuted and macerated with methanol (3 times $\times$ 8 h) in a water-bath not exceeding 40 °C (3 $\times$ 150 mL) while mixing at 300 rpm with the use of a mechanical mixer. Conjoined aerial parts, roots, flowers and fruits extracts were filtered and concentrated up to dryness using a rotating evaporator (Heidolph VV2000, Schwabach, Germany). After that the residue was dissolved in methanol:water (1:9) and subjected to three further fractionation steps with 150 mL of *n*-hexane, dichloromethane, ethyl acetate and *n*-butanol, respectively. The weights of the comminuted parts of *Zosima absinthifolia* and extracts/fractions obtained are indicated in Table 6.

Table 6. Weights of the crushed plants and obtained extracts and fractions.

Species	Extracts/Fractions	Aerial Part	Root	Flower	Fruit
	MeOH (g)	25.01	29.88	23.92	85.98
	Hexane (g)	3.28	4.05	2.98	11.88
Zosima	CH_2Cl_2 (g)	9.12	10.10	8.97	26.01
absinthifolia	EtOAc (g)	1.66	2.24	1.59	4.81
	BuOH (g)	4.92	5.86	4.77	18.57
	Aqueous residue (g)	5.02	3.22	4.98	6.96

The extraction, and identification of purified compounds from the CH_2Cl_2 fruit fraction was done according to [41]. The effective CH_2Cl_2 fraction of fruit was first applied to a silica gel column and eluted with a gradient of hexane:EtOAc (100:0 $\rightarrow$ 0:100, v/v) and EtOAc:MeOH (100:0 $\rightarrow$ 0:100, v/v), and three fractions (Fr. A–C) were acquired. Repetitive silica gel column chromatography with hexane:EtOAc (85:15 and 80:20) solvent systems on Fr. A gave compound **1**. Fr. B was applied to a silica gel column and eluted with hexane:EtOAc (75:25) and a Sephadex LH-20 column eluting with ethyl acetate to give compounds **2** and **3**. Elution with hexane:EtOAc (70:30) of a silica gel column of Fr. C gave compound **4**. The chemical structures of compounds **1–4** are presented in Figure 1.

4.3. Isolation of the Essential Oil, GC-FID and GC-MS Analyses

Isolation of the essential oils, GC-FID and GC-MS analyses processes were performed according to [42]. The crushed parts, essential oil % yields of the species and colours of essential oils are presented in Table 2.

4.4. Determination of Total Phenolics

The total phenolic content of specimens was evaluated utilising the Folin–Ciocalteu assay [43] with slight modifications [44]. The total phenolics absorbance was determined at 765 nm with the use of a Jenway UV/Vis 6405 spectrophotometer (Jenway, Chelmsford, UK). The findings are reported as gallic acid equivalents (GAE/g specimens).

4.5. 1,1-Diphenyl-2-picrylhydrazyl (DPPH) Radical Scavenging Capacity Assay

The previously detailed DPPH assay [45] was applied with slight alterations. Reagent stock solution (1 $\times$ 10^{-3} M) was prepared by dissolving 22 mg of DPPH in 50 mL of methanol. This solution was kept at 20 °C until used. Samples (0.02 g) were extracted in two steps: first, to the dry material in an Eppendorf tube was added 1 mL of distilled water. Specimens were heated at 95 °C during 15 min and further for 5 min centrifuged (12,000 rpm, 25 °C). The supernatant was transferred into a fresh tube. The supernatant with 1 mL of dist. water was diluted and the same heating and centrifugation procedure was repeated. The study solution (6×10^{-5} M) was prepared by mixing 100 mL of methanol with 6 mL of the stock solution. Then, 0.1 mL of each experimental solution was mixed in to react with 3.9 mL of the solution of DPPH, followed by vortexing for 30 s and a further reaction time of 30 min. The optical absorbance was gauged at 515 nm with the Jenway UV/Vis 6405 spectrophotometer. A sample without DPPH solution was utilized as a blank sample. The scavenging activity of DPPH was measured according to the formula below:

$$\text{Scavenging activity of DPPH (\%)} = [(A \text{ control} - A \text{ sample})/A\text{control}] \times 100$$

where A = absorbance at 515 nm.

4.6. Anti-Lipid Peroxidation Activity

The thiobarbituric acid (TBA) assay was utilised to assess the protective activity of samples on liposomes against lipid peroxidation [46]. Seven different sample concentrations (0.016–1 mg/mL) were studied in this test. Chlorogenic acid, rutin and propyl gallate were prepared as reference compounds at seven different concentrations (0.000064–1 mg/mL), and chlorogenic acid and rutin were utilised in the same concentration interval. Brain extract (0.2 mL), phosphate buffer (0.5 mL), ferric chloride (0.1 mL), ascorbic acid (0.1 mL) and the samples were mixed and incubated at 37 °C for 20 minutes. Next, 25% HCl (0.5 mL), 1% TBA (0.5 mL) and 2% BHT (0.1 mL) were added into the mixture, which was shaken and incubated at 85 °C for 30 minutes. The mixture was cooled and *n*-butanol (2.5 mL) was added. After centrifugation, the absorbance of the samples was recorded at 532 nm using the UV-1800 spectrophotometer. These tests were repeated four times. The IC_{50} values were established through linear regression analysis. A low IC_{50} value means that the antioxidant activity is high.

4.7. Determination of AChE and BuChE Inhibition Activities

The determination of AChE and BuChE inhibition activities of the samples were performed according to [41]. This process was repeated three times for each plate. All data were expressed as mean $\pm$ SE of three independent assays.

4.8. Microscopic Analysis

Materials (kept in 70% alcohol) from *Zosima absinthifolia* were assessed by light microscopy using Sartur and chloral hydrate reagents. In the light microscopy study, cross-sections of peduncles, rays, pedicels, and fruits from *Z. absinthifolia* were prepared manually. Images prepared with Sartur R [47,48] were recorded with a Zeiss 51425 camera attached to a light microscope (Zeiss 415500-1800-000). In the scanning electron microscopy (SEM) investigations, leaf, stem, pedicel and fruit parts were attached to aluminium stubs and covered with gold for 4 min in a sputter-coater. Morphological observations were done in a Jeol JSM 6490LV scanning electron microscope at the Turkish Petroleum International Company (TPAO) Research Centre SEM laboratory, Ankara.

4.9. Molecular Docking Studies

Umbelliferone was found to be an active compound against AChE and pimpinellin was found to be active against BuChE was found. These active compounds were docked at the binding sites of 1-EVE and 1-P0I.

4.10. Protein Preparation

The three-dimensional complex structures of AChE (PDB ID: 1EVE) and BuChE (PDB ID: 1P0I) were obtained from the Protein Data Bank [49,50]. The protein structures were prepared using the Protein Preparation Wizard panel tool of the Scrödinger software suite (Maestro 11.8). Firstly water molecules (>5Å radius) and other small molecules were removed from the crystal structures, hydrogen atoms were added and physiological pH was set at 7. Finally, the restrained minimization was performed with the added hydrogen atoms to OPLS3e.

4.11. Ligand Preparation

The ligands were prepared for docking with using the Ligand Preparation Panel in the programme. The grid files were created using the Receptor Grids Generation Panel. Finally the Glide Ligand Docking Panel was used for docking studies.

4.12. Statistical Analysis

All findings are stated as mean $\pm$ SE and variations between means were statistically analyzed through One-way analysis of ANOVA followed via Bonferroni's complementary analysis, with $p < 0.05$ considered to demonstrate statistical significancy.

5. Conclusions

The CH_2Cl_2 fraction of fruit from *Zosima absinthifolia* and umbelliferone had a remarkable antioxidant and anticholinesterase activities. The tested extracts and essential oils displayed high radical scavenging capacity (RSC), which was found to be in correlation to their content of phenolic compounds. Octyl acetate was the dominant component in the essential oils. Original information has been presented regarding the total phenolics content and high presence of chlorogenic acid and flavonoid rutin in the extracts and essential oils of plant *Z. absinthifolia*. Novel data of a comparative study on the essential oil constituents of aerial parts, roots, flowers and fruits of *Z. absinthfolia* has shown different phenolic compositions which can depend from the function of secretory canals of the various plant parts. Due to the remarkable presence of compounds with high anticholinesterase activities in the plant we presume that *Z. absinthifolia* could be utilised as an herbal alternative to synthetic drugs in the prophylaxis of AD.

Author Contributions: Conceptualization, S.K., B.D., N.M.P. and O.S.; methodology, S.K., M.K., B.D., N.M.P. and O.S.; software, S.K., M.K., B.D., N.M.P. and O.S.; validation, S.K., M.K., B.D., N.M.P. and O.S.; formal analysis, S.K., M.K., B.D., N.M.P. and O.S.; investigation, S.K., M.K., S.V.Y., K.Y., M.B., B.D., N.M.P. and O.S.; resources, S.K. and S.V.Y.; data curation, S.K., M.K., B.D., N.M.P., M.B. and O.S.; writing—original draft preparation, S.K., B.D., N.M.P. and O.S.; writing—review and editing, S.K. and O.S, M.B.; visualization, S.K. and O.S.; supervision, O.S.; project administration, S.K. and O.S.; funding acquisition, S.K. and O.S.

Funding: This work was supported by Research Fund of the Ataturk University.

Conflicts of Interest: The authors report no declarations of interest.

References

1. Warren, B.Z.; Ira, T.L. Alzheimer's Disease in Down Syndrome: Neurobiology And Risk. *Ment. Retard. Dev. Disabil. Res. Rev.* **2007**, *13*, 237–246.
2. Karch, S.; Broichhagen, J.; Schneider, J.; Böning, D.; Hartmann, S.; Schmid, B.; Tripal, P.; Palmisano, R.; Alzheimer, C.; Johnsson, K.; Huth, T. A New Fluorogenic Small-Molecule Labeling Tool for Surface Diffusion Analysis and Advanced Fluorescence Imaging of β-Site Amyloid Precursor Protein-Cleaving Enzyme 1 Based on Silicone Rhodamine: SiR-BACE1. *J. Med. Chem.* **2018**, *61*, 6121–6139. [CrossRef] [PubMed]
3. World Health Organization. *The Epidemiology and Impact of Dementia*; WHO: Geneva, Switzerland, 2015. Available online: http://www.who.int/mental_health/neurology/dementia/dementia_thematicbrief_epidemiology.pdf (accessed on 23 October 2018).

4. Sadaoui, N.; Bec, N.; Barragan-Montero, V.; Kadrie, N.; Cuisinier, F.; Larroque, C.; Arab, K.; Khettal, B. The essential oil of Algerian *Ammodaucus leucotrichus* Coss. & Dur. and its effect on the cholinesterase and monoamine oxidase activities. *Fitoterapia* **2018**, *130*, 1–5. [PubMed]

5. Ustun, O.; Senol, F.S.; Kurkcuoglu, M.; Orhan, I.E.; Kartal, M.; Baser, K.H.C. Investigation on chemical composition, anticholinesterase and antioxidant activities of extracts and essential oils of Turkish *Pinus* species and pycnogenol. *Ind. Crops Prod.* **2012**, *38*, 115–123. [CrossRef]

6. Kim, J.; Lee, H.J.; Lee, K.W. Naturally occurring phytochemicals for the prevention of Alzheimer's disease. *J. Neurochem.* **2010**, *112*, 1415–1430. [CrossRef] [PubMed]

7. Ferreira, A.; Proenc, C.; Serralheiro, M.L.M.; Araujo, M.E.M. The in vitro screening for acetylcholinesterase inhibition and antioxidant activity of medicinal plants from Portugal. *J. Ethnopharmacol.* **2006**, *108*, 3–37. [CrossRef]

8. Albanese, L.; Ciriminna, R.; Meneguzzo, F.; Pagliaro, M. Innovative beer-brewing of typical, old and healthy wheat varieties toboost their spreading. *J. Clean. Prod.* **2018**, *171*, 297–311. [CrossRef]

9. Luz, D.A.; Pinheiro, A.M.; Silva, M.L.; Monteiroa, M.C.; Predige, R.D.; Maiaa, C.S.F.; Andrad, E.; Júniora, F. Ethnobotany, phytochemistry and neuropharmacological effects of *Petiveria alliacea* L. (Phytolaccaceae): A review. *J. Ethnopharmacol.* **2016**, *185*, 182–201. [CrossRef]

10. Perry, E.K.; Pickering, A.T.; Wang, W.W.; Houghton, P.; Perry, N.L. Medicinal Plants and Alzheimer's Disease: Integrating Ethnobotanical and Contemporary Scientific Evidence. *J. Altern. Complement. Med.* **1998**, *4*, 419–428. [CrossRef]

11. Adsersen, A.; Gauguin, B.; Gudiksen, L.; Jäger, A.K. Screening of plants used in Danish folk medicine to treat memory dysfunction for acetylcholinesterase inhibitory activity. *J. Ethnopharmacol.* **2006**, *104*, 418–422. [CrossRef]

12. Huang, W.Y.; Cai, Y.Z.; Zhang, Y. Natural Phenolic Compounds from Medicinal Herbs and Dietary Plants: Potential Use for Cancer Prevention. *Nutr. Cancer* **2009**, *62*, 1–20. [CrossRef] [PubMed]

13. Anand, P.; Singh, B.; Singh, N. A review on coumarins as acetylcholinesterase inhibitors for Alzheimer's disease. *Bioorg. Med. Chem.* **2012**, *20*, 1175–1180. [CrossRef] [PubMed]

14. Ferreira, V.T.H.; Guimaraes, I.M.; Flavia, R.S.; Fabiola, M.R. Alzheimer's Disease: Targeting the Cholinergic System. *Curr. Neuropharmacol.* **2016**, *14*, 101–115. [CrossRef]

15. Ciriminna, R.; Meneguzzo, F.; Fidalgo, A.; Ilharco, L.M.; Pagliaro, M. Extraction, benefits and valorization of olive polyphenols. *Eur. J. Lipid Sci. Technol.* **2016**, *118*, 503–511. [CrossRef]

16. Mimica, N.D.; Bozin, B.; Sokovıc, M.; Simin, N. Antimicrobial and antioxidant activities of *Melissa officinalis* L. (Lamiaceae) Essential Oil. J. Agric. *Food Chem.* **2004**, *52*, 2485–2489. [CrossRef] [PubMed]

17. Bahadir, O.; Citoglu, G.S.; Ozbek, H.; Dall'Acqua, S.; Hosek, J.; Smejkal, K. Hepatoprotective and TNF-alpha inhibitory activity of *Zosima absinthifolia* extracts and coumarins. *Fitoterapia* **2011**, *82*, 454–459. [CrossRef] [PubMed]

18. Bahadir, Ö.; Saltan, Ç.G.; Özbek, H. Evaluation of anti-inflammatory effect of Zosima absinthifolia and deltoin. *J. Med. Plant. Res.* **2010**, *4*, 909–914.

19. Razavi, S.M.; Nejad-Ebrahimi, S. Chemical composition, allelopatic and antimicrobial potentials of the essential oil of Zosima absinthifolia (Vent.) Link fruits from Iran. *Nat. Prod. Res.* **2010**, *24*, 1125–1130. [CrossRef]

20. Razavi, S.M.; Ghasemiyan, A.; Salehi, S.; Zahri, F. Screening of biological activity of *Zosima absinthifolia* fruits extracts. *EurAsia J. BioSci.* **2009**, *4*, 25–28. [CrossRef]

21. Al-Shamma, A.; Mitscher, L.A. Comprehensive survey of indigenous Iraqi plants for potential economic value. 1. Screening results of 327 species for alkaloids and antimicrobial agents. *J. Nat. Prod.* **1979**, *42*, 633–642. [CrossRef]

22. Reed, M.W.; Moore, H.W. Efficient Synthesis of Furochromone and Furocoumarin Natural Products (Khellin, Pimpinellin, Isophellopterin) by Thermal Rearrangement of 4-Furyl-4-hydroxycyclobutenones. *J. Org. Chem.* **1988**, *53*, 4166–4171. [CrossRef]

23. Ellman, G.L.; Courtney, K.D.; Andresjr, V.; Featherstone, R.M. A new and rapid colorimetric determination of acetylcholinesterase activity. *Biochem. Pharmacol.* **1961**, *7*, 88–95. [CrossRef]

24. Yerdelen, K.Ö.; Tosun, E. Synthesis, docking and biological evaluation of oxamide and fumaramide analogs as potential AChE and BuChE inhibitors. *Med. Chem. Res.* **2015**, *24*, 588–602. [CrossRef]

25. Ali, Y.; Seong, H.; Reddy, M.R.; Seo, S.Y.; Choi, J.S.; Jung, H.A. Kinetics and Molecular Docking Studies of 6-Formyl Umbelliferone Isolated from Angelica decursiva as an Inhibitor of Cholinesterase and BACE1 Md. *Molecules* **2017**, *22*, 1604. [CrossRef] [PubMed]

26. Karakaya, S.; Gözcü, S.; Güvenalp, Z.; Özbek, H.; Yuca, H.; Dursunoğlu, B.; Kazaz, C.; Kılıç, C.S. The α-amylase and α-glucosidase inhibitory activities of the dichloromethane extracts and constituents of *Ferulago bracteata* roots. *Pharm. Biol.* **2018**, *56*, 18–24. [CrossRef] [PubMed]

27. Razavi, S.; Imanzadeh, G.; Jahed, F.S.; Zarrini, G. Pyranocoumarins from Zosima absinthifolia (Vent) Link roots. *Bioorg. Khim.* **2013**, *39*, 244–246. [CrossRef] [PubMed]

28. Sytar, O.; Bruckova, K.; Hunkova, E.; Zivcak, M.; Kiessoun, K.; Brestic, M. The application of Muliplex flourimetric sensor for analysis flavonoids content in the medical herbs family *Asteraceae, Lamiaceae, Rosaceae. Biol. Res.* **2015**, *48*, 48. [CrossRef]

29. Granato, D.; Shahidi, F.; Wrolstad, R.; Kilmartin, P.; Melton, L.D.; Hidalgo, F.J.; Miyashita, K.; Camp, J.; Alasalvar, C.; Ismail, A.B.; et al. Antioxidant activity, total phenolics and flavonoids contents: Should we ban in vitro screening methods? *Food Chem.* **2018**, *264*, 471–475. [CrossRef]

30. Đorđević, M.R.; Radulović, N.S.; Blagojević, P.D.; Pešić, M.S.; Akhlaghi, H. The essential oil of *Zosima absinthifolia* Link (Apiacee) from Iran: A rich source of lavandulyl esters. *Nat. Volatiles Essential Oils* **2017**, *4*, 99.

31. Kılıç, Ö. Essential Oil Composition of Two Apiacee Species from Bingol (Turkey). *Tr. J. Nat. Sci.* **2014**, *3*, 18–21.

32. Baser, K.H.C.; Ozek, T.; Demirci, B.; Kurkcuoglu, M.; Aytac, Z.; Duman, H. Composition of the essential oils of *Zosima absinthifolia* (Vent.) Link and Ferula elaeochytris Korovin from Turkey. *Flavour Frag J.* **2000**, *15*, 371–372. [CrossRef]

33. Karthikeyan, R.; Kanimozhi, G.; Prasad, N.R.; Agilan, B.; Ganesan, M.; Srithar, G. Alpha pinene modulates UVA-induced oxidative stress, DNA damage and apoptosis in human skin epidermal keratinocytes. *Life Sci.* **2018**, *212*, 150–158. [CrossRef] [PubMed]

34. Villa, N.R.; Pacheco-Hernández, Y.; Becerra-Martínez, E.; Zárate-Reyes, J.A.; Cruz-Duráne, R. Chemical profile and pharmacological effects of the resin and essential oil from Bursera slechtendalii: A medicinal "copal tree" of southern Mexico. *Fitoterapia* **2018**, *128*, 86–92. [CrossRef] [PubMed]

35. Weston, R.J. Composition of Essential Oils from the Leaves of Seven New Zealand Species of *Pittosporum* (*Pittosporaceae*). *J. Essential Oil Rese.* **2004**, *16*, 453–458. [CrossRef]

36. Chen, Y.; Zhou, C.; Ge, Z.; Liu, Y.; Liu, Y.; Feng, W.; Wei, T. Composition and potential anticancer activities of essential oils obtained from myrrh and frankincense. *Oncol. Lett.* **2013**, *6*, 1140–1146. [CrossRef] [PubMed]

37. Cheniclet, C.; Carde, J.P. Presence of Leucoplasts in Secretory Cells and of Monoterpenes in the Essential Oil: A Correlative Study. *Isr. J. Bot.* **1985**, *34*, 219–238.

38. Figueiredo, A.C.; Barroso, J.G.; Pedro, L.G.; Scheffer, J.J.C. Factors affecting secondary metabolite production in plants: Volatile components and essential oils Factors affecting volatile and essential oil production in plants. *Flavour Fragr. J.* **2008**, *23*, 213–226. [CrossRef]

39. Dickson, D.W. Neuropathological diagnosis of Alzheimer's disease: A perspective from longitudinal clinicopathological studies. *Neurobiol. Aging* **1997**, *18*, 21–26. [CrossRef]

40. Burcul, F.; Blazevic, I.; Radan, M.; Politeo, O. Terpenes, phenylpropanoids, sulfur and other essential oil constituents as inhibitors of cholinesterases. *Curr. Med. Chem.* **2018**. [CrossRef]

41. Karakaya, S.; Koca, M.; Kilic, C.S.; Coskun, M. Antioxidant and anticholinesterase activities of *Ferulago syriaca* Boiss. and *F. isaurica* Peşmen growing in Turkey. *Med. Chem. Res.* **2018**, *27*, 1843–1850. [CrossRef]

42. Karakaya, S.; Göger, G.; Kılıç, C.S.; Demirci, B. Composition of volatile oil of the aerial parts, flowers and roots of *Ferulago blancheana* Post. (Apiaceae) growing in Turkey and determination of their antimicrobial activities by bioautography method. *Turk. J. Pharm. Sci.* **2016**, *13*, 173–180.

43. Molyneux, P. The use of the stable free radical diphenylpicryl-hydrazyl (DPPH) for estimating antioxidant activity. *Songklanakarin J. Sci. Technol.* **2004**, *26*, 211–219.

44. Sytar, O.; Bośko, P.; Živčák, M.; Brestic, M.; Smetanska, I. Bioactive Phytochemicals and Antioxidant Properties of the Grains and Sprouts of Colored Wheat Genotypes. *Molecules* **2018**, *23*, 2282. [CrossRef] [PubMed]

45. Singleton, V.L.; Rossi, J.A. Colorimentry of total phenolics with phosphomolybdic-phosphotungstic acid reagents. *Am. J. Enol. Viticult.* **1965**, *16*, 144–158.

46. Dinis, T.C.P.; Madeira, V.M.C.; Almeida, L.M. Action of phenolic derivates (acetoaminophen, salycilate, and 5-aminosalycilate) as inhibitors of membrane lipid peroxidation and as peroxyl radical scavengers. *Arch. Biochem. Biophys.* **1994**, *315*, 161–169. [CrossRef] [PubMed]
47. Çelebioğlu, S.; Baytop, T. Bitkisel tozların tetkiki için yeni bir reaktif. *Farmakolog* **1949**, *19*, 301.
48. ÖZKAN, Y. *Türk Farmakopesi 2017*; Genel monograflar I, T.C. Sağlık Bakanlığı Yayın No: 1098, 1.; Baskı: Ankara, Turkey, 2018. Available online: https://www.titck.gov.tr/Dosyalar/Laboratuvar/T%C3%BCrkFarmakopeDergisi2.Cilt1.Say%C4%B1s%C4%B1.pdf (accessed on 4 May 2016).
49. Kryger, G.; Silman, I.; Sussman, J.L. Structure of acetylcholinesterase complexed with E2020 (Aricept): Implications for the design of new anti-Alzheimer drugs. *Structure* **1999**, *7*, 297–307. [CrossRef]
50. Nicolet, Y.; Lockridge, O.; Masson, P.; Fontecilla-Camps, J.C.; Nachon, F. Crystal structure of human butyrylcholinesterase and of its complexes with substrate and products. *J. Biol. Chem.* **2003**, *278*, 41141–41147. [CrossRef]

Sample Availability: Samples of the compounds are available from the authors.

Article

Changes of Phytochemical Components (Urushiols, Polyphenols, Gallotannins) and Antioxidant Capacity during *Fomitella fraxinea*–Mediated Fermentation of *Toxicodendron vernicifluum* Bark

Da-Ham Kim [1], Min-Ji Kim [1], Dae-Woon Kim [1], Gi-Yoon Kim [1], Jong-Kuk Kim [1], Yoseph Asmelash Gebru [1], Han-Seok Choi [2], Young-Hoi Kim [1] and Myung-Kon Kim [1,*]

[1] Department of Food Science and Biotechnology, Chonbuk National University, Jeonju 54896, Jeonbuk, Korea; dadaham@naver.com (D.-H.K.); kmj6202@hanmail.net (M.-J.K.); kdwoon1@naver.com (D.-W.K.); seokmin0000@naver.com (G.-Y.K.); rlawodrnr@naver.com (J.-K.K.); yagebru@gmail.com (Y.A.G.); yhoi1307@hanmail.net (Y.-H.K.)

[2] Department of Agriculture and Fisheries Processing, Korea National College of Agriculture and Fisheries, Jeonju 54874, Jeonbuk, Korea; coldstone@korea.kr

* Correspondence: kmyuko@jbnu.ac.kr; Tel.: +82-63-270-2551; Fax: +82-63-270-2572

Academic Editors: Marian Brestic, Marek Zivcak, Oksana Sytar and Marco Landi
Received: 21 December 2018; Accepted: 10 February 2019; Published: 14 February 2019

Abstract: The stem bark of *Toxicodendron vernicifluum* (TVSB) has been widely used as a traditional herbal medicine and food ingredients in Korea. However, its application has been restricted due to its potential to cause allergies. Moreover, there is limited data available on the qualitative and quantitative changes in the composition of its phytochemicals during fermentation. Although the *Formitella fraxinea*-mediated fermentation method has been reported as an effective detoxification tool, changes to its bioactive components and the antioxidant activity that takes place during its fermentation process have not yet been fully elucidated. This study aimed to investigate the dynamic changes of urushiols, bioactive compounds, and antioxidant properties during the fermentation of TVSB by mushroom *F. fraxinea*. The contents of urushiols, total polyphenols, and individual flavonoids (fisetin, fustin, sulfuretin, and butein) and 1,2,3,4,6-penta-*O*-galloyl-β-D-glucose (PGG) significantly decreased during the first 10 days of fermentation, with only a slight decrease thereafter until 22 days. Free radical scavenging activities using 2,2-diphenyl-1-picrylhydrazyl (DPPH), 2,2′-azino-bis(3-ethylbenzothiazoline-6- sulfonic acid) (ABTS), and ferric reducing/antioxidant power (FRAP) as an antioxidant function also decreased significantly during the first six to nine days of fermentation followed by a gentle decrease up until 22 days. These findings can be helpful in optimizing the *F. fraxinea*–mediated fermentation process of TVSB and developing functional foods with reduced allergy using fermented TVSB.

Keywords: *Toxicodendron vernicifluum*; *Fomitella fraxinea*; fermentation; urushiols; polyphenols; antioxidant activity

---

1. Introduction

Toxicodendron vernicifluum (TVSB, Stokes) F. Barkley (formerly known as *Rhus verniciflua* Stokes) belongs to the Anacardiaceae family [1], and is widely distributed in China, Japan, and Korea [2]. It is commonly known as the lacquer tree or vanish tree (also known as sumac). The stem and bark of *T. vernicifluum* have traditionally been used as folk medicines for treating blood disorders, hepatic disorders, gastric disorders, inflammatory diseases, anti-aging, paralysis, hypertension, and various cancers, as well as materials for lacquered chicken and duck soup recipes in Korea [3–5].

The xylem and bark of *T. vernicifluum* have also been reported to possess strong antioxidant [6,7], antiplatelet [8,9], immune-enhancing [10], neuroprotective [11,12], anti-inflammatory [12,13], and anti-cancer activities [4,14,15]. These health benefits were found to be attributed to the presence of flavonoids, gallotannins, and phenolic acids [3,16]. However, urushiol congeners found in the plant can cause allergic contact dermatitis with irritation, inflammation, and blistering in sensitive individuals [17,18]. Various methods have been attempted to remove or reduce urushiol congeners from the xylem and bark of *T. vernicifluum*, including physicochemical treatments such as solvent extraction, pyrolysis at high temperature, and enzymatic methods. Although some of these methods are effective at removing urushiols, they have limitations such as the generation of harmful substances, high cost, process complexity, and low efficiency [19].

A new biological method has been introduced to remove urushiols through fermentation with Basidiomycete *Formitella fraxinea* as an alternative method to solve these problems. This method was able to efficiently decrease more than 90% of the urushiols in TVSB [19]. Recently, an aqueous extract of urushiol-free fermented TVSB by *F. fraxinea* has been approved for use in soy sauce, fermented vinegar, and some alcoholic beverages by the Korea Food and Drug Administration (KFDA) [20], and its extracts are currently being marketed as functional health foods in Korea. Many studies on the biological activities of TVSB fermented by *F. fraxinea* have been reported, and most of these results were obtained by fermentation with *F. fraxinea* for 20–28 days [5,7,11,21,22]. While the fermentation method using *F. fraxinea* is being accepted as an effective tool to remove urushiols in TVSB, the dynamic changes of its bioactive components such as total polyphenols, individual phenolic compounds, gallotannins, and antioxidant activity during the fermentation process have not yet been fully elucidated. Flavonoids (fustin, fisetin, sulfuretin, and butein), PGG, and gallic acid as a gallotannin group are considered the major active constituents that are responsible for the various biological effects of TVSB [9,23–25]. Therefore, the objective of this study was to investigate dynamic changes in the polyphenols, individual phenolic compounds, gallotannins, and antioxidant activity during the fermentation of TVSB by mushroom *F. fraxinea*.

In this study, we demonstrated the optimization of an effective detoxification of urushiol congeners in TVSB through monitoring fermentation time while simultaneously intending to achieve a minimized loss of other useful phenolic compounds. The optimal fermentation period to achieve the desired detoxification effect was found to be 13–16 days. This finding provides valuable information to reconsider the common practice of fermenting TVSB extracts with *F. fraxinea* for 20–28 days prior to their applications in Korea.

2. Results and Discussion

2.1. Changes of Urushiols

Although TVSB has attracted much attention since ancient times in Korea due to its beneficial properties to human health, its allergenic urushiol congeners have limited its application in the food and pharmaceutical industries. Moreover, the contents of urushiols in *T. verniciflua* are higher in the bark than in its xylem. Therefore, it is necessary to remove its urushiol congeners during food or pharmaceutical applications. It has been reported that fermentation with *F. fraxinea* is an effective way to remove toxic urushiols from TVSB, and that mushroom laccases play an important role in the detoxification [19]. In this study, changes of urushiol congeners during *F. fraxinea*-mediated fermentation were analyzed by HPLC. As shown in Figure 1A, pentadecatrienylcatechol ($C_{15:3}$) (671.08 $\pm$ 25.43 µg/g dry weight (DW)) was the most abundant, followed by pentadecenylcatechol ($C_{15:1}$) (461.80 $\pm$ 4.95 µg/g DW) and pentadecadienylcatechol ($C_{15:2}$) (95.77 $\pm$ 6.43 µg/g DW) in unfermented TVSB (UTVSB). The content of $C_{15:3}$ rapidly decreased from 671.08 $\pm$ 25.43 µg/g DW in UTVSB to 115.22 $\pm$ 11.14 µg/g DW (82.8% decrease) and 48.61 $\pm$ 6.51 µg/g DW (92.8% decrease) after 10 and 13 days of fermentation, respectively. Thereafter, it showed only a slow decrease, reaching 30.66 $\pm$ 2.70 µg/g DW (95.4% decrease) at the end of the fermentation process (22 days). The $C_{15:2}$ and $C_{15:1}$

contents also decreased in a similar pattern to that of $C_{15:3}$. The control was incubated for the same time intervals as the fermented sample. However, HPLC analysis of the controls was carried out only for 0 and 22-day time points, because the thin-layer chromatography (TLC) results (Supplementary Data Figures S1–S4) showed no difference among all of the matched controls. When the TVSB was treated under the sample conditions, but without *F. fraxinea* inoculation as the matched control, no meaningful changes were observed, even after 22 days compared to zero days (Supplementary Data, Figures S1−S4). These results suggest that the gradual decreases of urushiols during fermentation are due to the action of related enzymes secreted from *F. fraxinea*.

The total content of urushiol congeners decreased from 1228.7 $\pm$ 34.6 µg/g DW in UTVSB to 186.02 $\pm$ 24.3 µg/g DW (84.9%), 78.77 $\pm$ 11.2 µg/g DW (93.6%), and 47.75 $\pm$ 9.5 µg/g DW (96.1% decrease) at the end of 10, 13, and 22 days of fermentations, respectively (Figure 1B). Considering these results, the optimal fermentation period for the removal or reduction of urushiol congeners from TVSB by *F. fraxinea* is suggested to be between 13–16 days of fermentation. Choi et al. [19], who attempted a similar biological method for the reduction of urushiols, reported that the content of $C_{15:3}$ decreased rapidly in the first five days of fermentation, followed by a gradual decrease, reaching the lowest levels after 15 days of fermentation.

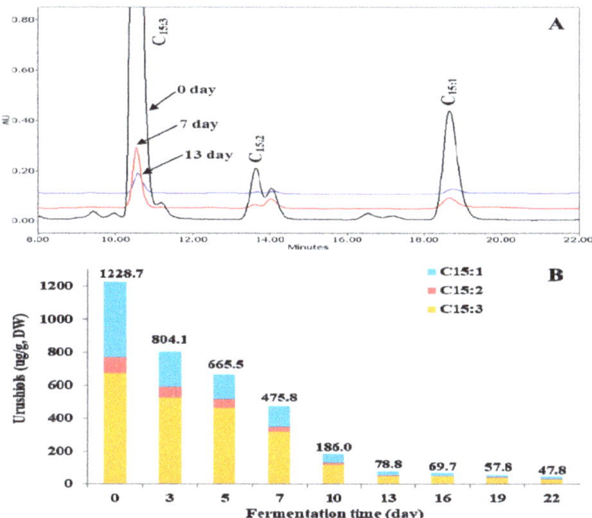

Figure 1. Changes of urushiol congeners during fermentation of *Toxicodendron vernicifluum* stem bark (TVSB) by *F. fraxinea.* Representative HPLC chromatograms (**A**) and contents (**B**).

The degradation of urushiols is closely associated with the laccase that is secreted from *F. fraxinea* mycelia. Laccase is a type of copper-containing polyphenol oxidase (*p*-diphenol oxidase or benzenediol; oxygen oxidoreductase, EC 1.10.3.2). More than 100 fungal laccases have been purified and characterized from *Basidiomycetes* and *Ascomycetes* [26]. Since these enzymes can catalyze the oxidation of phenolic compounds such as polyphenols, methoxy-substituted phenols, *o*-diphenols, *p*-diphenols, aromatic amines, and syringaldazine, it is somewhat difficult to categorize laccases based on the type of substrates, because a wide range of substrates can be catabolized [26–28]. However, laccase can oxidize urushiols and result in the formation of semiquinone radicals under aerobic conditions. This is a typically unstable product, and it undergoes subsequent changes either through an attack on the urushiol nucleus, forming biphenyl compounds, or a disproportionation reaction to give urushiol quinone [29]. It may also undergo polymerization reactions, which results in the formation of a conjugated insoluble product. All of these mechanisms lead to the reduction of the active site of urushiols, thereby resulting in its detoxification [29,30].

2.2. Change of Total Phenol, Total Flavonoid, and Individual Flavonoid

The diverse pharmacological activities of *T. vernicifluum* are believed to be mainly attributed to the presence of phenolic compounds such as fustin, fisetin, sulfuretin, quercetin, taxifolin, garbanzol, butein, gallic acid, dihydroxybenzoic acids, and other phenolic compounds [24,25]. It is known that fustin, fisetin, sulfuretin, and butein are abundant in the bark and xylem of *T. vernicifluum* [12,13]. Therefore, it is necessary to minimize the degradation of these valuable bioactive components during fermentation. Changes to the total phenol and total flavonoid contents in TVSB during *F. fraxinea*-mediated fermentation were investigated.

The total phenol content decreased sharply from 2927.6 ± 228.8 mg GAE/100 g DW in UTVSB to 859.2 ± 156.6 mg GAE/100 g DW after 10 days of fermentation, reaching 599.3 ± 78.1 mg GAE/100 g DW at the end of the fermentation (Figure 2A). The total flavonoid content decreased from 1365.4 ± 130.6 mg RE/100 g DW in UTVSB to 598.3 ± 64.5 mg RE/100 g DW and 272.8 ± 39.6 mg RE/100 g DW after 10 and 22 days of fermentation, respectively (Figure 2B). The reduction of total phenol and total flavonoid during fermentation is closely associated with the polyphenol oxidases that are secreted from *F. fraxinea* mycelia. Polyphenol oxidases (EC 1.14.18.1) that are known as tyrosinase, laccase, catechol oxidase, catecholase, phenolase, cresolase, and urushiol oxidase based on substrate specificity are known to have considerable overlap in substrate affinities [31,32]. These enzymes are distributed in a wide range of fungi, higher plants, and mammals. They can catalyze the oxidation of a broad range of phenolic compounds, including phenols, phenolic acids, flavonoids, tyrosine, L-3,4-dihydroxyphenylalanine (L-DOPA), naphthols, bisphenols, and other phenolic compounds [28,31,33].

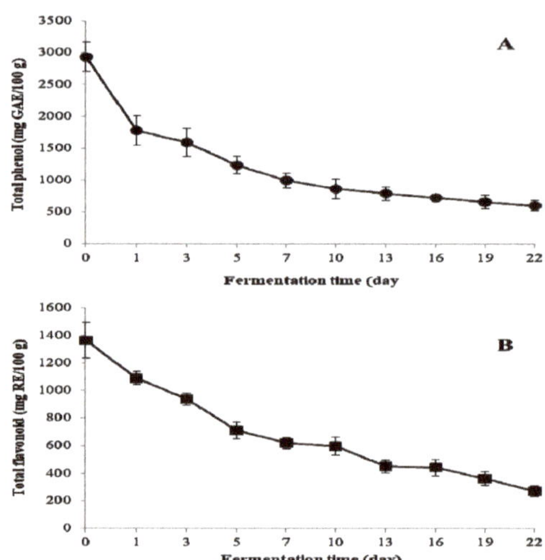

Figure 2. Changes in the contents of total phenol (**A**) and total flavonoid (**B**) during the fermentation of TVSB by *F. fraxinea*. Error bars are the standard deviations of triplicate measurements.

In this study, changes of individual flavonoids during fermentation were also monitored by HPLC analysis (Figure 3A). The contents of major flavonoids such as fustin, fisetin, sulfuretin, and butein showed noticeable decreases at the initial stage (five days) of fermentation, followed by gradual decreases up until 22 days (Figure 3B). It has been previously reported that basidiomycete polyphenol oxidases can effectively utilize flavonoids such as quercetin, fisetin, rutin, luteolin, myricetin, and kaempferol, as well as phenolic acids as substrates [28,31,33,34]. Therefore, the reduction of flavonoids that was observed during the fermentation of the TVSB extract is suggested to be mediated by polyphenol oxidases that are secreted from *F. fraxinea*.

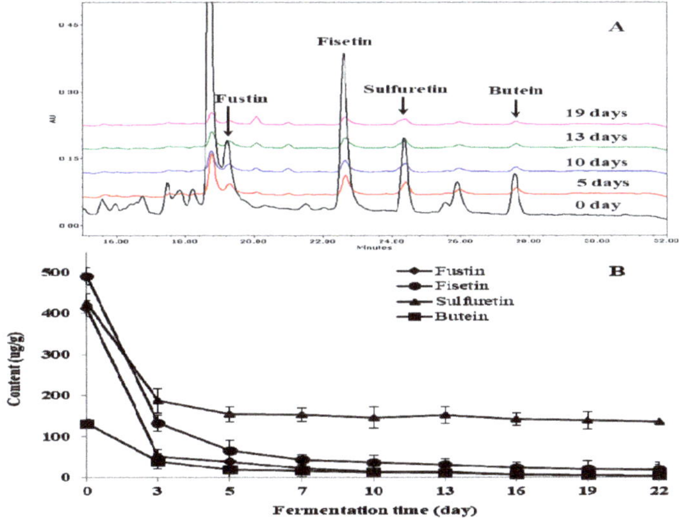

Figure 3. Changes of individual flavonoid compounds during the fermentation of TVSB by *F. fraxinea*. HPLC chromatograms (**A**), contents of individual flavonoids (**B**). Error bars are the standard deviations of triplicate measurements.

2.3. Changes of PGG

Gallotannin appears to have important biological roles including antioxidant, antibacterial, anti-inflammatory, anti-hypoglycemic, anti-angiogenic, and anti-cancer activities. 1,2,3,4,6-penta-O-galloyl-β-D-glucose (PGG), which is a gallotannin, are widely distributed in fruit, berries, and woody plants. PGG and its hydrolysates are among the main groups of antioxidant polyphenols in UTVSB. They have attracted attention in recent years due to their beneficial properties for human health [16,23]. In this study, PGG, gallic acid, and methyl gallate were detected as gallotannins in UTVSB.

PGG content decreased from 5590.0 ± 324.5 µg/g DW in UTVSB to 1580.0 ± 282.2 µg/g DW (71.7% decrease) and 1452.9 ± 211.9 µg/g DW (74.0% decrease) after 10 and 13 days of fermentation, respectively. Only slight changes in PGG content were observed from day 13 to day 22 of fermentation (Figure 4). Gallic acid and methyl gallate contents also showed noticeable decreases within seven days from the start of fermentation.

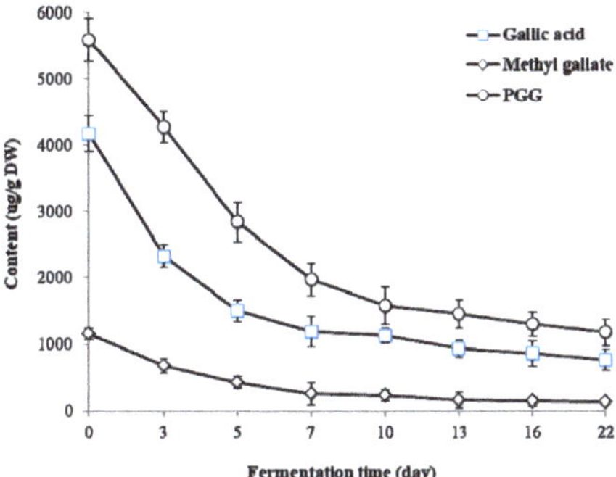

Figure 4. Changes of 1,2,3,4,6-penta-O-galloyl-β-D-glucose (PGG) and its metabolites gallic acid and methyl gallate during the fermentation of TVSB by *F. fraxinea*. Error bars are the standard deviations of triplicate measurements.

2.4. Characterization of PGG Hydrolysates by HPLC-MS

Gallotannins are important bioactive compounds in medicinal plants. PGG is especially abundant in the TVSB [25]. This compound showed a consistent decrease during the initial stage of fermentation, as shown in Figure 4. HPLC analysis (Figure 5B) revealed the formation of new peaks in the 80% methanol extract of fermented TVSB that were not detected in the similar extract of UTVSB (Figure 5A). These peaks were then tentatively identified by HPLC-MS analysis. Total ion current (TIC) chromatograms of the 80% methanol extract from 10-day fermented TVSB and the hydrolysate of PGG by crude enzyme preparation isolated from fermented TVSB are shown in Figure 6. Forty-two compounds in the fermented TVSB and 40 compounds in PGG hydrolysate were identified tentatively (Table 1).

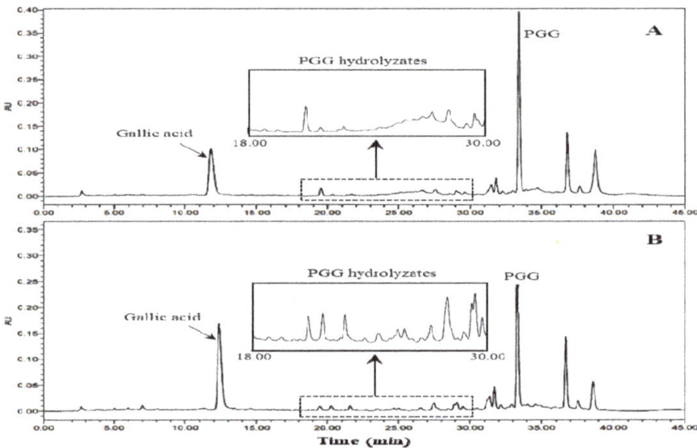

Figure 5. HPLC profiles (at 310 nm) of 80% methanol extracts of the unfermented *Toxicodendron vernicifluum* stem bark (UTVSB) (**A**) and 10-day fermented TVSB (**B**).

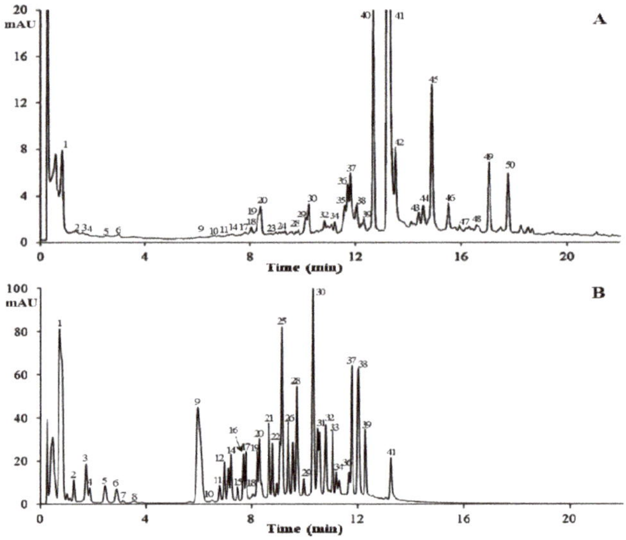

Figure 6. Total ion current (TIC) chromatograms of 10-day fermented TVSB (**A**) and PGG hydrolysate (**B**) by a crude enzyme preparation isolated from fermented TVSB.

The peak **1** at 0.714 min was identified as gallic acid based on a quasimolecular ion at m/z 169.0 [M − H]$^-$. Peaks **2** and **4**, which were detected in both samples, showed the same quasimolecular ions at m/z 331.1 [M − H]$^-$ in the negative electrospray ionization mass spectrometry (ESI-MS) spectra and λ_{max} at 276 nm. These compounds possessed one hexosyl moiety (162 amu) more than the fragment of peak **1** (gallic acid). Thus, these compounds were characterized as monogalloylglucose (MGG) isomers. Thirteen isomers (peaks **3, 5, 6, 11, 12, 13, 16, 17,** and **19–23**) detected in fermented TVSB showed a quasimolecular ion at 483.1 [M − H]$^-$ with λ_{max} at 273–276 nm.

Table 1. Compounds detected in fermented TVSB and PGG hydrolysate by crude enzyme preparation isolated from fermented TVSB.

Peak No.	tR (min)	UV (λ_{max}, nm)	[M − H]$^-$ (m/z)	Other Fragment (m/z)	Identification	Detection FTVSB	Detection PGGH
1	0.714	272	169.0	331.1	Gallic acid	O	X
2	1.264	276	331.1	445.1, 271.0	MGG	O	O
3	1.699	276	483.1	597.0, 224.9	DGG	O	O
4	1.890	276	331.1	445.1, 270.9	MGG (isomer)	O	O
5	2.371	276	483.1	597.0	DGG (isomer)	O	O
6	2.781	273	483.1	181.0, 597.0	DGG (isomer)	O	O
7	3.141		255.1	123.0	Unidentified	X	O
8	3.500		301.0	415.0, 197.8	Unidentified	O	O
9	5.848	276	183.1	375.1	Methyl gallate	O	O
10	6.464		283.0	396.8	Unidentified	O	O
11	6.744	276	483.1	241.0, 596.9	DGG (isomer)	O	O
12	6.950	275	483.1	597.1, 241.1	DGG (isomer)	X	O
13	7.118	273	483.1	597.0, 241.0	DGG (isomer)	O	O
14	7.234	276	225.1	483.1, 339.0	Unidentified	O	O
15	7.449		283.1	397.0, 575.0	Unidentified	X	O
16	7.692	275	483.1	241.1, 597.0	DGG (isomer)	X	O
17	7.799	275	483.1	597.0, 241.1	DGG (isomer)	O	O
18	7.991		283.1	397.1	Unidentified	O	O
19	8.215	277	483.1	241.0, 597.0	DGG (isomer)	O	O
20	8.313	277	483.1	597.0, 241.1	DGG (isomer)	O	O
21	8.658	216, 273	483.1	597.1, 241.0	DGG (isomer)	X	O
22	8.803	216, 273	483.2	597.0, 295.0	DGG (isomer)	X	O
23	8.962	275	483.2	597.1, 241.0	DGG (isomer)	O	O
24	9.097	276	635.1	749.1, 317.1	TGG	O	O
25	9.111	216, 279	431.2	499.1, 563.1	Unidentified	X	O
26	9.386	216, 276	635.1	749.1, 317.2	TGG (isomer)	O	O
27	9.578	217, 276	635.1	317.1	TGG (isomer)	O	O
28	9.708	216, 276	635.1	749.1, 317.1	TGG (isomer)	O	O
29	9.979	216, 276	635.1	317.0, 169.0	TGG (isomer)	O	O
30	10.315	217, 278	635.1	317.1, 169.0	TGG (isomer)	O	O
31	10.502	217, 278	635.1	317.2	TGG (isomer)	O	O
32	10.777	217, 278	787.1	635.1, 393.1	TeGG	O	O
33	11.057	217, 278	635.1	317.1, 749.1	TGG (isomer)	O	O
34	11.225	218, 277	787.1	393.2, 309.0	TeGG (isomer)	O	O
35	11.328	217, 278	635.1	317.2, 749.0	TGG (isomer)	X	O
36	11.688	217, 238	301.0	610.9	Ellagic acid	O	O
37	11.790	217, 278	787.1	393.2, 301.1	TeGG (isomer)	O	O
38	12.005	217, 279	787.1	393.2	TeGG (isomer)	O	O
39	12.280	217, 279	787.1	393.1	TeGG (isomer)	O	O
40	13.195	217, 279	787.1	393.1	TeGG (isomer)	O	X
41	13.256	217, 279	939.1	469.1	PGG	O	O
42	13.503	316, 360	285.1	113.0	Fisetin	O	X
43	14.344		255.1	432.9	Unidentified	O	X
44	14.549		401.0	287.2, 723.4	Unidentified	O	X
45	14.806		209.1	539.1	Unidentified	O	X
46	15.441	265, 264	301.1	415.1, 603.0	Quercetin	O	X
47	16.430		423.1	271.1, 536.9	Unidentified	O	X
48	16.995	261, 379	271.1	551.1, 385.1	Butein	O	X
49	21.014		417.1	531.0	Unidentified	O	X
50	23.105		653.0	539.0, 518.7	Unidentified	O	X

DGG: digalloylglucose; FTVSB, 10-days fermented TVSB; MGG: monogalloylglucose; PGGH, PGG hydrolysate; O, detected; TeGG: tetragalloylglucose; TGG: trigalloylglucose; X, not detected.

These compounds possessed one galloyl moiety (152 amu) more than monogalloylglucose. Accordingly, these compounds were tentatively characterized as digalloylglucose (DGG) isomers. In a similar manner, nine isomers (peaks **24, 26–31, 33,** and **35**) showed quasimolecular ions at *m/z* 635.1 [M − H]⁻ and similar UV profiles. They possessed one galloyl moiety (152 amu) more than DGG. These compounds were characterized as trigalloylglucose (TGG) isomers. Six isomers at peaks **32, 34,** and **37–40** were also suggested as tetragalloylglucose (TeGG) isomers by their quasimolecular ion at *m/z* 787.1 [M − H]⁻ and UV profiles (λ_{max} at 273–276 nm). The mass spectra of oligomeric galloylglucoses that were identified in both samples are presented in Figure 7.

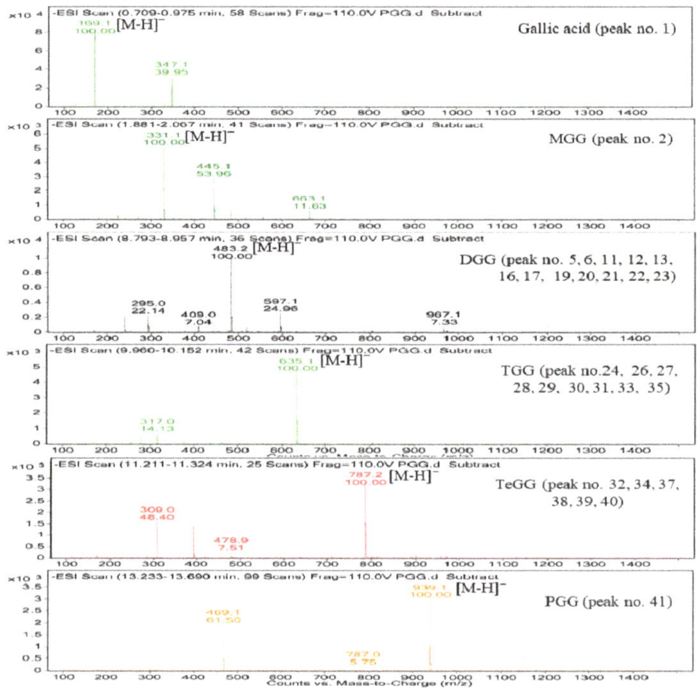

Figure 7. LC-MS spectra of PGG and its hydrolysates identified in fermented TVSB.

Some of these PGG hydrolysates have been previously found in *Euvcalyptus nitens* wood [35], mango [36], *Rhus coriaria* [37], *Psidium guineense, Syzygium cumini, Pouteria macrophylla* [38], and the degradation products of tannin by *Aspergillus niger* [39]. These compounds can affect the expression of biological activities of target plants. In addition, galloylglucose oligomers may have more potent antioxidant activities than PGG or ascorbic acid [40]. Generally, microbial tannase (E.C.3.1.1.20) hydrolyses the ester and depside bonds of gallotannins and PGG to produce gallic acid and methyl gallate as end products [41,42]. However, the PGG-degrading enzyme from *F. fraxinea* in this study catalyzed the hydrolysis of the bonds that were present in the molecules of PGG to produce oligomeric galloylglucoses—mainly DGG, TGG, and TeGG—instead of gallic acid. In addition, peak **36** was tentatively identified as ellagic acid based on its quasimolecular ion at 301.0 [M − H]⁻ and its UV profile. Peaks **42, 46,** and **48** with quasimolecular ions [M − H]⁻ of *m/z* 285.1, *m/z* 301.1, and *m/z* 271.1 were identified as fisetin, quercetin, and butein, respectively. These compounds have been previously reported as the major bioactive components of *T. vernicifluum* [24,25].

2.5. Changes of Antioxidant Activities

Due to the differences in the theoretical bases of different antioxidant measurements, a single antioxidant property model can hardly reflect the antioxidant capacity of the samples [43]. For this reason, three model systems, i.e., 2,2-diphenyl-1-picrylhydrazyl (DPPH) radical scavenging activity, 2,2′-azino-bis(3-ethylbenzothiazoline-6- sulfonic acid) (ABTS) radical scavenging activity, and ferric reducing/antioxidant power (FRAP) were used to evaluate the antioxidant properties of fermented TVSB. DPPH and ABTS rely on the reaction of radicals and cation radicals, respectively, and the FRAP method relies on the reduction by the antioxidant components of complex ferric ion–TPTZ (2,4,6-tri(2-pyridyl)-*s*-triazine). A study by Gorinstein et al. [44] observed a high correlation of antioxidant capacities measured by ABTS, DPPH, and FRAP assays in some fruits, while Pellegrini et al. [45] reported a weak correlation on antioxidant capacity measured by FRAP and ABTS assays in some vegetables and beverages. The results of changes in antioxidant capacities are presented in Figure 8. DPPH radical scavenging activity rapidly decreased from $82.45 \pm 1.41\%$ in unfermented TVSB (day 0) to $21.31 \pm 1.35\%$ after nine days and $15.88 \pm 0.85\%$ after 12 days of fermentation, while no significant change was observed thereafter until 22 days. The antioxidant properties of *T. vernicifluum* stem and bark can be explained by the higher contents of phenolic compounds such as fustin, fisetin, sulfuretin, quercetin, gallotannins, and urushiols [6,7].

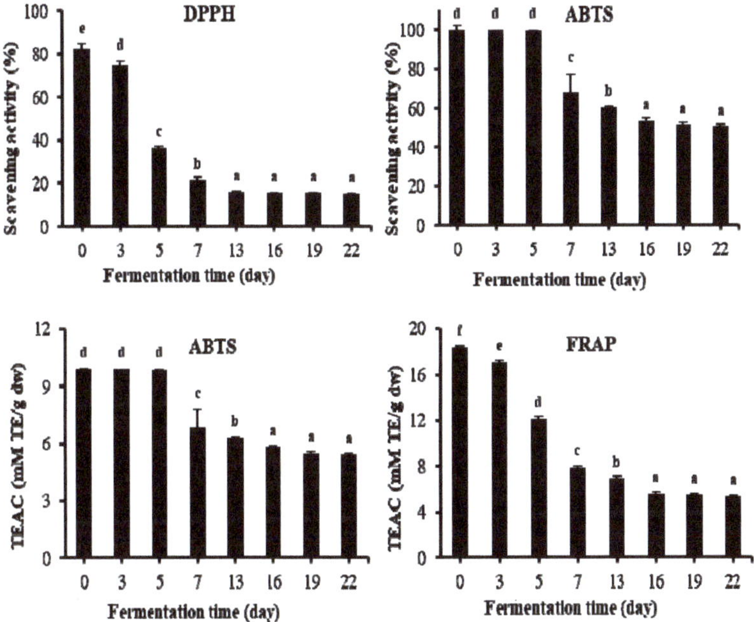

Figure 8. Changes of antioxidant activities during fermentation of TVSB by *F. fraxinea*. Different letters indicate values that are significantly different by Duncan's multiple range test at 5% level. TEAC, Trolox equivalent antioxidant capacity. DPPH, 2,2-diphenyl-1-picrylhydrazyl; ABTS, 2,2′-azino-bis-3-ethylbenzothiazoline-6-sulfonic acid; FRAP, ferric reducing antioxidant power. Error bars are the standard deviations of triplicate measurements.

ABTS radical scavenging activity as an inhibition percentage (%) showed a significant decrease ($p < 0.05$) from $99.83 \pm 1.96\%$ in the unfermented sample to $67.85 \pm 9.40\%$ and $53.04 \pm 2.11\%$ after nine and 15 days of fermentation, respectively. No meaningful change was observed thereafter until 22 days. Trolox equivalent antioxidant capacity (TEAC) is a method that provides information on the overall status of antioxidants within a test sample, and has proven to be a useful indicator for determining the ability of an organism to mitigate the potential damage caused by reactive oxygen species. It has been widely used for studying the antioxidant capacity of phytochemicals and biological samples [46]. In this study, the TEAC value also decreased from 9.89 ± 0.06 mM TE/g DW in the unfermented sample to 6.82 ± 0.03 mM TE/g DW (31% decrease) and 6.29 ± 0.03 mM TE/g DW (36.4% decrease) after nine and 15 days of fermentation, respectively, with no significant change until 22 days. Similarly, the FRAP value decreased significantly ($p < 0.05$) during the first 12 days of fermentation. It decreased by 57.3% (from 18.39 ± 0.10 to 7.85 ± 0.21 mM/g DW), with no significant decrease observed thereafter until 22 days.

Phenolic compounds can be enzymatically oxidized and polymerized by polyphenol oxidases such as laccases or tyrosinases. Previous studies have reported that oxidized products, polymers, and aggregates exhibit stronger antioxidant activities than their corresponding monomeric flavonoids [34,47–49]. However, it can be clearly observed that fermented TVSB still retained antioxidant activities, although it was relatively weaker compared to that of UTVSB. Kim et al. [7] reported that the methanol and ethyl acetate fractions of fermented TVSB by *F. fraxinea* showed higher radical scavenging and reducing power activities than those of synthetic antioxidant butylated hydroxyanisole (BHA) and butylated hydroxytoluene (BHT). Therefore, fermented TVSB might be more preferable than the above synthetic antioxidant products. Oligomeric galloylglucoses that originated from PGG such as MGG, DGG, TGG, and TeGG may also have contributed to the antioxidant activities of fermented TVSB.

3. Materials and Methods

3.1. Plant Material

TVSB was collected from a 10-year-old tree that had been cultivated in Imsil-Gun, Chonbuk Province, Republic of Korea in October 2015. The plant material was authenticated by one of the authors (M.K. Kim). A voucher was deposited at the Fermentation Laboratory, Department of Food Science and Biotechnology, Chonbuk National University, Jeonju 54896, Jeonbuk, Republic of Korea. The fresh TVSB was air-dried (60 °C) for 24 hours and kept at room temperature (20 °C) until used for further experiments within one week.

3.2. Chemicals

Urushiol standards ($C_{15:3}$, $C_{15:2}$, $C_{15:1}$) were purchased from Phytolab GmbH and Co. (Dutendorfer Straße, Vestenbergsgreuth, Germany). Fustin, fisetin, sulfuretin, and butein were purchased from Chromadex Co. (Irvine, CA, USA). Gallic acid, methyl gallate, quercetin, taxifolin, PGG, Folin-Ciocalteu reagent, 2,2-diphenyl-1-picrylhydrazyl (DPPH), 2,2′-azino -bis(3-ethylbenzothiazoline-6-sulfonic acid), diammonium salt (ABTS), 6-hydroxy-2,5,7,8 -tetramethylchromane-2-carboxylic acid (Trolox), 2,4,6-tri(2-pyridyl)-*s*-triazine (TPTZ), potassium persulfate, tannic acid, bovine serum albumin (BSA), and sodium dodecyl sulfate (SDS) were purchased from Sigma-Aldrich (St. Louis, MO, USA). HPLC grade methanol, acetonitrile, and deionized water were purchased from J.T. Baker Co. (Phillipsburg, NJ, USA). All of the other reagents were of analytical grade.

3.3. Microorganism and Culture Conditions

The strain of *Fomitella fraxinea* (Bull.) Imazeki (KACC 42289) was kindly donated by the Korean Agricultural Culture Collection (KACC) of the Rural Development Administration (RDA), Wanju, Jeonbuk, Republic of Korea.

3.4. Fermentation of TVSB by F. fraxinea

The strain was preincubated onto potato dextrose agar (Becton, Dickinson and Company, Sparks, MD, USA) for six days at 25 °C. Sterilization of the culture media was performed at 121 °C for 30 min. The preincubated strain was inoculated into germinated-malt medium (11 Brix°) saccharified at 65 °C with four-fold tap water (*v/v*) for eight hours, and then cultured for two weeks at 25–26 °C with gentle shaking (120 rpm) using an orbital shaker (model SK-600, Jeiotech Co., LTD. Daejon, Korea). The air-dried TVSB of crushed coarse powder was placed in an Erlenmeyer flask (500 mL). The moisture content of each powdered sample (100 g) was then adjusted to approximately 65% with tap water. Each flask was sterilized at 121 °C for 30 min. The day-0 (control) sample was freeze-dried immediately and stored at −20 °C until further analysis. The remaining samples were inoculated with five mL of *F. fraxinea* liquid culture except for the unfermented (matched controls) and incubated at room temperature (25–26 °C) with gentle shaking (120 rpm) for each required time frame before further analysis. All of the fermentations were set up on the same day, and each sample for the specific time point was retrieved with its matched control at the same time. Each sample was then freeze-dried at the end of each time frame and stored until use.

3.5. Extraction and HPLC Analysis of Urushiols.

3.5.1. Extraction

A powdered sample (1.0 g) of UTVSB and fermented TVSB was extracted twice with 20 mL of acetone in an ultrasonic bath (Hwa Shin Instrument Co., Ltd., Seoul, Korea) for 20 min at room temperature and centrifuged at 5000 rpm for 15 min. Supernatants were combined and concentrated under vacuum at 45 °C, and the residue was dissolved in acetone (five mL).

3.5.2. Thin-Layer Chromatography (TLC) Analysis

TLC was carried out for the extracted samples as follows. The developing solvent was a mixture of chloroform–methanol–water (65:35:10, lower phase). The spots were detected either under UV (254 nm) or by spraying a 10% $CuSO_4$ solution in an 8% H_2SO_4 solution followed by heating at 110 °C for 10 min.

3.5.3. HPLC Analysis

The standard stock solution of urushiols was prepared at a concentration of 1000 µg/mL in acetone. The stock solution was serially diluted with acetone to obtain a calibration curve at seven concentration levels (10–500 µg/mL) for C_{15} and $C_{15:1}$ and at 5–250 µg/mL for $C_{15:2}$. HPLC analysis was performed using an HPLC system (Waters, Milford, MA, USA) equipped with a 2690 separation module and 996 photodiode array (PDA) detector with a YMC-Pak Pro C18 column (4.6 mm × 250 mm, five µm; YMC Co., LTD, Tokyo, Japan). The mobile phase was 90% MeOH in deionized water at a flow rate of 1.0 mL/min (isocratic) with an injection volume of 10 µL. The UV detection wavelength was set at 273 nm. The concentration of each constituent was calculated using the calibration curve by plotting the peak area of the corresponding substance against the concentration (in µg/mL) of the standard substance.

3.6. Extraction for Polyphenols and Gallotannins

A powdered sample (1.0 g) of UTVSB and fermented TVSB was extracted twice with 20 mL of 80% aqueous MeOH using an ultrasonic bath for 30 min and centrifuged at 5000 rpm for 15 min. The supernatants was concentrated at 45 °C under reduced pressure, and the residue was dissolved in 80% methanol (five mL) for an analysis of total flavonoids, individual flavonoids, and gallotannins by HPLC, and for antioxidant activity assay.

3.7. Total Phenol

The total phenolic content of the samples was determined according to a method described by Chandra et al. [50] with some modifications. Briefly, an 80% MeOH extract (20 μL) of each sample was mixed with a 50% Folin–Ciocalteu phenol reagent (20 μL) in 96-well plates. After five minutes, one N of sodium carbonate (20 μL) was added to the mixture, and distilled water was added to reach a final volume of 200 μL. After incubation at room temperature in the dark for 30 min, the absorbance of a test sample against a blank was measured at a wavelength of 725 nm using a VersaMax ELISA microplate leader (Molecular Devices, LLC, CA, USA). The phenolic content was calculated based on a calibration curve of gallic acid. The result was expressed as mg of gallic acid equivalent (GAE) per 100 g of the dried sample.

3.8. Total Flavonoid

The total flavonoid content was measured according to a method described by Zhishen et al. [51] with some modifications. Briefly, an 80% MeOH extract (30 μL) of each sample was mixed with 30 μL of 5% sodium nitrite solution. After five minutes of reaction, 300 μL of 5% aluminum chloride was added. Then, 200 μL of one N of NaOH was added six minutes later, and the total volume was adjusted to one mL with distilled water. The absorbance of the test sample against a blank was measured at a wavelength of 510 nm with a Shimadzu UV-1601 spectrophotometer (Kyoto, Japan). The flavonoid content was calculated using a calibration curve of rutin. The result was expressed as mg rutin equivalent (RE) per 100 g of the dried sample.

3.9. HPLC Analysis of Individual Flavonoids

The stock solution was serially diluted with methanol to obtain seven concentration levels (12.5–750 μg/mL) for each analyte to prepare a calibration curve. HPLC analysis was performed using an HPLC system (Waters, Milford, MA, USA) equipped with a 2690 separation module and a 996 photodiode array (PDA) detector with a ZORBAX Eclipse XDB C18 column (4.6 mm × 250 mm, five μm; Agilent Technologies, Technologies, Inc., Santa Clara, CA, USA). The mobile phase was 0.1% formic acid in ionized water (A) and 90% acetonitrile in water (B). The ratio of the mobile phase was A:B mixed at 90:10 (zero to one minute), 20:80 (one to 15 min), and 90:10 (15–25 min) at a flow rate of 1.0 mL/min. The UV detection wavelength was set at 280 nm. The concentration of each constituent was calculated using a calibration curve by plotting the peak area of the corresponding substance against the concentration (μg/mL) of the standard substance.

3.10. Hydrolysis of PGG by Tannases

3.10.1. Preparation of PGG

PGG was prepared from tannic acid according to the method of Chen and Hagerman [52] with slight modification. A sample of 5.0 g of tannic acid was methanolyzed in 100 mL of 70% aqueous methanol in 0.1 M of sodium acetate (pH 5.00 at 65 °C) After methanolysis for 15 h, the pH of the reaction mixture was adjusted to 6.0, and methanol was removed by evaporation under reduced pressure below 35 °C. Water was added to maintain the volume, replacing evaporated methanol. The resulting aqueous solution was sequentially extracted with three volumes of diethyl ether and ethyl acetate. The ethyl acetate fraction was combined and evaporated under reduced pressure, and then, distilled water was added to the ethyl acetate extract. The resulting milky suspension was centrifuged for 15 min at 4500 rpm (model VS-550, Vision Scientific Co., LTD, Daejon, Korea). The precipitate containing PGG was washed twice with 20 mL of ice-cold 2% aqueous methanol. In the last step, the final material was obtained by freeze-drying. The material obtained was identified as PGG by LC-MS and NMR spectroscopy (JEOL JNM-ECA 600 FT-NMR, Akishima, Tokyo, Japan) as follows: UV, λ_{max} 280 nm (methanol); ESI-LC-MS, m/z 939.2 [M − H]$^-$ ($C_{41}H_{32}O_{26}$); ^{1}H−NMR (600 MHz, methanol-d_4), δ 7.01 (2H, s), 6.95 (2H, s), 6.87 (2H, s), 6.84 (2H, s), 6.79 (2H, s), 6.13 (1H, d,

J = 8.25 Hz), 5.81 (1H, t, J = 9.62 Hz), 5.51 (1H. dd, J = 9.62 Hz), 5.50 (1H, t, J = 9.62 Hz), 4.42 (1H, d, J = 11.68 Hz), 4.30 (2H, m); $^{13}C-$NMR (150 Hz, methanol-d_4), δ 168.05, 167.41, 167.14, 167.04, 166.33, 146.66 (x2), 146.58, 146.54, 146.48, 146.39, 140.87, 140.47 (x2) 140.41(x2), 140.24 (x2), 140.12 (x2), 121.15, 120.47, 120.34, 120.31, 119.84, 110.73 (x2), 110.57 (x2), 110.51 (x2), 110.48 (x2), 110.44 (x2), 93.93, 74.53, 74.22, 72.30, 69.91, 63.22.

3.10.2. Preparation of Crude Enzymes from Fermented TVSB

All of the procedures for the preparation of crude enzyme were carried out at 4 °C unless otherwise indicated. TVSB (50 g) that had been fermented for 10 days with *F. fraxinea* was mixed with 500 mL of 0.1 M of sodium phosphate buffer (pH 5.5) with gentle stirring for four hours, and then homogenized with an Omni mixer homogenizer (Omni International, Kennesaw, GA, USA) for one minute. The homogenate was squeezed through a cheese cloth, and the filtrate was centrifuged at 10000 $\times$ g for 20 min. The crude enzyme containing tannase was prepared by solid ammonium sulfate (30–80%) saturation. After centrifugation at 10000 $\times$ g for 20 min, the precipitate was dissolved in 10 mM of sodium acetate buffer (pH 5.5). After dialysis for 24 hours, the solution was centrifuged at 10000 $\times$ g for 20 min, and the supernatant was lyophilized.

3.10.3. Tannase Assay

Lyophilized crude enzyme preparation (50 mg) was dissolved in 10 mL of 0.1 M of sodium acetate buffer (pH 5.5). Tannase activity was measured using the method of Mondal et al. [53] with modifications. Briefly, the solution (150 µL) containing three mM of tannic acid (0.2 M of acetate buffer, pH 5.5) as a substrate was added to 200 µL of crude enzyme, and the reaction solution was incubated at 40 °C for 30 min. Subsequently, a BSA solution (1 mg/mL) was added to the flask and then kept at room temperature for 15 min. The reaction solution was centrifuged (5000 rpm, 20 min) to remove the supernatant, and 1.5 mL of SDS–triethanolamine solution was added to the precipitate. To stabilize the color, 0.5 mL of 0.16% $FeCl_2$ (0.01 N HCl) was added, and the mixture was allowed to stand at room temperature for 15 min. Its absorbance was then measured at a wavelength of 530 nm. The calibration curve was prepared using tannic acid under the same conditions. One unit (U) of tannase was defined as the amount of enzyme that was required to hydrolyze 1.0 µM of tannic acid per minute under specified conditions.

3.10.4. Enzymatic Hydrolysis of PGG

The reaction mixture (50 mL) containing 0.5 g of PGG in 2.5 mL of methanol and crude enzyme preparation (30–80% ammonium sulfate precipitate) isolated from fermented TVSB or *Asp. oryzae* tannase containing 30 U of tannase activity in 47.5 mL of 0.1 M sodium acetate buffer (pH 5.5) were incubated for 12 hours at 45 °C with gentle shaking. After the reaction mixture was kept in a boiling water bath for 10 min, it was subjected to HPLC and LC-MS analysis.

3.11. Analysis of PGG Hydrolysates in Fermented TVSB

3.11.1. HPLC

The standard solution was serially diluted with 80% methanol to obtain seven concentration levels (12.5–750 µg/mL) for each compound to prepare a calibration curve. HPLC analysis was performed using an HPLC system equipped with a 2690 separation module and a 996 photodiode array (PDA) detector with a ZORBAX Eclipse XDB$-C_{18}$ column. The mobile phase consisted of 0.1% formic acid in ionized water (A) and 90 % acetonitrile in water (B). The ratio of A:B as the mobile phase was maintained at 95:5 (zero to two minutes), 45:55 (two to 25 min), 40:60 (25 to 30 min), and 95:5 (30 to 40 min) at a flow rate of 0.8 mL/min. The UV detection wavelength was set at 310 nm.

3.11.2. HPLC-MS

The powder of fermented TVSB (1.0 g) was extracted with 20 mL of acetone in an ultrasonic bath (Hwa Shin Instrument Co., Ltd., Seoul, Korea) for 20 min at room temperature and centrifuged at 4500 rpm for 20 min. The residue was extracted one more time using the same solvent. The supernatant was concentrated under vacuum at 45 °C, and the residue was dissolved in five mL of acetone. A standard stock solution of urushiols was prepared at a concentration of 1000 μg/mL in acetone. The stock solution was serially diluted with acetone to obtain a calibration curve at seven concentration levels (10–500 μg/mL) for C_{15} and $C_{15:1}$, and at 5–250 μg/mL for $C_{15:2}$.

HPLC-MS analysis was performed on an Agilent chromatographic system 1100 series (Agilent Technologies, Santa Clara, CA, USA) equipped with a G1379A degasser, a G1312A binary pump, a 1329A autosampler, and a G1316A column thermostat. The LC was coupled with an Agilent ion trap 1100 SL mass detector. Separation was performed using a reverse phase column Kinetex C_{18} (50 × 2.1 mm i. d., 2.6 μm, Phenomenex, Torrance, CA, USA). The mobile phase consisted of 0.1% formic acid in water (A) and 0.1% formic acid in 90% acetonitrile with the following gradient program: zero to four minutes, 90% A; four to 25 min, 90–40% A; 25 to 30 min, 40–90% A; and 30 to 35 min, 90% A. The flow rate was set at 0.5 mL/min, and the temperature was set at 40 °C. The injection volume was two μL. The MS was equipped with an electrospray ionization (ESI) interface in negative ion mode. The ESI parameters were set as follows: ion source temperature, 350 °C; gas flow rate (nitrogen), 15 L/min; nebulizer pressure, 40 psi pressure; capillary voltage, 3000 V; fragmentation voltage, 400 V; collision energy, 0 V; and full-scan data acquisition, 50–1200 *m/z*.

3.12. Antioxidant Activity

3.12.1. DPPH Free Radical Scavenging Activity

The DPPH radical scavenging activity of the sample was determined according to the method described by Thaipong et al. [54] with some modifications. Briefly, the extract (20 μL) was added to an 80-μL DPPH solution (500 μM) and 100-μL Tris-HCl buffer (0.1 M). The mixture was incubated at room temperature in the dark for 20 min. As a blank, the test was repeated using a buffer instead of a sample. Absorbance was measured at a wavelength of 517 nm using a microplate leader, and the scavenging activity of the extract was calculated against a blank as follows.

$$\text{DPPH radical scavenging activity (\%)} = (1 - A_0/A_1) \times 100 \tag{1}$$

where A_0 and A_1 are the absorbance values of the test sample and control, respectively.

3.12.2. ABTS Free Radical Scavenging Activity

ABTS free radical scavenging activity was determined by the methods described by Thaipong et al. [54] with some modifications. Briefly, a mixture of ABTS (7.4 mM) solution and potassium persulfate (2.6 mM) solution in equal volumes was kept for 12 hours at room temperature in a dark to form an ABTS cation. The solution was diluted by mixing with methanol to obtain an absorbance of 1.0 ± 0.02 at 734 nm using a UV-Vis spectrophotometer. Then, 100 μL of the extract was added to 1400 μL of the diluted ABTS solution, and the mixture was incubated at room temperature for one hour in the dark. After the reaction, its absorbance was measured at a wavelength of 734 nm. The calibration curve was linear between 25–400 μM Trolox. Results were expressed as μM Trolox equivalent (TE) g dry weight, and the ABTS radical scavenging activity (%) was also calculated with the following equation:

$$\text{ABTS radical scavenging activity (\%)} = (1 - A_0/A_1) \times 100 \tag{2}$$

where A_0 and A_1 are the absorbance values of the test sample and control, respectively.

3.12.3. Ferric Reducing/Antioxidant Power (FRAP)

Ferric reducing power was determined using FRAP assay [55] with some modification. The working solution for FRAP assay was prepared by mixing 10 volumes of 300 mM of acetate buffer, pH 3.6, with one volume of 10 mM of TPTZ in 40 mM of HCl, and with one volume of 20 mM of ferric chloride. All of the required solutions were freshly prepared before their uses. A sample extract (80 µL) was added to 1420 µL of FRAP reagent. The reaction mixture was incubated at room temperature for 30 min in the dark. Then, the absorbance of the samples was measured at 593 nm. The calibration curve was linear between 25–600 µM Trolox. Results were expressed as µM Trolox equivalent (TE)/g dry weight.

3.13. Statistical Analysis

All of the experiments were performed in triplicate. All of the values were expressed as mean $\pm$ standard deviation (SD). All of the statistical analyses were performed with SPSS (ver. 10.1) for Windows. One-way analysis of variance (ANOVA) and Duncan's multiple range test was carried out to test any significant difference among various treatments. Significant differences were determined at $p < 0.05$.

4. Conclusions

Dynamic changes in urushiol congeners, polyphenols, gallotannins, and antioxidant activities during the fermentation of TVSB by mushroom *F. fraxinea* to reduce or detoxify urushiols were investigated. The content of urushiol congeners noticeably decreased within 10 days of fermentation with a slow decrease thereafter until 22 days. The contents of total phenol, total flavonoid, and individual flavonoids also decreased in similar patterns as those of the urushiols. PGG was hydrolyzed during the fermentation process, resulting in the formation of a number of oligomeric galloylglucoses isomers. These results indicate that PGG was mainly hydrolyzed into oligomeric galloylglucoses (TeGG, TGG, DGG, and MGG) by tannin-hydrolyzing enzymes secreted from *F. fraxinea* mycelia during fermentation. The overall decrease in antioxidant activity during the fermentation of TVSB may be associated with a decrease in phenolic compounds. Nonetheless, the fermented TVSB has to some extent retained strong antioxidant activity, although it is relatively weaker than that of UTVSB. Oligomeric galloylglucoses such as MGG, DGG, TGG, and TeGG, which originated from PGG may also have contributed to the antioxidant activity of fermented TVSB. Although TVSB extracts are fermented for 20–28 days using *F. fraxinea* to be marketed as functional food raw materials in Korea, the present study suggests that the optimal fermentation period to remove or reduce toxic urushiols is between 13–16 days, with a minimized reduction of useful constituents such as flavonoids and gallotannins. The findings in the present work also provide clear evidence for the need of additional studies on the enzymatic characteristics of PGG-hydrolyzing enzymes secreted from *F. fraxinea* during the fermentation of TVSB.

Supplementary Materials: Supplementary materials are available online. Figures S1–S4.

Author Contributions: M.-K.K. and H.-S.C. designed overall experiments. Preparation of plant materials and fermentation work were performed by D.-W.K., M.-J.K. and D.-H.K. HPLC analysis were performed by M.-J.K. G.-Y.K. and J.-K.K. M.-J.K. and D.-H.K. performed antioxidant activity and enzyme assays. Y.-H.K. performed structural elucidation of LC-MS data. M.-K.K. and Y.-H.K. drafted the manuscript. Y.A.G. organized new supplementary data and carried out major revision of the manuscript. All authors read and approved the final manuscript.

Funding: This work was supported by 2017 Regional Food Strategy Development Project (Grant number: PJ0610240-000), Imsil-gun, Jeonbuk, Republic of Korea.

Acknowledgments: The authors would like to thank Center for University–Wide Research Facilities, Chonbuk National University, for providing help with LC-MS analysis.

Conflicts of Interest: The authors declare no conflict of interest.

References

1. Wiart, C. A note on *Toxicodendron vernicifluum* (Stokes) F.A. Barkley. *Food Chem. Toxicol.* **2014**, *64*, 410. [CrossRef] [PubMed]
2. Zhao, M.; Liu, C.; Zheng, G.; Wei, S.; Hua, Z. Comparative studies of bark structure, lacquer yield and urushiol content of cultivated *Toxicodendron vernicifluum* varieties. *N. Z. J. Bot.* **2013**, *51*, 13–21. [CrossRef]
3. Kim, J.H.; Shin, Y.C.; Ko, S.G. Integrating traditional medicine into modern inflammatory diseases care: Multitargeting by *Rhus verniciflua* Stokes. *Mediat. Inflamm.* **2014**. [CrossRef] [PubMed]
4. Choi, W.C.; Jung, H.S.; Kim, K.S.; Lee, S.K.; Yoon, S.W.; Park, J.H.; Kim, S.H.; Cheon, S.H.; Eo, W.K.; Lee, S.H. *Rhus verniciflua* Stokes against advanced cancer: A perspective from the Korean integrative cancer center. *J. Biomed. Biotechnol.* **2012**. [CrossRef]
5. Liu, C.S.; Nam, T.G.; Han, M.W.; Ahn, S.M.; Choi, H.S.; Kim, T.Y.; Chun, O.K.; Koo, I.K.; Kim, D.O. Protective effect of detoxified *Rhus verniciflua* Stokes on human keratinocytes and dermal fibroblasts against oxidative stress and identification of the bioactive phenolics. *Biosci. Biotechnol. Biochem.* **2013**, *77*, 1682–1688. [CrossRef]
6. Lim, K.T.; Hu, C.; Kitts, D.D. Antioxidant activity of a *Rhus verniciflua* Stokes ethanol extract. *Food Chem. Toxicol.* **2001**, *39*, 229–237. [CrossRef]
7. Kim, M.O.; Yang, J.F.; Kwon, Y.S.; Kim, M.J. Antioxidant and anticancer effects of fermented *Rhus verniciflua* stem bark extracts in HCT-116 cells. *ScienceAsia* **2015**, *41*, 322–328. [CrossRef]
8. Jeon, W.K.; Lee, J.H.; Kim, H.K.; Lee, A.Y.; Lee, S.O.; Kim, Y.S.; Ryu, S.Y.; Kim, S.Y.; Lee, Y.J.; Ko, B.S. Anti-platelet effects of bioactive compounds isolated from the bark of *Rhus verniciflua* Stokes. *J. Ethnopharmacol.* **2006**, *106*, 62–69. [CrossRef]
9. Lee, J.H.; Kim, M.; Chang, K.H.; Hong, C.Y.; Na, C.S.; Dong, M.S.; Lee, D.; Lee, M.Y. Antiplatelet effects of Rhus verniciflua stokes heartwood and its active constituents–fisetin, butein, and sulfuretin–in rats. *J. Med. Food.* **2015**, *18*, 21–30.
10. Lee, E.J.; Lee, G.H.; Sohn, S.H.; Bae, H.S. Extract of *Rhus verniciflua* Stokes enhances Th1 response and NK cell activity. *Mol. Cell. Toxicol.* **2016**, *12*, 399–407. [CrossRef]
11. Byun, J.S.; Han, Y.H.; Hong, S.J.; Hwang, S.M.; Kwon, Y.S.; Lee, H.J.; Kim, S.S.; Kim, M.J.; Chun, W.J. Bark constituents from mushroom-detoxified *Rhus verniciflua* suppress kainic acid-induced neuronal cell death in mouse hippocampus. *Korean J. Physiol. Pharmacol.* **2010**, *14*, 279–283. [CrossRef]
12. Cho, N.K.; Choi, J.H.; Yang, H.J.; Jeong, E.J.; Lee, K.Y.; Kim, Y.C.; Sung, S.H. Neuroprotective and anti-inflammatory effects of flavonoids isolated from *Rhus verniciflua* in neuronal HT22 and microglial BV2 cell lines. *Food Chem. Toxicol.* **2012**, *50*, 1940–1945. [CrossRef]
13. Kim, K.H.; Moon, E.J.; Choi, S.U.; Pang, C.H.; Kim, S.Y.; Lee, K.R. Identification of cytotoxic and anti-inflammatory constituents from the bark of *Toxicodendron vernicifluum* (Stokes) F. A. Barkley. *J. Ethnopharmacol.* **2015**, *162*, 231–237. [CrossRef]
14. Lee, J.C.; Lee, K.Y.; Kim, J.; Na, C.S.; Jung, N.C.; Chung, G.H.; Jang, Y.S. Extract from *Rhus verniciflua* Stokes is capable of inhibiting the growth of human lymphoma cells. *Food Chem. Toxicol.* **2004**, *42*, 1383–1388. [CrossRef]
15. Lee, S.H.; Choi, W.C.; Yoon, S.W. Impact of standardized *Rhus verniciflua* Stokes extract as complementary therapy on metastatic colorectal cancer: A Korean single-center experience. *Integr. Cancer Ther.* **2009**, *8*, 148–152. [CrossRef]
16. Rayne, S.; Mazza, G. Biological activities of extracts from Sumac (*Rhus* spp.): A review. *Plant Foods Hum. Nutr.* **2007**, *62*, 165–175. [CrossRef]
17. Gross, M.; Baer, H.; Fales, H.M. Urushiols of poisonous Anacardiaceae. *Phytochem.* **1975**, *14*, 2263–2266. [CrossRef]
18. Ma, X.M.; Lu, R.; Miyakoshi, T. Recent advances in research on lacquer allergy. *Allergol. Int.* **2012**, *61*, 45–50. [CrossRef]
19. Choi, H.S.; Kim, M.K.; Park, H.S.; Yun, S.E.; Mun, S.P.; Kim, J.S.; Sapkota, K.; Kim, S.; Kim, T.Y.; Kim, S.J. Biological detoxification of lacquer tree (*Rhus verniciflua* Stokes) stem bark by mushroom species. *Food Sci. Biotechnol.* **2007**, *16*, 935–942.
20. Cheong, S.H.; Choi, Y.W.; Min, B.S.; Choi, H.Y. Polymerized urushiol of the commercially available *Rhus* product in Korea. *Ann. Dermatol.* **2010**, *22*, 16–20. [CrossRef]

21. Sapkota, K.; Kim, S.; Kim, M.K.; Kim, S.J. A detoxified extract of *Rhus verniciflua* Stokes upregulated the expression of BDNF and GDNF in the rat brain and the human dopaminergic cell line SH-SY5Y. *Biosci. Biotechnol. Biochem.* **2010**, *74*, 1997–2004. [CrossRef]

22. Shin, S.H.; Koo, K.H.; Bae, J.S.; Cha, S.B.; Kang, I.S.; Kang, M.S.; Kim, H.S.; Heo, H.S.; Park, M.S.; Gil, G.H.; et al. Single and 90-day repeated oral dose toxicity studies of fermented *Rhus verniciflua* stem bark extract in Sprague–Dawley rats. *Food Chem. Toxicol.* **2013**, *55*, 617–626. [CrossRef]

23. Djakpo, O.; Yao, W. *Rhus chinensis* and *Galla chinensis*—Folklore to modern evidence: Review. *Phytother. Res.* **2010**, *24*, 1739–1747. [CrossRef]

24. Jin, M.J.; Kim, I.S.; Park, J.S.; Dong, M.S.; Na, C.S.; Yoo, H.H. Pharmacokinetic profile of eight phenolic compounds and their conjugated metabolites after oral administration of *Rhus verniciflua* extracts in rats. *J. Agric. Food Chem.* **2015**, *63*, 5410–5416. [CrossRef]

25. Jang, J.Y.; Shin, H.; Lim, J.W.; Ahn, J.H.; Jo, Y.H.; Lee, K.Y.; Hwang, B.Y.; Jung, S.J.; Kang, S.Y.; Lee, M.K. Comparison of antibacterial activity and phenolic constituents of bark, lignum, leaves and fruit of *Rhus verniciflua*. *PLoS ONE* **2018**, *13*, e0200257.

26. Rivera-Hoyos, C.M.; Morales-Álvarez, E.D.; Poutou-Piñales, R.A.; Pedroza-Rodríguez, A.M.; Rodríguez-Vázquez, R.; Delgado-Boada, J.M. Fungal laccases. *Fungal Biol. Rev.* **2013**, *7*, 67–82. [CrossRef]

27. Minussi, R.C.; Pastore, G.A.; Durán, N. Potential applications of laccase in the food industry. *Trends Food Sci. Technol.* **2002**, *13*, 205–216. [CrossRef]

28. Ghidouche, S.; Es-Safi, N.E.; Ducrot, P.H. *Trtametes versicolor* laccase mediated oxidation of flavonoids. Influence of the hydroxylation pattern of ring B of flavonols. *Amer. J. Food Technol.* **2007**, *2*, 630–640.

29. Kumanotani, J. Enzyme catalyzed durable and authentic oriental lacquer: A natural microgel-printable coating by polysaccharide–glycoprotein–phenolic lipid complexes. *Prog. Org. Coat.* **1998**, *34*, 135–146. [CrossRef]

30. Lu, R.; Miyakoshi, T. Studies on acetone powder and purified *Rhus* laccase immobilized on zirconium chloride for oxidation of phenols. *Enzyme Res.* **2012**, *2012*, 375309. [CrossRef]

31. Motoda, S. Purification and characterization of polyphenol oxidase from *Trametes* sp. MS39401. *J. Biosci. Bioeng.* **1999**, *87*, 137–143. [CrossRef]

32. Burke, R.M.; Cairney, J.W.G. Laccases and other polyphenol oxidases in ecto- and ericoid mycorrhizal fungi. *Mycorrhiza* **2002**, *12*, 105–116. [CrossRef]

33. Martínková, L.; Kotik, M.; Marková, E.; Homolka, L. Biodegradation of phenolic compounds by Basidiomycota and its phenol oxidases: A review. *Chemosphere* **2016**, *149*, 373–382. [CrossRef]

34. Kubo, I.; Nihei, K.I.; Shimizu, K. Oxidation products of quercetin catalyzed by mushroom tyrosinase. *Bioorg. Med. Chem.* **2004**, *12*, 5343–5347. [CrossRef]

35. Barry, K.M.; Davies, N.W.; Mohammed, C.L. Identification of hydrolysable tannins in the reaction zone of *Eucalyptus nitens* wood by high performance liquid chromatography–electrospray ionisation mass spectrometry. *Phytochem. Anal.* **2001**, *12*, 120–127. [CrossRef]

36. Gómez-Caravaca, A.M.; López-Cobo, A.; Verardo, V.; Segura-Carretero, A.; Fernádez-Gutíerrez, A. HPLC-DAD-q-TOF-MS as a powerful platform for the determination of phenolic and other polar compounds in the edible part of mango and its by-products (peel, seed, and seed husk). *Electrophoresis* **2016**, *37*, 1072–1084. [CrossRef]

37. Abu-Reidah, I.M.; Ali-Shtayeh, M.S.; Jamous, R.M.; Román, D.A.; Segura-Carretero, A. HPLC−DAD−ESI-MS/MS screening of bioactive components from *Rhus coriaria* L. (Sumac) fruits. *Food Chem.* **2015**, *166*, 179–191. [CrossRef]

38. Gordon, A.; Jungfer, E.; da Silva, B.A.; Maia, J.G.S.; Marx, F. Phenolic constituents and antioxidant capacity of four underutilized fruits from the Amazon region. *J. Agric. Food Chem.* **2011**, *59*, 7688–7699. [CrossRef]

39. Abdel-Nabey, M.A.; Sherief, A.A.; El-Tanash, A.B. Tannin biodegradation and some factors affecting tannase production by two *Aspergillus* sp. *Biotechnol.* **2011**, *10*, 149–158. [CrossRef]

40. Sugimoto, K.; Nakagawa, K.; Hayashi, S.; Amakura, Y.; Yoshimura, M.; Yoshida, T.; Yamaji, R.; Nakano, Y.; Inui, H. Hydrolysable tannins as antioxidants in the leaf extract of *Eucalyptus globulus* possessing tyrosinase and hyaluronidase inhibitory activities. *Food Sci. Technol. Res.* **2009**, *15*, 331–336. [CrossRef]

41. Yao, J.; Guo, G.S.; Ren, G.H.; Liu, Y.H. Production, characterization and applications of tannase. *J. Mol. Catal. B: Enzym.* **2014**, *101*, 137–147. [CrossRef]

42. Govindarajan, R.K.; Revathi, S.; Rameshkumar, N.; Krishnan, M.; Kayalvizhi, N. Microbial tannase: Current perspectives and biotechnological advances. *Biocatal. Agric. Biotechnol.* **2016**, *6*, 168–175. [CrossRef]

43. Xiao, Y.; Xing, G.; Rui, X.; Li, W.; Chen, X.; Jiang, M.; Dong, M. Enhancement of the antioxidant capacity of chickpeas by solid state fermentation with *Cordyceps militaris* SN-18. *J. Funct. Foods* **2014**, *10*, 210–222. [CrossRef]

44. Gorinstein, S.; Haruenkit, R.; Poovarodom, S.; Vearasilp, S.; Ruamsuke, P.; Namiesnik, J.; Leontowicz, M.; Leontowicz, H.; Suhaj, M.; ShengSome, G.P. Analytical assays for the determination of bioactivity of exotic fruits. *Phytochem. Anal.* **2010**, 355–362. [CrossRef]

45. Pellegrini, N.; Serafini, M.; Colombi, B.; Del Rio, D.; Salvatore, S.; Bianchi, M.; Brighenti, F. Total antioxidant capacity of plant foods, beverages and oils consumed in Italy assessed by three different in vitro assays. *J. Nutr.* **2003**, 2812–2819. [CrossRef]

46. Shama, P.; Singh, R.P. Evaluation of antioxidant activity in foods with special reference to TEAC method. *Am. J. Food Technol.* **2013**, *8*, 83–101.

47. Kurisawa, M.; Chung, J.E.; Uyama, H.; Kobayashi, S.; Uyama, H.; Kobayashi, S. Enzymatic synthesis and antioxidant properties of poly(rutin). *Biomacromolecules* **2003**, *4*, 1394–1399. [CrossRef]

48. Riebel, M.; Sabel, A.; Claus, H.; Fronk, P.; Xia, N.; Li, H.; König, H.; Decker, H. Influence of laccase and tyrosinase on the antioxidant capacity of selected phenolic compounds on human cell lines. *Molecules* **2015**, *20*, 17194–17207. [CrossRef]

49. Sánchez-Mundo, M.L.; Escobedo-Crisantes, V.M.; Mendoza-Arvizu, S.; Jaramillo-Flores, M.E. Polymerization of phenolic compounds by polyphenol oxidase from bell pepper with increase in their antioxidant capacity. *CYTA J. Food* **2016**, *14*, 594–603.

50. Chandra, S.; Khan, S.; Avula, B.; Lata, H.; Yang, M.H.; ElSohly, M.A.; Khan, I.A. Assessment of total phenolic and flavonoid content, antioxidant properties, and yield of aeroponically and conventionally grown leafy vegetables and fruit crops: A comparative study. *Evid. Based Complement. Alternat. Med.* **2014**. [CrossRef]

51. Zhishen, J.; Mengcheng, T.; Jianming, W. The determination of flavonoid contents in mulberry and their scavenging effects on superoxide radicals. *Food Chem.* **1999**, *64*, 555–559. [CrossRef]

52. Chen, Y.; Hagerman, A.E. Characterization of soluble non-covalent complexes between bovine serum albumin and β-1,2,3,4,6-penta-*O*-galloyl-D-glucopyranose by MALDI-TOF MS. *J. Agric. Food Chem.* **2004**, *52*, 4008–4011. [CrossRef]

53. Mondal, K.C.; Banerjee, D.; Jana, M.; Pati, B.R. Colorimetric assay method for determination of the tannase activity. *Anal. Biochem.* **2001**, *295*, 168–171. [CrossRef]

54. Thaipong, K.; Boonprakob, U.; Crosb, K.; Cisneros-Zevallos, L.; Byrne, D.H. Comparison of ABTS, DPPH, FRAP, and ORAC assays for estimating antioxidant activity from guava fruit extracts. *J. Food Composit. Anal* **2006**, *19*, 669–675. [CrossRef]

55. Benzie, I.F.F.; Strain, J.J. The ferric reducing ability of plasma (FRAP) as a measure of "antioxidant power": The FRAP assay. *Anal. Biochem.* **1996**, *239*, 70–76. [CrossRef]

Sample Availability: Samples are available from the authors.

Article

Study of Interactions between Amlodipine and Quercetin on Human Serum Albumin: Spectroscopic and Modeling Approaches

Zuzana Vaneková [1,*], Lukáš Hubčík [2], José Luis Toca-Herrera [3], Paul Georg Furtműller [4], Jindra Valentová [5], Pavel Mučaji [1] and Milan Nagy [1]

[1] Department of Pharmacognosy and Botany, Faculty of Pharmacy, Comenius University in Bratislava, 83232 Bratislava, Slovakia; mucaji@fpharm.uniba.sk (P.M.); nagy@fpharm.uniba.sk (M.N.)

[2] Department of Physical Chemistry of Drugs, Faculty of Pharmacy, Comenius University in Bratislava, 83232 Bratislava, Slovakia; hubcik@fpharm.uniba.sk

[3] Department of Nanobiotechnology, University of Natural Resources and Life Sciences, 1190 Vienna, Austria; jose.toca-herrera@boku.ac.at

[4] Department of Chemistry, University of Natural Resources and Life Sciences, 1190 Vienna, Austria; paul.furtmueller@boku.ac.at

[5] Department of Chemical Theory of Drugs, Faculty of Pharmacy, Comenius University in Bratislava, 83232 Bratislava, Slovakia; valentova@fpharm.uniba.sk

* Correspondence: vanekova27@uniba.sk

Academic Editors: Marian Brestic, Marek Zivcak, Oksana Sytar and Marco Landi
Received: 14 December 2018; Accepted: 29 January 2019; Published: 30 January 2019

Abstract: The aim of this study was to analyze the binding interactions between a common antihypertensive drug (amlodipine besylate—AML) and the widely distributed plant flavonoid quercetin (Q), in the presence of human serum albumin (HSA). Fluorescence analysis was implemented to investigate the effect of ligands on albumin intrinsic fluorescence and to define the binding and quenching properties. Further methods, such as circular dichroism and FT-IR, were used to obtain more details. The data show that both of these compounds bind to Sudlow's Site 1 on HSA and that there exists a competitive interaction between them. Q is able to displace AML from its binding site and the presence of AML makes it easier for Q to bind. AML binds with the lower affinity and if the binding site is already occupied by Q, it binds to the secondary binding site inside the same hydrophobic pocket of Sudlow's Site 1, with exactly the same affinity. Experimental data were complemented with molecular docking studies. The obtained results provide useful information about possible pharmacokinetic interactions upon simultaneous co-administration of the food/dietary supplement and the antihypertensive drug.

Keywords: human serum albumin; amlodipine; quercetin; fluorescence; circular dichroism; FT-IR; molecular modeling

---

1. Introduction

Human serum albumin (HSA) is the most prevalent protein in human plasma, constituting ~60% of the total plasma protein content. Its structure and purpose have been studied in great detail and it offers a wide variety of applications in research as well as in clinical medicine [1].

HSA and its ability to bind and transport a wide range of molecules (e.g., drugs, metabolites, fatty acids, etc.) play a key role in drug distribution. The most well-known binding sites are Sudlow's Site 1 (located at subdomain IIA and containing a tryptophan residue) and Site 2 (located on subdomain IIIA) [2–4].

HSA in the presence of multiple ligands creates a complex system, where several molecules may or may not compete to bind to the same binding site. Since only the free fraction of the total amount of drug is responsible for the therapeutic effect, these competitive displacement interactions should be considered when administering multiple drugs [5].

The novel aspect of pharmacokinetic interaction which has arisen in the recent years is the displacement of drugs from their binding sites by plant metabolites. Phenolic compounds like flavonoids or tannins naturally occur in all plant materials, whether their use is dietary (the daily intake of fresh fruit and vegetables) or medicinal (herbal teas and supplements). Several studies have explored these interactions between selected drugs and various plant polyphenols, such as warfarin with quercetin [6] and multiple other flavone, flavonol, and flavanone aglycones [7]; nifedipine with rutin and baicalin [8]; ticagrelor with quercetin; rutin with baicalin [9]; propranolol with quercetin [10]; and gliclazide with quercetin [11]. All of these studies came to a common conclusion that flavonoids bind with high affinity to HSA and are able to alter the binding of the other molecule.

Amlodipine (AML; Figure 1a) is a long-acting calcium channel blocker belonging to the dihydropyridine derivatives group. Members of this group selectively inhibit the transmembrane influx of calcium ions into vascular smooth muscle which causes a peripheral vasodilatation. AML is widely used for the treatment of hypertension in the form of a besylate salt, often in combination with other types of antihypertensives [12]. In vitro studies have shown that approximately 93% of circulating amlodipine is bound to plasma proteins [13]. Other studies extensively proved that AML binds to the Site 1 on HSA [14–17]. Its (*S*)-enantiomer binds to a higher extent than the (*R*)-enantiomer [18].

Quercetin (Q; Figure 1b) is one of the most commonly occurring flavonoids in any plant material and one of the most abundant dietary flavonoids with an average daily consumption of 25–50 mg [19]. It is known for its strong antioxidant, anti-inflammatory, and venoprotective properties. After consumption, quercetin undergoes an extensive glucuronide conjugation in liver. Four of the five hydroxyl groups can be conjugated with glucuronic acid (with the exception of 5-hydroxyl group) and the most common metabolite is quercetin-3-glucuronide [19,20]. Quercetin and its metabolites strongly bind to HSA (99.4% for quercetin) and the main binding site is Sudlow's Site 1 [6,9–11,21,22].

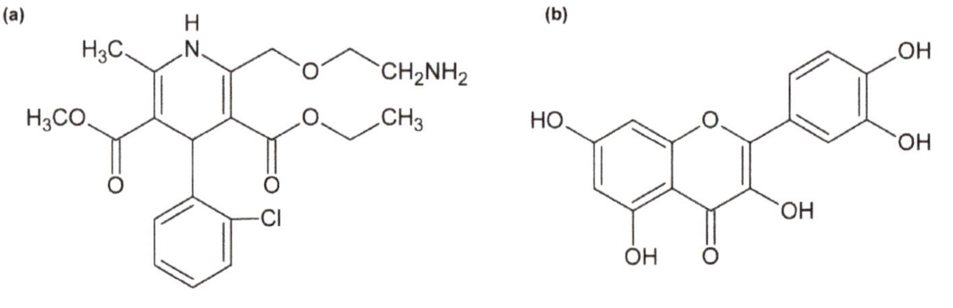

Figure 1. Structures of (**a**) amlodipine and (**b**) quercetin.

The aim of this study was to investigate a possibility of binding interactions between AML and Q in the presence of HSA which, to the best of our knowledge, has not yet been studied. The competitive binding was investigated first by fluorescence spectroscopy. A more detailed view on conformational alterations of HSA was obtained by implementing circular dichroism and FT-IR techniques. Molecular docking studies were used to complement the experimental data.

2. Results

2.1. Fluorescence Quenching and Enhancement

Fluorescence spectroscopy is one of the most widely used methods for study of protein–ligand interactions due to its convenience, high reliability, and sensitivity. Intrinsic fluorescence of proteins is caused by aromatic amino acid residues, namely tryptophan (Trp), tyrosine (Tyr), and phenylalanine (Phe). Trp fluorescence is usually dominant in the protein emission spectrum, with excitation maximum wavelength at 295 nm and emission maximum wavelength at 350 nm. Tyrosine has a quantum yield similar to tryptophan, but its emission spectrum is more narrowly distributed on the wavelength scale, occurs at shorter wavelengths and is generally quenched in the protein matrix. Fluorescence spectra are highly sensitive to changes on the environment around amino acid residues: excited-state reactions, molecular rearrangements, changes in solvent polarity, energy transfer, ground-state complex formation, and collisional quenching. These result in changes of the fluorescence intensity (quenching or enhancement) and shifts in maximum wavelengths [23].

HSA structure consists of 585 amino acids with 18 Tyr residues and one Trp residue (located at position 214), which is conveniently placed inside the hydrophobic pocket of the Sudlow's Site 1. HSA in solution presented a strong fluorescence emission with a peak maximum at 347 nm when excitation wavelength was 295 nm. The studied drugs had no intrinsic fluorescence at these wavelengths. However, increasing concentrations of AML and Q in the HSA solution result in large-scale quenching of HSA intrinsic fluorescence (Figure 2), which indicates that both drugs bind into the close proximity of Trp^{214} in Site 1. There is also a blue shift in the Q + HSA system spectra, indicating that the polarity around Trp residue decreased. The AML + HSA system does not show any considerable shift in maximum wavelengths.

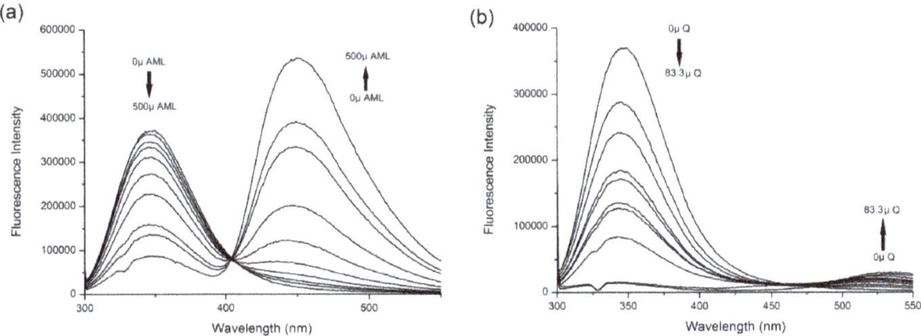

Figure 2. The fluorescence emission spectra of (**a**) amlodipine besylate (AML) + human serum albumin (HSA) and (**b**) Q + HSA systems at excitation λ = 295 nm. Conditions: T = 310.15 K, pH = 7.4. The HSA concentration was 5 μM; AML concentrations were 0, 3.9, 7.8, 15.6, 31.25, 62.5, 125, 250, 333.3, and 500 μM; Q concentrations were 0, 1.95, 3.9, 7.8, 10.4, 15.6, 20.8, 31.25, 41.6, 62.5 and 83.3 μM.

Both studied drugs contain conjugated double bond systems in their structures (Figure 1). AML shows observable fluorescence emission at λ_{EX} = 370 nm and λ_{EM} = 450 nm and Q shows weak fluorescence emission at λ_{EX} = 375 nm and λ_{EM} = 540 nm. HSA does not show any intrinsic fluorescence at these wavelengths. Therefore, we were also able to observe changes in ligand fluorescence emission. Upon addition of HSA increasing concentrations into ligand solutions, we observed significant increase of the ligand fluorescence for both AML and Q (Figure 3).

Quercetin fluorescence shows an unusual dual behavior, as described by Sengupta & Sengupta [22]. At λ_{EX} = 375 nm we observed an enhancement of intrinsic fluorescence with a significant blue shift at λ_{EM} = 525 nm, and in samples with higher concentrations of HSA, a second peak with emission maximum at λ_{EM} = 450 nm appears, which is characteristic of its parent compound, 3-hydroxyflavone (Figure 3b). If we use a different excitation wavelength, λ_{EX} = 450 nm, which is a selective excitation wavelength for Q + HSA ground-state complex, a single emission band appears with a maximum at λ_{EM} = 525 nm (Figure 3c).

Fluorescence resonance energy transfer (FRET) occurs when the donor (in this case Trp^{214} of HSA) and acceptor (in this case AML or Q) exist in a close proximity and the donor emission spectrum overlaps with the absorption spectrum of the acceptor [23]. Figure 4 shows that both studied compounds provide an unusually strong spectral overlap. In Figure 2, both secondary mixtures (AML + HSA and Q + HSA) show bands at λ_{EM} = 450 nm for AML and λ_{EM} = 530 nm for Q, respectively. This behavior further supports that both compounds bind to the close proximity of Trp[214].

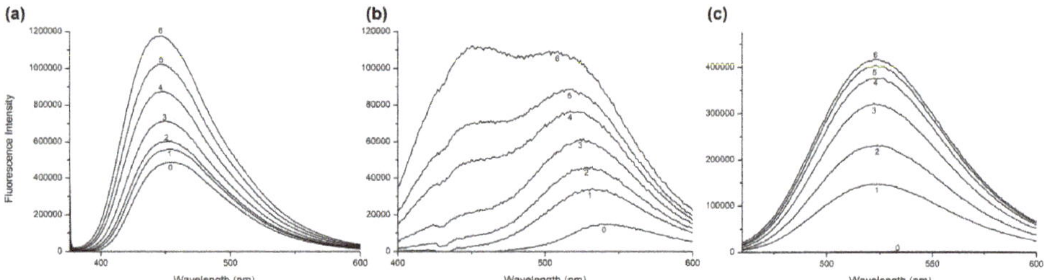

Figure 3. The fluorescence emission spectra of ligand + HSA systems for (**a**) AML at λ_{EX} = 370 nm, (**b**) Q at λ_{EX} = 375 nm, and (**c**) Q at λ_{EX} = 450 nm. In all experiments, the ligand concentration was 7.5 µM and HSA concentrations were 0 (line 0), 1.87 (line 1), 3.75 (line 2), 7.5 (line 3), 15 (line 4), 20 (line 5), and 30 (line 6) µM. Conditions: T = 310.15 K, pH = 7.4.

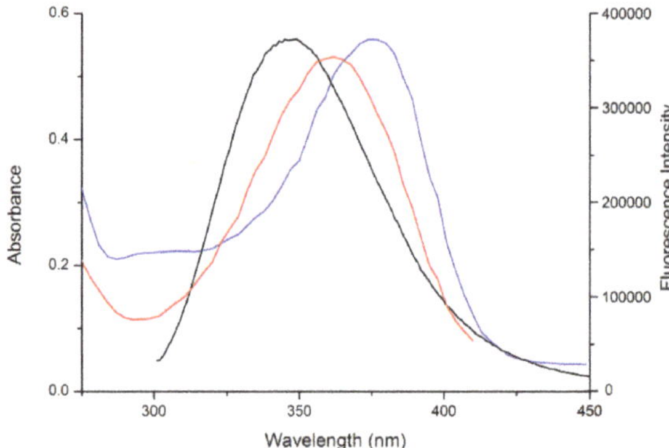

Figure 4. Comparison of absorption spectra of AML (**red line**), Q (**blue line**), and fluorescence emission spectra of HSA (**black line**).

In order to find out if there was any interaction between the two studied compounds and the binding site on HSA, we performed an analysis of ternary mixtures, [(Q + HSA) + AML] and [(AML + HSA) + Q], respectively. The comparison of quenching curves in the binary and ternary systems provides general information concerning the change of affinity of one drug to serum albumin in the presence of another drug.

Figure 5 shows the quenching curves comparison of binary and ternary systems. In the case of the AML binding, there was no discernible difference between the binary and the ternary systems where HSA was preincubated with Q. This means that the presence or absence of Q did not influence the ability of AML to bind to HSA. However, Figure 5b shows a case of a cooperative binding. When AML is bound first, Q causes stronger fluorescence quenching. This could be explained by allosteric modulation: the presence of AML inside the Sudlow's Site 1 deforms the protein molecule in a way that allows for better access and stronger binding of Q.

In order to investigate if any displacement interaction was present, it was necessary to also observe the behavior of the firstly-added drug in ternary mixtures. Figure 6 shows the behavior of the fluorescence enhancement bands (characteristic for ligand + HSA) of the firstly added ligand in ternary mixtures. Upon addition of AML into Q + HSA system, we could see the band for Q + HSA sample stays almost intact and there was only a slight displacement at the highest AML concentrations.

On the contrary, upon addition of Q into AML + HSA system, the enhanced fluorescence band of AML + HSA gradually decreased and that at the highest Q concentrations it was completely replaced by the characteristic double fluorescence band of Q + HSA system as seen in Figure 3. Therefore we can assume that Q was able to displace AML from its binding site, but in the case when Q was bound first it was not displaced and both drugs were binding at the same time to Sudlow's Site 1.

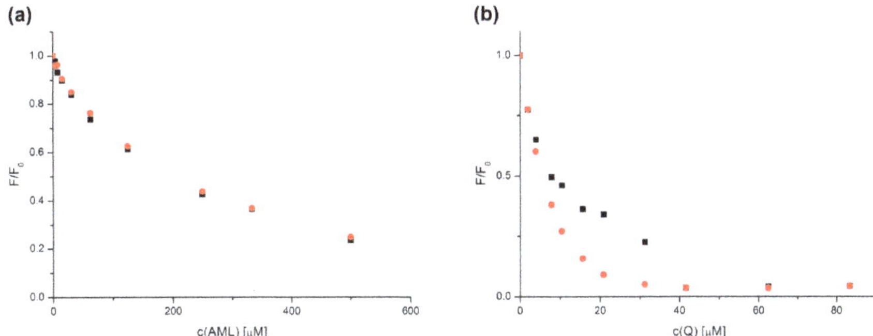

Figure 5. Quenching curves of HSA with AML (**a**) and with Q (**b**), respectively, for binary (black squares) and ternary (red dots) systems. Conditions: T = 310.15 K, pH = 7.4. The HSA concentration was 5 µM; AML and Q concentrations increased from 0 to 500 µM and from 0 to 83.3 µM, respectively.

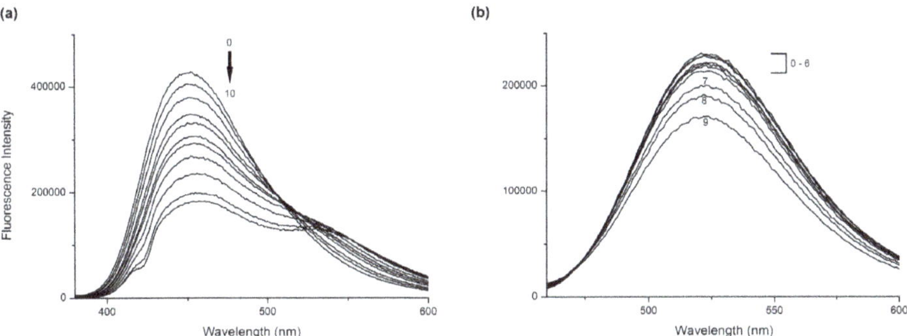

Figure 6. The fluorescence emission spectra of ternary systems (**a**) (AML + HSA) + Q at λ_{EX} = 370 nm and (**b**) (Q + HSA) + AML at λ_{EX} = 450 nm. Conditions: T = 310.15 K, pH = 7.4. The concentration of HSA and firstly-added drug was 5 µM; Q and AML concentrations increased from 0 (line 0) to 83.3 µM (line 10) and from 0 (line 0) to 500 µM (line 9), respectively.

2.2. Stern–Volmer Analysis

Fluorescence quenching can be described by the well-known Stern–Volmer equation:

$$F_0/F = 1 + K_q\,[Q]$$

where F_0 and F are the fluorescence intensities before and after a quencher addition, K_q is the Stern–Volmer quenching constant, and [Q] is the quencher's concentration [23].

It is possible to determine the type of fluorescence quenching by obtaining the K_q value at various temperatures. For AML the difference between Stern–Volmer plots at different temperatures is almost nonexistent and the plots display a slight upward curvature, concave towards the y-axis (Figure 7a). This is characteristic for the combination of static and dynamic quenching and the decrease of K_q with the increasing temperature is compensated by the increase of K_S (an association constant of static quenching). However, in high molar ratios, such as in this experiment, the upward curvature is rather caused by the extreme concentrations of the quencher molecule. As the quencher concentration increases, the probability increases that a quencher is within the first solvent shell of the fluorophore at the moment of excitation. These closely spaced fluorophore–quencher pairs are immediately quenched upon excitation, and thus appear to be dark complexes [23]. Still, from the fact that the K_q value stays the same for all temperatures, we can conclude the quenching mechanism is a combination of static and dynamic quenching.

Figure 7b shows that for Q the K_q value significantly decreases with the increasing temperature, therefore the type of fluorescence quenching present is the static one characteristic by a fluorophore–ligand complex formation. This is in an agreement with the fluorescence enhancement experiments where we observed a specific emission band for Q + HSA ground-state complex. Q + HSA system similarly suffered from an extreme upward curvature at the highest concentrations (data not shown). Therefore, for the K_q calculations we used the first seven samples from each dataset starting from the lowest quencher concentration to avoid the distortion caused by extreme concentrations.

From K_q we were able to determine the accessibility of fluorophores for quenching using the following equation.

$$K_q = k_q \times \tau_0$$

where k_q is the bimolecular quenching constant and τ_0 is the average fluorescence lifetime of the fluorophore without quencher (10^{-8} s for Trp) [23]. Table 1 is listing the K_q and k_q values for all binary and ternary systems at 310.15 K. The K_q values further support the behavior described by Figure 5. Values of k_q are larger than the maximum scattering collision quenching constant k_q (2.0×10^{10} L mol^{-1} s^{-1}) for all systems, however the values for AML binding in both secondary and ternary systems are one order of magnitude lower than those of Q which might support the combination of static and dynamic quenching.

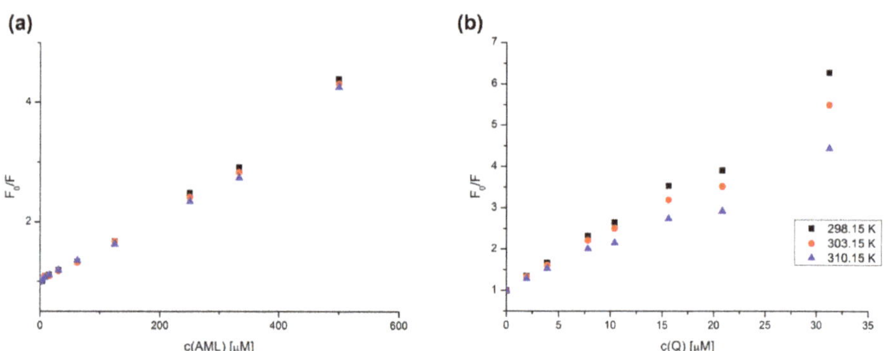

Figure 7. Stern–Volmer plots for (**a**) AML + HSA and (**b**) Q + HSA systems at different temperatures and λ_{EX} = 295 nm. The HSA concentration was 5 µM; AML and Q concentrations increased from 0 to 500 µM and from 0 to 31.25 µM, respectively.

Table 1. The values of Stern–Volmer quenching constants (K_q), bimolecular quenching rate constants (k_q), correlation coefficients of Stern–Volmer curves (R), binding constants (K_D), and Gibbs free energy (ΔG) for the binary and ternary systems at λ_{EX} = 295 nm and T = 310.15 K.

System	K_q [10^3 M^{-1}]	k_q [10^{11} M^{-1}s^{-1}]	R	K_D [µM]	ΔG [kJ mol^{-1}]
AML + HSA	5.35 ± 0.09	5.35 ± 0.09	0.9998	183.77 ± 25.28	-22.181
Q + HSA	15.23 ± 1.89	15.23 ± 1.89	0.9756	6.48 ± 2.01	-30.807
(Q +HSA) +AML	5.07 ± 0.07	5.07 ± 0.07	0.9998	192.81 ± 24.09	-22.057
(AML + HSA) + Q	76.02 ± 1.35	76.02 ± 1.35	0.9989	2.39 ± 0.32	-33.379

2.3. UV Absorption Measurements

Careful examination of the absorption spectra of the fluorophore is an additional method to distinguish static and dynamic quenching. Collisional quenching affects the excited states of the fluorophores only and thus no changes in the absorption spectra are expected. In contrast, ground-state complex formation will frequently result in perturbation of the absorption spectrum of the fluorophore [23].

In order to investigate this possibility, we performed UV absorption measurements of the systems from Figure 3 and focused on ligand absorption. In Figure 8 we see that the AML absorption band undergoes some minor changes in the intensity and no changes in the band shape. On the contrary, the UV absorption spectra of Q changes dramatically, with changes in the shape and the band maximum shifts from 375 nm (pure quercetin) to 407 nm (Q + HSA ground-state complex). These findings further confirm the results of Stern–Volmer analysis: Q forms a ground-state complex with Trp^{214} and AML quenches the HSA fluorescence by combination of static and dynamic quenching.

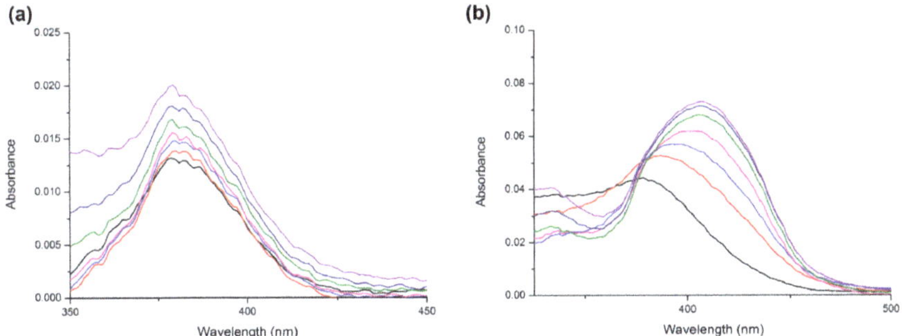

Figure 8. The UV absorption spectra of ligand + HSA systems for (**a**) AML and (**b**) Q. In all experiments, ligand concentration was 7.5 μM and HSA concentrations were 0 (black line), 1.87 (red line), 3.75 (blue line), 7.5 (pink line), 15 (green line), 20 (dark blue line), and 30 (purple line) μM. Conditions: T = 310.15 K, pH = 7.4.

2.4. Binding Constant Analysis

Analysis of binding constants and thermodynamic parameters plays an important role in pharmacokinetic research. According to Klotz [24], widely used linear expression of fluorescence data (i.e., Scatchard plot) is not suitable for obtaining both the K_D value and the number of binding sites from a single dataset since in order to do that, at least two different measurements have to be performed. Moreover, all linear transformations of binding data suffer from similar limitations: minor nonlinearities are very difficult to detect and evaluate, point-spread masks the inaccuracies caused by titration errors, but most importantly, all linear transformations of binding data invalidate the rules for applying a linear transformation to a dataset.

Lissi et al. [25] also confirmed that values for n ("number of binding sites") obtained from the Scatchard plot were significantly different from values obtained by ultracentrifugation experiments. Suspiciously, all values of n from several dozens of published research articles are very close to one and much smaller than values obtained by ultracentrifugation experiments. There is also a common discrepancy in the actual meaning of n in this equation—where most research articles label it as "number of binding sites", it in fact represents the stoichiometry of the binding step.

Therefore we used a more simplistic approach by using the DynaFit software script. Data input consisted of [L_{total}] (total ligand concentrations) as the variable reactant concentrations, [P_{total}] (total protein concentration) as the fixed reactant concentration, and F/F_0 as the experimentally observed value. The software creates a semilogarithmic plot of F/F_0 vs [L_{free}] (free ligand concentration) and determines the K_D value by finding the halfway point in the sigmoid-shaped curve.

The results are displayed in Table 1. We can see that AML binding is two factors of ten weaker than that of Q. This result supports the observations that Q forms a ground-state complex and is able to displace AML from its binding site. From the comparison of K_D values for binary and ternary mixtures we can (in correlation with Figure 5) similarly conclude that the presence of Q does not change the binding affinity of AML and the presence of AML significantly increased the binding affinity of Q.

From K_D values we were also able to calculate the values of ΔG (Gibbs free energy) using the following equation.

$$\Delta G = -RT \ln (c^\theta / K_D)$$

In which, R is the ideal gas constant, T the temperature, and the standard reference concentration $c^\theta = 1$ mol/L. All ΔG values for all systems were negative. Therefore, it can be inferred that the binding interaction of HSA with all systems is spontaneous and enthalpy driven.

2.5. Circular Dichroism Measurements

Circular dichroism spectroscopy is a valuable technique for detecting changes in the protein secondary and tertiary structure. In order to obtain an insight into the HSA structure, the far-UV (200–260 nm) and the near-UV (250–340 nm) CD spectra were recorded in the presence or absence of drugs in six molar ratios of ligands to protein (L/P) (0, 0.5, 1, 2, 3, and 4) for both binary and ternary systems.

The far-UV CD spectrum of HSA showed two negative minima in the UV region at 208 nm and 222 nm, which is characteristic of the α-helical structure of the protein. The spectral profiles of the binary systems shown in Figure 9 indicate that the binding of Q or AML induced a minor perturbation in the HSA secondary structure. Table 2 lists the α-helix percentage values for all molar ratios. We can see that upon binding of AML, the protein reaches the maximum α-helix content at molar ratio 1:1 and then decreases again. The sample with Q, similarly, reaches maximum α-helix content at molar ratio 1:1 and the value continues to fluctuate around the similar percentage upon adding more Q into the solution.

The near-UV CD spectrum is a useful tool for observation of the protein tertiary structure. Figure 10 shows the spectra for all binary and ternary systems. We can see that AML caused minimal changes in the CD spectrum. On the contrary, we see that upon addition of Q, the spectrum of HSA shows major perturbations of the tertiary structure, both in Tyr and Trp range (275–287 nm and 285–305 nm, respectively) [26] which is in agreement with the study by Zsila et al. [27]. In ternary mixtures we can see that upon addition of AML into Q + HSA system there was a small interaction in the Trp range which supports the claim that in the presence of Q, AML binds to a slightly different binding site near Trp[214]. In the vice versa situation, preincubation with AML caused a small decrease of quercetin's Trp-range CD band and a small increase of Tyr-range CD band.

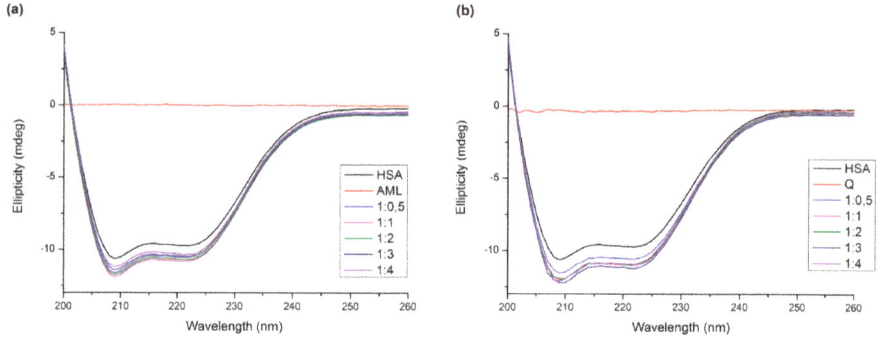

Figure 9. Far-UV CD spectra of (a) AML + HSA and (b) Q + HSA systems. Conditions: T = 310.15 K, pH = 7.4. The HSA concentration was 1 μM; AML or Q concentrations increased from 0 to 4 μM.

Table 2. The values of α-helix percentages for binary systems AML + HSA and Q + HSA.

System	Ratio Protein:Ligand	α-Helix Content [%]
AML + HSA	1:0	59.1 ± 1.0
	1:0.5	63.6 ± 0.3
	1:1	64.7 ± 0.2
	1:2	64.1 ± 0.3
	1:3	62.5 ± 0.1
	1:4	61.5 ± 0.7
Q + HSA	1:0	59.1 ± 1.0
	1:0.5	63.0 ± 2.0
	1:1	65.4 ± 0.5
	1:2	65.2 ± 0.9
	1:3	66.2 ± 0.7
	1:4	66.3 ± 1.5

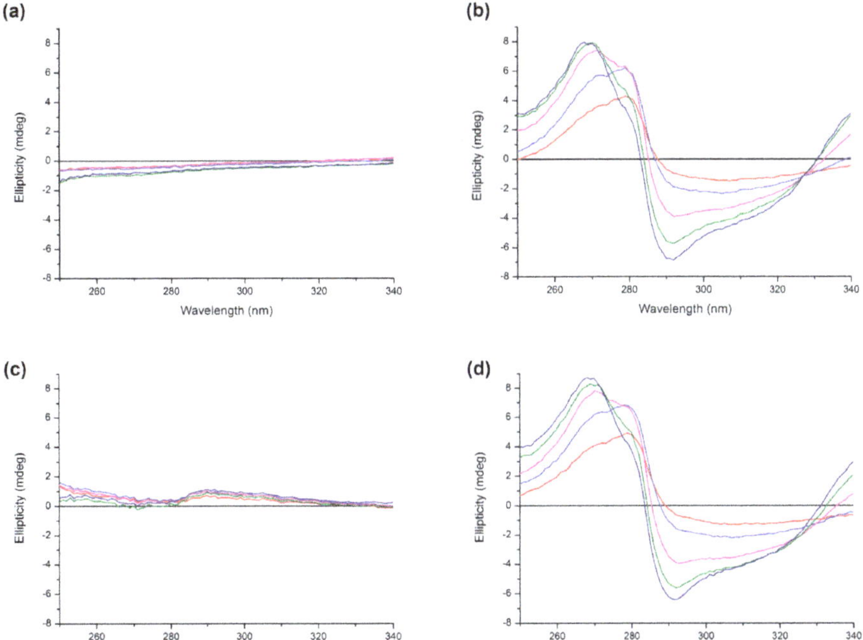

Figure 10. Near-UV CD spectra of (**a**) AML + HSA, (**b**) Q + HSA, (**c**) (Q + HSA) + AML, and (**d**) (AML + HSA) + Q. Conditions: T = 310.15 K, pH = 7.4. The HSA concentration was 15 μM while the concentrations of AML or Q were 7.5 μM (red line), 15 μM (blue line), 30 μM (pink line), 45 μM (green line), and 60 μM (dark blue line).

2.6. FT-IR

The secondary structure of HSA and its dynamics can be effectively studied by Fourier transform infrared spectroscopy. All proteins, including HSA, possess the protein amide I band at 1650–1654 cm^{-1} (C=O stretching) and amide II band at 1540–1560 cm^{-1} (C–N stretch coupled with N–H bending mode), which are related to the protein secondary structure [28]. The changes of the amide I and amide II bands observed in the FT-IR spectra are related to the interaction of protein with the analyzed drug. As shown in Figure 11, amide I band maxima for all studied samples are almost identical.

That means binding AML or Q did not result in any major change in the secondary structure of HSA. However, amide II bands of most ligand + HSA samples are slightly downshifted. This suggests slightly different binding site near Trp[214] for AML and Q, respectively. This supports our conclusions based on the CD spectra analysis. Subtle changes in the local environment of both ligands could be confirmed by deconvoluted band I analysis of all samples (spectra not shown), too. In short, three main bands (in decreasing intensity of order, after rounding the wavelength values) for free HSA are at 1651 cm^{-1} (random coil), 1660 cm^{-1} (α-helix), and 1667 cm^{-1} (β-turn); for the binary system Q + HSA we found bands at 1652 cm^{-1} (α-helix), 1648 cm^{-1} (random coil), and 1656 cm^{-1} (α-helix); for the binary system AML + HSA bands exist at 1660 cm^{-1} (α-helix), 1651 cm^{-1} (random coil), and 1667 cm^{-1} (β-turn); for the ternary system (Q + HSA) + AML we found bands at 1652 cm^{-1} (random coil), 1656 cm^{-1} (α-helix), and 1661 cm^{-1} (β-turn); and for the ternary system (AML + HSA) + Q we found bands at 1652 cm^{-1} (random coil), 1657 cm^{-1} (α-helix), and 1661 cm^{-1} (β-turn), respectively [29]. This confirms slightly different binding locations of quercetin and amlodipine in Site I of HSA as proposed by our CD spectra analysis and docking results.

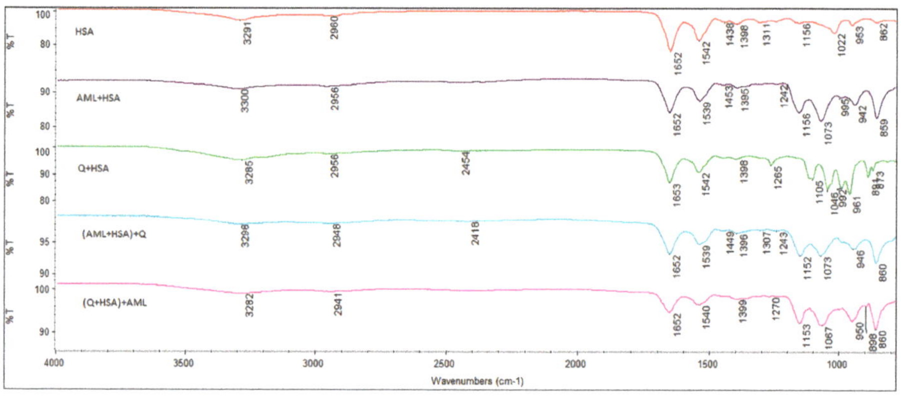

Figure 11. FT-IR spectra of HSA (red), AML+HSA (violet), Q+HSA (green), (AML + HSA) + Q (aquamarine), and (Q + HSA) + AML (magenta).

2.7. Molecular Docking Study

Both ligands (Q and AML) were separately docked and the same binding Site I was found. However, some difference in concrete position was discovered as shown by a superposition of best docking results (Figure 12a,b). The space occupied by AML is defined by the following amino acids; Tyr[150], Glu[153], Ser[192], Lys[195], Lys[199], Trp[214], Arg[222], Leu[238], His[242], and Glu[292]. For Q, amino acids Tyr[150], Ser[192], Lys[195], Lys[199], Trp[214], Arg[222], Leu[238], Leu[260], and Ala[291] define its placement on HSA.

Different positioning of both ligands in binding Site I is fixed by different hydrogen bonds: AML interacts through the oxygen (bearing an ethylamino group) with the amino group of Lys[199]; the length of this hydrogen bond is 2.97 Å (Figure 13a). For Q, three hydrogen bonds were found (Figure 13b) between

- - the carbonyl oxygen and the amino group of guanidine part of Arg[222] (distance 3.46 Å),
- - the hydroxyl group on the C-3 and the amino group of Lys[199] (3.43 Å), and
- - the hydroxyl group on the C-4' and the hydroxyl group of Ser[192] (3.35 Å)

These results support the interpretation of all our spectroscopic measurements.

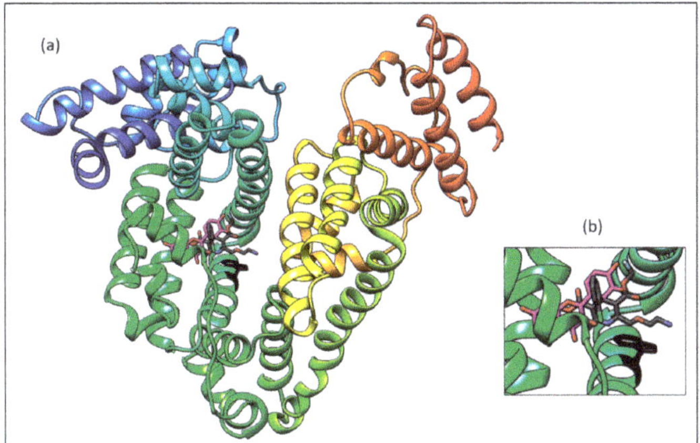

Figure 12. Superposition of the best docking placement of quercetin (magenta) and amlodipine (grey) in site I of HSA (**a**) and a detailed view (**b**); always in the proximity of Trp[214] (black).

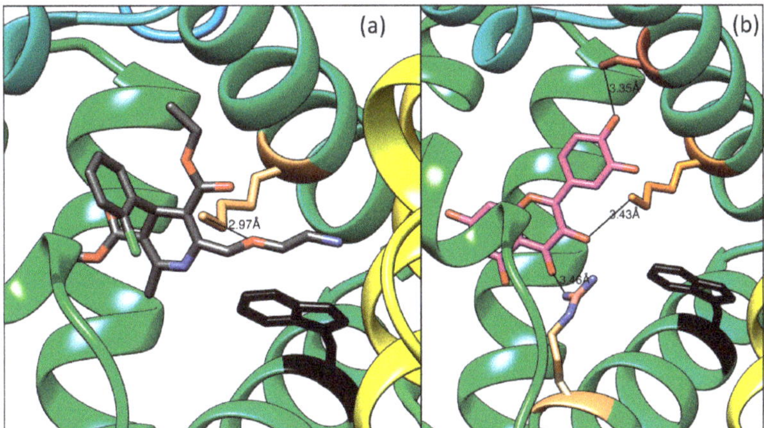

Figure 13. Detailed view on hydrogen bonds between (**a**) AML and Lys[199] or (**b**) Q and Ser[192], Lys[199] and Arg[222], respectively, in HSA.

3. Discussion

Our results demonstrate for the first time, to the best of our knowledge, the binding interactions between quercetin and amlodipine on human serum albumin. From the results shown above it can be concluded that both drugs bind to the same primary binding site localized inside the Sudlow's Site 1 on HSA and there exist a competitive interaction between them. Quercetin is binding with the higher affinity and is able to displace amlodipine from its binding site. Amlodipine binds with the lower affinity and if the binding site is already occupied by quercetin, it binds with the same affinity to the secondary binding site inside the same hydrophobic pocket of Sudlow's Site 1. The displacement of amlodipine by quercetin may elevate the concentration of the unbound amlodipine in the serum which might cause fluctuations in patient's blood pressure or elevated risk of adverse side effects. However, more studies, particularly in vivo monitoring of the free plasma levels of drugs, should be performed to evaluate the magnitude and severity of this interaction.

These results help further the knowledge of binding interactions and are helpful for understanding the interactions between plant compounds and drugs. In concurrence with other studies mentioned in this work, the plant compound–drug interactions are known to cause adverse side effects and therefore should be treated with similar caution as any other drug–drug interaction.

4. Materials and Methods

4.1. Materials

Human serum albumin (recombinant, expressed in rice), amlodipine besylate (pharmaceutical secondary standard), quercetin ($\geq$95%, HPLC), and dimethylsulfoxide (DMSO; $\geq$99.5% (GC) plant cell culture tested) were purchased from Sigma-Aldrich. Phosphate buffer (50 mM, pH 7.4) was prepared from $Na_2HPO_4 \times 12\,H_2O$ and $NaH_2PO_4 \times 2\,H_2O$ (p.a., Centralchem, Slovakia). The HSA stock solutions were prepared by dissolving an appropriate amount in phosphate buffer. AML stock solutions were prepared by mixing an appropriate amount of the substance with phosphate buffer and then heating the mixture to approx. 60 °C until completely dissolved. Quercetin stock solutions were prepared by dissolving the substance in DMSO and then diluting in phosphate buffer to the required concentration. DMSO concentration in final mixtures did not exceed 1% (v/v). Milli-Q water was used for all the measurements.

4.2. Methods

4.2.1. Fluorescence Measurements

Fluorescence spectra were measured in triplicates on a FluoroMax 4 spectrofluorimeter (Horiba Jobin Yvon Scientific, Edison, NJ, USA), equipped with a 1.0 cm path length quartz cell. The slit widths for the excitation and emission were 3.0 nm for all measurements. The temperatures used for the measurements were 298.15 K, 303.15 K, and 310.15 K, respectively. Samples were incubated for 2 min. Buffer background was subtracted from the raw spectra. Fluorescence intensities were corrected for the absorption of excitation light and reabsorption of emitted light to decrease the inner filter using the following relationship [30]:

$$F_{cor} = F_{obs} \times 10^{(A_{ex}+A_{em})/2}$$

where F_{cor} and F_{obs} are the corrected and observed fluorescence intensities, respectively. A_{ex} and A_{em} are the absorbance values at excitation and emission wavelengths, respectively.

4.2.2. Binding Constant Analysis

Fluorescence spectral data after correction were used to calculate the dissociation constant (K_D) and Gibbs free energy (ΔG) for all studied systems. The 10 nm wide section around the fluorescence maximum was selected and the fluorescence intensity values were added up to minimize the influence of signal noise.

For evaluation and logarithmic plot fitting we used DynaFit software (DynaFit 4; BioKin, Ltd.: Watertown, MA, USA, 2015) using a custom-written script.

4.2.3. UV Absorption Measurements

UV absorption spectra were performed on Infinite M200 Tecan (Männedorf, Switzerland) using Sarstedt TC Plate 96 Well, Standard, F. The temperature was 310.15 K. Plates were incubated for 2 min. Buffer background was subtracted from the raw spectra.

4.2.4. Circular Dichroism (CD) Measurements

The isothermal wavelength scan studies of HSA in the absence or the presence of Q and/or AML were carried out in triplicates using a Chirascan CD spectrophotometer equipped with a Peltier type temperature controller (Applied Photophysics Ltd., Leatherhead, UK). The instrument was flushed with nitrogen with a flow rate of 5 L per minute, the path length was 1 mm, spectral bandwidth was set to 1 nm, the scan time per point to 5 s, and the temperature was set to 310.15 K. Buffer background was subtracted from the raw spectra.

For the far-ultraviolet (far-UV) CD spectra (200–260 nm) the HSA concentration was 1 µM. For the near-UV CD (250–340 nm) spectra an HSA concentration of 15 µM was used. Six molar ratios of ligands to protein (L/P) (0, 0.5, 1, 2, 3, and 4) were investigated for both binary and ternary systems.

The far-UV spectral data were used to calculate the α-helix percentage using the following equation [31].

$$\alpha\ helix\ \% = \frac{-MRE_{208\ nm} - MRE_{\beta\text{-}sheets}}{-MRE_{\alpha\text{-}helix} - MRE_{\beta\text{-}sheets}} \times 100$$

where $MRE_{208\ nm}$ is the mean residue ellipticity value of the sample at the excitation wavelength of 208 nm, $MRE_{\alpha\text{-}helix}$ is the standardized value for a protein with 100% content of α-*helixes* and is equal to 33,000, and $MRE_{\beta\text{-}sheets}$ is the standardized value for a protein with 100% content of β-*sheets* and is equal to 4000.

The near-UV spectral data were evaluated by an empiric method described by Zsila et al. [26,32] The spectra were brought down to a common baseline by subtracting the spectrum of pure HSA.

4.2.5. FT-IR

For this method, purified lyophilized crystallic samples were prepared. The concentrations of HSA and ligands for the FT-IR spectra analyses were 10 µM and 80 µM, respectively. The solutions were mixed and incubated for 15 min, then filtered through 30 kDa ultracentrifugation filters Amicon® Ultra 0.5 mL using Hettich Universal 320 laboratory centrifuge. The filtrate was freeze-dried and used as a FT-IR sample.

ATR-FT-IR spectra of all samples were recorded on a Nicolet 6700 FTIR spectrometer. All spectra were taken on a germanium crystal with a resolution of 4 cm^{-1} and using 32 scans at 298 K. The number, position and width of component bands were estimated by performing a Fourier self-deconvolution to the protein infrared amide I band after subtraction of the free HSA spectrum from the sample ones using Omnic 9 software (Thermo Fisher Scientific Inc.: Waltham, MA, USA). The featureless original HSA spectrum between 2200 and 1800 cm^{-1} was the subtraction criterion.

4.2.6. Docking Study

Human serum albumin structure from the Protein Data Bank (PDB ID: 1E78A) was used for calculations. Quercetin and amlodipine structures were created in ChemSketch software (ChemSketch, version 12.01, Advanced Chemistry Development, Inc., Toronto, ON, Canada) and converted from a *.mol file format to a *.pdb one by OpenBabel 2.3.2 and used without any optimization. To be in line with conditions in fluorescence and CD experiments (pH = 7.4) a corresponding protonized amlodipine molecule (based on ACD/ADME Suite version 5, Build 1339 prediction; Advanced Chemistry Development, Inc.: Toronto, ON, Canada) was used for docking. A PatchDock web server (http://bioinfo3d.cs.tau.ac.il/PatchDock/index.html) was used to dock (complex type: protein–small ligand, clustering RMSD = 4.0). Best 10 docking solutions were evaluated for both ligands. Molecular graphics images were produced using the UCSF Chimera 1.13 package (Resource for Biocomputing, Visualization, and Informatics at the University of California, San Francisco, CA, USA). Hydrogen bonds were calculated using this package using relax constraints of 0.4 Å and 20.0 degrees, respectively.

Author Contributions: Fluorescence, UV absorption, CD experiments, and their evaluation, Z.V.; Fluorescence supervision, L.H.; CD spectroscopy supervision, P.G.F.; FT-IR and molecular docking and their evaluation, M.N.; FT-IR Supervision, J.V.; Writing—Original Draft Preparation, Z.V.; Writing—Review and Editing, M.N., L.H., P.G.F., and J.L.T.-H.; Project Administration, M.N. and P.M.; Funding Acquisition, Z.V., M.N. and P.M.

Funding: The research leading to these results has received funding from Comenius University in Bratislava, grant numbers UK/134/2017 and FaF/32/2018, OeAD, grant number ICM-2017-06608, and Slovak Ministry of Education, grant number VEGA 1/0359/18.

Acknowledgments: The first author (Z.V.) would like to express her biggest gratitude to the tutors at FEBS 2018 Ligand Binding Theory and Practice Advanced Course for valuable consulting and guidance.

Conflicts of Interest: The authors declare no conflicts of interest. The funders had no role in the design of the study; in the collection, analyses, or interpretation of data; in the writing of the manuscript, or in the decision to publish the results.

References

1. Peters, T., Jr. *All About Albumin*, 1st ed.; Academic Press: Cambridge, MA, USA, 1995.
2. Fanali, G.; di Masi, A.; Trezza, V.; Marino, M.; Fasano, M.; Ascenzi, P. Human serum albumin: From bench to bedside. *Mol. Asp. Med.* **2012**, *33*, 209–290. [CrossRef] [PubMed]
3. Sudlow, G.; Birkett, D.J.; Wade, D.N. Spectroscopic techniques in the study of protein binding. A fluorescence technique for the evaluation of the albumin binding and displacement of warfarin and warfarin-alcohol. *Clin. Exp. Pharmacol. Physiol.* **1975**, *2*, 129–140. [CrossRef] [PubMed]
4. He, X.M.; Carter, D.C. Atomic structure and chemistry of human serum albumin. *Nature* **1992**, *358*, 209–215. [CrossRef] [PubMed]
5. Ni, Y.; Su, S.; Kokot, S. Spectrofluorimetric studies on the binding of salicylic acid to bovine serum albumin using warfarin and ibuprofen as site markers with the aid of parallel factor analysis. *Anal. Chim. Acta* **2006**, *580*, 206–215. [CrossRef] [PubMed]
6. Poór, M.; Boda, G.; Needs, P.W.; Kroon, P.A.; Lemlic, B.; Bencsik, T. Interaction of quercetin and its metabolites with warfarin: Displacement of warfarin from serum albumin and inhibition of CYP2C9 enzyme. *Biomed. Pharmacother.* **2017**, *88*, 574–581. [CrossRef]
7. Poór, M.; Li, Y.; Kunsági-Máté, S. Molecular displacement of warfarin from human serum albumin by flavonoid aglycones. *J. Lumin.* **2013**, *142*, 122–127. [CrossRef]
8. Wang, X.; Liu, Y.; He, L.; Liu, B.; Zhang, S.-Y.; Ye, X.; Jing, J.-J.; Zhang, J.-F.; Gao, M.; Wang, X. Spectroscopic investigation on the food components–drug interaction The influence of flavonoids on the affinity of nifedipine to human serum albumin. *Food Chem. Toxicol.* **2015**, *78*, 42–51. [CrossRef]
9. Liu, B.-M.; Zhang, J.; Bai, C.-L.; Wang, X.; Qiu, X.-Z.; Wang, X.-L.; Ji, H.; Liu, B. Spectroscopic study on flavonoid–drug interactions: Competitive binding for human serum albumin between three flavonoid compounds and ticagrelor, a new antiplatelet drug. *J. Lumin.* **2015**, *168*, 69–76. [CrossRef]
10. Mohseni-Shahri, F.S.; Housaindokht, M.R.; Bozorgmehr, M.R.; Moosavi-Movahedi, A.A. The influence of the flavonoid quercetin on the interaction of propranolol with human serum albumin: Experimental and theoretical approaches. *J. Lumin.* **2014**, *154*, 229–240. [CrossRef]
11. Kameníková, M.; Furtmüller, P.G.; Klacsová, M.; Lopez-Guzman, A.; Toca-Herrera, J.L.; Vitkovská, A.; Devínsky, F.; Mučaji, P.; Nagy, M. Influence of quercetin on the interaction of gliclazide with human serum albumin-spectroscopic and docking approaches. *Luminescence* **2017**, *32*, 1203–1211. [CrossRef] [PubMed]
12. Ritter, J.; Flower, R.; Henderson, G.; Rang, H. *Rang & Dale's Pharmacology*, 8th ed.; Elsevier: Amsterdam, The Netherlands, 2016.
13. NORVASC—Amlodipine Besylate Tablet. Available online: http://labeling.pfizer.com/ShowLabeling.aspx?id=562 (accessed on 2 December 2018).
14. Shahri, P.A.; Rad, A.S.; Beigoli, S.; Saberi, M.R.; Chamani, J. Human serum albumin–amlodipine binding studied by multi-spectroscopic, zeta-potential, and molecular modeling techniques. *J. Iran. Chem. Soc.* **2018**, *15*, 223–243. [CrossRef]
15. Housaindokht, M.R.; Zaeri, Z.R.; Bahrololoom, M.; Chamanic, J.; Bozorgmehrd, M.R. Investigation of the behavior of HSA upon binding to amlodipine and propranolol: Spectroscopic and molecular modeling approaches. *Spectrochim. Acta A* **2012**, *85*, 79–84. [CrossRef] [PubMed]
16. Abdollahpour, N.; Soheili, V.; Saberi, M.R.; Chamani, J. Investigation of the Interaction between Human Serum Albumin and Two Drugs as Binary and Ternary Systems. *Eur. J. Drug Metab. Pharmacokinet.* **2016**, *41*, 705–721. [CrossRef] [PubMed]
17. Abdollahpour, N.; Asoodeh, A.; Saberi, M.R.; Chamani, J.K. Separate and simultaneous binding effects of aspirin and amlodipine to human serum albumin based on fluorescence spectroscopic and molecular modeling characterizations: A mechanistic insight for determining usage drugs doses. *J. Lumin.* **2011**, *131*, 1885–1899. [CrossRef]

18. Maddi, S.; Yamsani, M.R.; Seeling, A.; Scriba, G.K.E. Stereoselective Plasma Protein Binding of Amlodipine. *Chiralit* **2010**, *22*, 262–266. [CrossRef] [PubMed]
19. Wang, W.; Sun, C.; Mao, L.; Ma, P.; Liu, F.; Yang, J.; Gao, Y. The biological activities, chemical stability, metabolism and delivery systems of quercetin. A review. *Trends Food Sci. Technol.* **2016**, *56*, 21–38. [CrossRef]
20. Formica, J.V.; Regelson, W. Review of the Biology of Quercetin and Related Bioflavonoids. Food and Chemical Toxicology. *Food Chem. Toxicol.* **1995**, *33*, 1061–1080. [CrossRef]
21. Boulton, D.W.; Walle, U.K.; Walle, T. Extensive Binding of the Bioflavonoid Quercetin to Human Plasma Proteins. *J. Pharm. Pharmacol.* **1998**, *50*, 243–249. [CrossRef]
22. Sengupta, B.; Sengupta, P.K. Binding of Quercetin with Human Serum Albumin: A Critical Spectroscopic Study. *Biopolymers* **2003**, *72*, 427–434. [CrossRef]
23. Lakowicz, J.R. *Principles of Fluorescence Spectroscopy*, 3rd ed.; Springer Science+Business Media, LLC: New York, NY, USA, 2006.
24. Klotz, I.M. *Ligand-Receptor Energetics: A Guide for the Perplexed*, 1st ed.; John Wiley & Sons: Hoboken, NJ, USA, 1997.
25. Lissi, E.; Calderón, C.; Campos, A. Evaluation of the number of binding sites in proteins from their intrinsic fluorescence: Limitations and pitfalls. *Photochem. Photobiol.* **2013**, *89*, 1413–1416. [CrossRef]
26. Zsila, F. Circular Dichroism Spectroscopic Detection of Ligand Binding Induced Subdomain IB Specific Structural Adjustment of Human Serum Albumin. *J. Phys. Chem. B* **2013**, *117*, 10798–10806. [CrossRef] [PubMed]
27. Zsila, F.; Bikádi, Z.; Simonyi, M. Probing the binding of the flavonoid, quercetin to human serum albumin by circular dichroism, electronic absorption spectroscopy and molecular modelling methods. *Biochem Pharmacol.* **2003**, *65*, 447–456. [CrossRef]
28. Byler, D.M.; Susi, H. Examination of the secondary structure of proteins by deconvolved FTIR spectra. *Biopolymers* **1986**, *25*, 469–487. [CrossRef] [PubMed]
29. Ionescu, S.; Matei, I.; Tablet, C.; Hillebrand, M. New insights on flavonoid-serum albumin interactions from concerted spectroscopic methods and molecular modeling. *Curr. Drug MeTable* **2013**, *14*, 474–490. [CrossRef]
30. Chi, Z.; Liu, R. Phenotypic characterization of the binding of tetracycline to human serum albumin. *Biomacromolecules* **2011**, *12*, 203–209. [CrossRef] [PubMed]
31. Zaccai, N.R.; Serdyuk, I.N.; Zaccai, J. *Methods in Molecular Biophysics. Structure, Dynamics, Function for Biology and Medicine*, 2nd ed.; Cambridge University Press: Cambridge, UK, 2007.
32. Zsila, F. Subdomain IB Is the Third Major Drug Binding Region of Human Serum Albumin: Toward the Three-Sites Model. *Mol. Pharm.* **2013**, *10*, 1668–1682. [CrossRef] [PubMed]

Sample Availability: Samples of the compounds Human serum albumin, Amlodipine and Quercetin are available from the authors.

Communication

Methyl Salicylate Enhances Flavonoid Biosynthesis in Tea Leaves by Stimulating the Phenylpropanoid Pathway

Xin Li [1,†], Li-Ping Zhang [1,†], Lan Zhang [1], Peng Yan [1], Golam Jalal Ahammed [2,*] and Wen-Yan Han [1,*]

[1] Key Laboratory of Tea Quality and Safety Control, Ministry of Agriculture, Tea Research Institute, Chinese Academy of Agricultural Sciences, 9 Meiling Road, Hangzhou 310008, China; lixin@tricaas.com (X.L.); lpzhang8263@163.com (L.-P.Z.); zhanglan@tricaas.com (L.Z.); yanpengzn@tricaas.com (P.Y.)

[2] College of Forestry, Henan University of Science and Technology, Luoyang 471023, China

* Correspondence: ahammed@haust.edu.cn (G.J.A.); hanwy@tricaas.com (W.-Y.H.); Tel.: +86-571-86650413 (W.-Y.H.)

† These authors contributed equally to this work.

Academic Editor: Marian Brestic
Received: 15 November 2018; Accepted: 18 January 2019; Published: 21 January 2019

Abstract: The phytohormone salicylic acid (SA) is a secondary metabolite that regulates plant growth, development and responses to stress. However, the role of SA in the biosynthesis of flavonoids (a large class of secondary metabolites) in tea (*Camellia sinensis* L.) remains largely unknown. Here, we show that exogenous methyl salicylate (MeSA, the methyl ester of SA) increased flavonoid concentration in tea leaves in a dose-dependent manner. While a moderate concentration of MeSA (1 mM) resulted in the highest increase in flavonoid concentration, a high concentration of MeSA (5 mM) decreased flavonoid concentration in tea leaves. A time-course of flavonoid concentration following 1 mM MeSA application showed that flavonoid concentration peaked at 2 days after treatment and then gradually declined, reaching a concentration lower than that of control after 6 days. Consistent with the time course of flavonoid concentration, MeSA enhanced the activity of phenylalanine ammonia-lyase (PAL, a key enzyme for the biosynthesis of flavonoids) as early as 12 h after the treatment, which peaked after 1 day and then gradually declined upto 6 days. qRT-PCR analysis of the genes involved in flavonoid biosynthesis revealed that exogenous MeSA upregulated the expression of genes such as *CsPAL*, *CsC4H*, *Cs4CL*, *CsCHS*, *CsCHI*, *CsF3H*, *CsDFR*, *CsANS* and *CsUFGT* in tea leaves. These results suggest a role for MeSA in modulating the flavonoid biosynthesis in green tea leaves, which might have potential implications in manipulating the tea quality and stress tolerance in tea plants.

Keywords: salicylic acid; flavonoids; phenylpropanoid pathway; phenylalanine ammonia-lyase (PAL); tea quality

1. Introduction

Tea is the most popular non-alcoholic beverage consumed across the seven continents [1]. However, tea cultivation is mostly confined to Asia and Africa [2]. Green tea, which is produced from the young shoots of *Camellia sinensis* (L.) Kuntze has a range of human health benefits, such as anti-cancer, anti-inflammatory, anti-allergic and anti-obesity effects [3,4]. Flavonoids are the major antioxidative constituents in tea leaves that function against cancer, cardiovascular disease, diabetes, obesity and metabolic syndrome [5,6]. Flavan-3-ol type flavonoids, i.e., catechin compounds, impart the characteristic astringency and bitterness to green tea infusions [7,8]. Catechins act as precursors of theaflavins and thearubigins that are developed during black tea processing [2]. Therefore, flavonoids are considered as an important group of constituents that largely determine the quality of tea.

In plants, flavonoids biosynthesis occurs in the endoplasmic reticulum, from where they are transported to different cellular compartments for specific functions [9,10]. Flavonoids are a large and diverse group of secondary metabolites that are synthesized through a specific branch of phenylpropanoid pathway [11,12]. Phenylalanine is the initial substrate of this pathway, which is deaminated by the catalysis of phenylalanine ammonia-lyase (PAL), the first and a rate-limiting enzyme that regulates overall carbon flux into phenylpropanoid metabolism due to its unique metabolic position [13]. Stress conditions trigger flavonoid biosynthesis, possibly to scavenge overproduced reactive oxygen species (ROS) [14]. In addition to the ability of flavonoids to scavenge ROS, role of flavonoids as "signaling molecules" or "developmental regulators" have been revealed in plants [15]. Some flavonoids interact with hormone signaling and regulate plant organ development [16].

Plant hormones are important endogenous signal molecules that regulate a plethora of metabolic processes and responses to stress [17]. Studies on the effect of plant hormones on flavonoid concentration in tea revealed a largely hormone-specific response [11]. For example, exogenous gibberellins (GA₃) and abscisic acid (ABA) decrease flavonoid concentration [13,18], whereas brassnosteroid application enhances endogenous flavonoid levels in tea leaves [7,8]. Salicylic acid (SA) is an important plant hormone that primarily functions in the immune response [19]. Nonetheless, roles of SA in plant growth, development and stress tolerance have also been revealed. Like flavonoids, SA is also synthesized from phenylalanine via cinnamic acid [20]. Since both SA and flavonoids are phenylpropanoid derivatives and have antioxidant capacity, SA application affects flavonoid biosynthesis [20]. Nevertheless, the effect of methyl salicylate (MeSA, the methyl ester of SA) on flavonoid biosynthesis in tea is largely unknown. In the present study, we show that exogenous MeSA could increase flavonoid concentration in tea leaves, which is associated with the MeSA-induced enhanced activity of PAL and transcriptional upregulation of genes involved in flavonoid biosynthesis. Our results suggest that MeSA has a significant stimulatory effect on flavonoid biosynthetic pathway which could be exploited to manipulate tea quality and stress tolerance.

2. Results

2.1. MeSA Increases Flavonoid Content in a Concentration-Dependent Manner

To assess whether exogenous MeSA could alter flavonoid levels in tea leaves, we analyzed the flavonoid concentration after application of different concentrations of MeSA. The results showed that moderate concentrations of MeSA (0.5–1 mM) increased flavonoid concentrations in tea leaves, reaching the highest value with 1 mM MeSA (27.78% compared to control). However, higher concentrations of MeSA either had no effect or negatively influenced flavonoid concentration in tea leaves (Figure 1). A time course of flavonoid concentrations after 1 mM MeSA treatment showed that flavonoid concentration began to rise after MeSA application, reaching the maximum level at 2 days post-treatment. For instance, flavonoid concentration increased by 30.64% and 29.91% after 1 and 2 days post treatment with MeSA, respectively as compared with that of control. Afterward, flavonoid concentrations gradually decreased.

After 6 days, the level of flavonoids in MeSA-treated leaves decreased by 29.69% compared with that of control (Figure 2). We also analyzed the total and individual catechin concentrations at 2 days post treatment with 1 mM MeSA. The results showed that MeSA treatment significantly increased the total catechin concentration in tea leaves which was attributed to significant increases in (−)-epigallocatechin-3-gallate (EGCG), epicatechins gallate (ECG) and (−)-epigallocatechin (EGC). However, (−)-catechin (C) and epicatechins (EC) concentrations were not altered by MeSA treatment (Supplementary Figure S1).

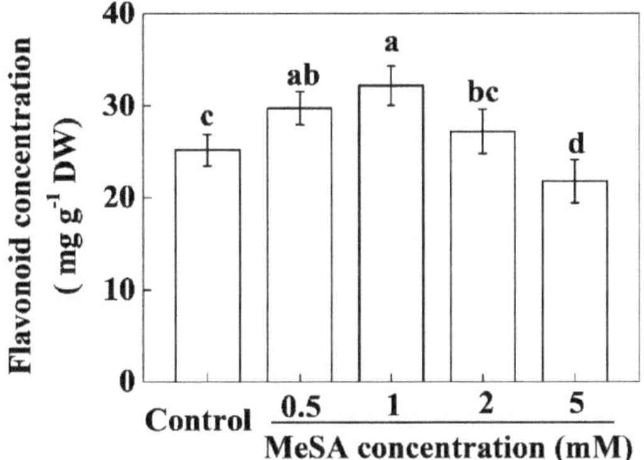

Figure 1. Effect of different concentrations of methyl salicylate (MeSA) on flavonoid concentrations in tea leaves. Tea bushes were sprayed with different concentrations of MeSA (0, 0.5, 1, 2 and 5 mM) and samples were harvested after 1 day for the biochemical analysis. The data of flavonoid concentrations were expressed as the mean values $\pm$ SD, $n = 6$. Mean denoted by the different letters indicate significant differences between the treatments ($p < 0.05$).

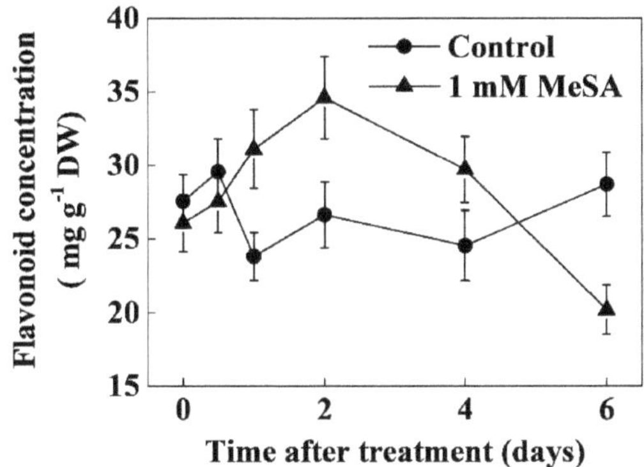

Figure 2. Time course of flavonoid concentration as influenced by exogenous methyl salicylate (MeSA) as foliar spray. Leaf samples were harvested at indicated time-points following foliar spray with 1 mM MeSA. The data of flavonoid concentrations were expressed as the mean values $\pm$ SD, $n = 6$.

2.2. Changes in PAL Activity after MeSA Application

Next, we analyzed the activity of PAL, the first enzyme of the phenylpropanoid pathway. Consistent with the time-course of flavonoid concentration, the PAL activity increased gradually after MeSA treatment, which peaked at 1 day after MeSA treatment (Figure 3). More precisely, the PAL activity increased by 66.58% compared with that of control at 1 day after MeSA treatment. Then the PAL activity in MeSA-treated leaves declined, but remained 36.70% and 23.90% higher than that of control at 2 and 4 days post treatment, respectively, before reaching the level close to the control at 6 days post treatment (Figure 3). Although the PAL activity was differentially regulated by exogenous MeSA at different time points, total soluble protein concentration was not remarkably affected by MeSA in tea leaves (Supplementary Figure S2).

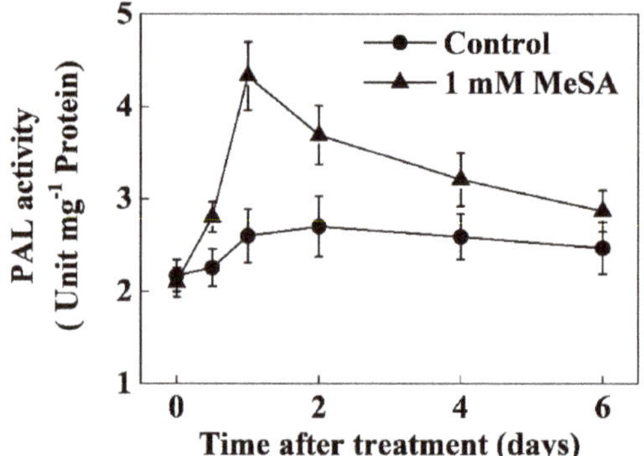

Figure 3. Time course of the phenylalanine ammonia-lyase (PAL) activity as influenced by exogenous methyl salicylate (MeSA). Tea bushes were sprayed with 1 mM MeSA. The data of PAL activity were expressed as the mean values $\pm$ SD, $n = 6$.

2.3. MeSA Modulates Transcript Levels of Flavonoid Biosynthetic Genes

To clarify whether the MeSA-induced changes in flavonoid concentration were attributed to concomitant changes in flavonoids biosynthesis, we analyzed the transcript levels of nine key genes involved in flavonoid biosynthetic pathway (Figure 4a), such as *PHENYLALANINE AMMONIA-LYASE (PAL), CINNAMATE 4-HYDROXYLASE (C4H), p-COUMARATE:COA LIGASE (4CL), CHALCONE SYNTHASE (CHS), CHALCONE ISOMERASE (CHI), FLAVANONE 3-HYDROXYLASE (F3H), DIHYDROFLAVONOL 4-REDUCTASE (DFR), ANTHOCYANIDIN SYNTHASE (ANS)* and *UDP-GLUCOSE FLAVONOID 3-O-GLUCOSYL TRANSFERASE (UFGT).*

(a)　　　　　　　　　　　　　　　　　**(b)**

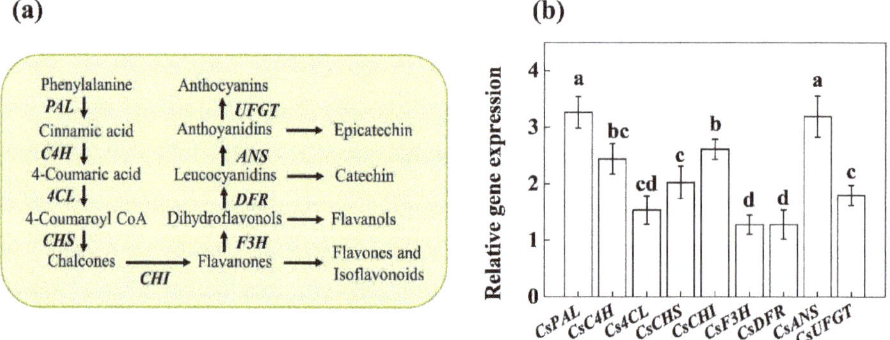

Figure 4. Effect of methyl salicylate on flavonoid biosynthetic pathway in tea leaves. (**a**) Nine key genes involved in flavonoid biosynthesis are marked in bold letters in italic. Adopted and redrawn from Li et al. [12]. (**b**) the expression of flavonoid biosynthetic genes in tea leaves. Transcript levels of the genes were analyzed by qRT-PCR using gene-specific primer pairs (Supplementary Table S1) and expressed as fold change relative to the control. Tea bushes were sprayed with 1 mM methyl salicylate (MeSA) and samples were harvested after 1 day for qRT-PCR assay. The data are mean of 3 biological replicates. Bars denoted by the different letters indicate significant differences between different expression levels of flavonoid biosynthetic genes ($p < 0.05$). *PHENYLALANINE AMMONIA-LYASE (PAL), CINNAMATE 4-HYDROXYLASE (C4H), p-COUMARATE:COA LIGASE (4CL), CHALCONE SYNTHASE (CHS), CHALCONE ISOMERASE (CHI), FLAVANONE 3-HYDROXYLASE (F3H), DIHYDROFLAVONOL 4-REDUCTASE (DFR), ANTHOCYANIDIN SYNTHASE (ANS)* and *UDP-GLUCOSE FLAVONOID 3-O-GLUCOSYL TRANSFERASE (UFGT).*

qRT-PCR analysis showed that exogenous MeSA caused approximately 3-fold increase in transcript levels of *CsPAL* which was more or less consistent with the activity of PAL (Figure 3 or Figure 4b). Similarly, MeSA treatment increased the transcript levels of the rest of the genes, such as *CsC4H* (2.4-fold), *Cs4CL* (1.5-fold), *CsCHS* (2.1-fold), *CsCHI* (2.6-fold), *CsF3H* (1.3-fold), *CsDFR* (1.2-fold), *CsANS* (3.2-fold) and *CsUFGT* (1.8-fold) compared to that of control. These findings suggest that exogenous MeSA differentially regulates transcription of different genes in flavonoid biosynthetic pathway to increase flavonoid concentrations in tea leaves.

3. Discussion

Flavonoids are the key secondary metabolites that contribute to the value of plant products from agronomic, industrial, and nutritional points of view [11]. Particularly, in case of green tea, flavonoids impact not only the tea taste (astringency) but also the health benefits and economic value [21]. Different hormone signaling pathways differentially regulate flavonoid biosynthesis, which is highly species-specific [8,11,13,18,22]. The role of methyl salycilate (MeSA) in flavonoid biosynthesis in tea leaves remained largely unknown. Here we found that an optimal dose of MeSA could increase flavonoid concentration in tea leaves (Figure 1). Time course analysis revealed that MeSA-induced enhancement in flavonoid concentration peaked after 1–2 days, which was consistent with the increased activity of PAL in tea leaves (Figures 2 and 3). Gene expression analysis relating to flavonoid biosynthesis suggested that exogenously applied MeSA stimulated transcriptional machinery causing differential upregulation in the respective transcripts (Figure 4). Therefore, MeSA application at 1 day prior to harvesting may potentially increase flavonoid concentrations in green tea.

Flavonoids are low molecular weight antioxidants that serve as major defense compounds against abiotic and biotic stressors in plants [11,15,20,23]. It is believed that flavonoids play a major role in scavenging ROS when antioxidant enzymes are depleted under stress [9,23]. In the presence of light, SA treatment significantly increases flavonoid content in *Ginkgo biloba* leaves [24], which is in agreement with the current study (Figure 1). Similarly, MeJA increases flavonoid accumulation in citrus leaves in the first 12 h after treatment [25], which is slightly different from our observation, possibly due to the differences in elicitors and plant species. Furthermore, we found that exogenous MeSA rapidly and transiently increased PAL activity in tea leaves, which was consistent with the concentrations of flavonoids (Figures 2 and 3). PAL is the key enzyme in the first step of the phenylpropanoid pathway which regulates biosynthesis of thousands of phenylpropanoids [18]. PAL links the secondary metabolism to primary metabolism and maintains the metabolic flow of carbon into the phenylpropanoid pathway in plants [13]. In the present study, exogenous MeSA increased both the activity and transcription of PAL gene (*CsPAL*) in tea leaves (Figures 3 and 4). MeSA-induced promotion in PAL activity perhaps stimulated subsequent reactions in phenylpropanoid pathway, leading to an enhanced production of phenylpropanoid derivatives including flavonoids (Figures 1–3). Our results are consistent with a previous report that showed that root application of SA in hydroponics increases PAL activity and specific flavonoid content in wheat leaves [20].

In the flavonoid biosynthesis pathway, *CHS*, *CHI* and *F3H* are termed as early biosynthetic genes, whereas downstream genes such as *DFR*, *ANS* and *UFGT* are named late biosynthetic genes [11,26]. In the current study, exogenous MeSA upregulated the transcript levels of both early and late biosynthetic genes in tea leaves (Figure 4). Different hormones differentially modulate the expression of flavonoid biosynthesis genes [8,11,22,26]. For instance, ABA and GA_3 down-regulate *CsPAL*, *CsC4H*, *CsF3H* and *CsANR* expression, leading to a decreased catechin content in tea leaves [13,18]. However, exogenous brassinosteroids increase *CsPAL* expression with concomitant increase in flavonoid concentration in tea leaves [7,8]. In *Vitis vinifera*, PAL expression reaches the peak at 3 h post-elicitation with MeJA, which gradually declines returning to the basal levels at 48 h post-treatment in the presence of light [26]. Consistent with this, jasmonic acid and brassinosteroids also enhance transcript levels of late biosynthetic genes and myeloblastosis (MYB) transcription factor in Arabidopsis seedlings [22]. Notably, at transcriptional levels, flavonoid biosynthesis is regulated by the MYB transcription factors [11], in which SA can induce MYBs to regulate specific phenylpropanoid (capsaicinoid) biosynthesis [27]. Therefore, such regulatory mechanism in response to exogenous MeSA might also function in tea leaves, however, this interpretation demands further in-depth investigation in current direction.

To sum up, in the current study, we found that: (1) a moderate dose of exogenous MeSA (1 mM) increased flavonoid concentration, but a high dose of MeSA (5 mM) showed an opposite effect, (2) 1 mM MeSA appeared to be the best concentration to simulate endogenous flavonoid accumulation in tea leaves, (3) MeSA-induced increase in flavonoid concentration was maximized after 1–2 day followed by gradual decline over time, (4) MeSA-induced increase in flavonoid concentration was associated with the simultaneous increase in the activity of PAL and transcription of early and late biosynthetic genes in flavonoid biosynthetic pathway. These results suggest that 1–2 day prior application of MeSA can be an effective method to manipulate flavonoid concentration in tea leaves. Moreover, MeSA-induced flavonoid biosynthesis may enhance plant tolerance to biotic and/or abiotic stressors.

4. Materials and Methods

4.1. Plant Materials and Growth Conditions

In the current experiment, "Longjing 43" tea (*Camellia sinensis* L.) cultivar was used as plant materials and the study was conducted at tea garden of the Tea Research Institute, Chinese Academy of Agricultural Sciences, Hangzhou, China. Foliar portion of tea bushes was sprayed with a series of freshly prepared methyl salicylate (MeSA) solutions (0.5, 1, 2 and 5 mM). Each treatment comprises 4 replicates, while each replicate represents an area of 10 m^2 consisting of 20 tea bushes.

4.2. Determination of Flavonoid Concentration

For the determination of flavonoid concentration, leaf samples were extracted in 70% ethanol (v/v) at 100 °C, and the concentration of total flavonoids was measured in the aqueous extract following AlCl$_3$ method as described previously [28]. Absorbance at 510 nm was recorded for the determination and rutin was used as the standard. Total and individual catechin concentrations in tea leaves were determined with a Waters 590 HPLC system (Waters Corp., Milford, MA, USA) equipped with a Thermo Scientific™ Hypersil™ ODS-2 C18 column (5 μm particle size, 4.6 mm × 250 mm, Thermo Fisher Scientific Inc., Waltham, MA, USA) at 280 nm as previously described [12].

4.3. Assay of Phenylalanine Ammonia-Lyase (PAL) Enzyme Activity

A tea sample (0.3 g) was homogenized in 3 mL 50 mM potassium phosphate buffer (pH 8.8, containing 2 mM EDTA, 2% PVPP, and 0.1% mercaptoethanol). The resulting homogenate was centrifuged at 15,000 rpm for 20 min at 4 °C and the crude enzyme extract was obtained as the supernatant. L-phenylalanine was used as substrate to assay the PAL activity based on the yield of cinnamic acid. The change in absorbance at 290 nm was monitored to determine the PAL activity as described previously [29].

4.4. Total RNA Extraction and Gene Expression Analysis

For gene expression analysis, tea leaf samples were collected at 1 day after MeSA treatment and immediately frozen into liquid nitrogen and kept at −80 °C until RNA isolation. Total RNA was extracted using an RNA extraction kit (Tiangen Biotech, Beijing, China) and reverse transcribed using a ReverTra Ace qPCR RT kit (Toyobo, Osaka, Japan) following the manufacturer's instructions. Gene-specific primers were designed based on their cDNA sequences (Supplementary Table S1). Quantitative real-time PCR (qRT-PCR) was performed on the ABI 7500 Real-Time PCR system (Applied Biosystems, Foster City, CA, USA) using SYBR Green PCR Master Mix (Takara, Shiga, Japan). The qRT-PCR cycling conditions were as follows: 95 °C for 30 s, and 40 cycles of 95 °C for 5 s and 60 °C for 34 s. Relative gene expression was calculated according to previously described method using *CsPTB* as the internal reference gene [30].

4.5. Statistical Analysis

The data were statistically analyzed using SAS 8.1 software package (SAS Institute Inc., Cary, NC, USA). Differences between treatments means were separated by the Tukey' test at a significance level of $p < 0.05$.

Supplementary Materials: The following is available online. Table S1: Primers used for qRT-PCR assays in tea leaves, Figure S1: Total and individual catechins concentrations in tea leaves as influenced by exogenous methyl salicylate as foliar spray, Figure S2: Time course of total soluble protein concentrations as influenced by exogenous methyl salicylate.

Author Contributions: Conceptualization, X.L. and W.-Y.H.; Formal analysis, Methodology, X.L., L.-P.Z., L.Z., P.Y. and G.J.A.; Writing—original draft, Writing—review & editing, X.L. and G.J.A.; Resources, Supervision, W.-Y.H.

Funding: This work was supported by the National Key R&D Program of China (2017YFE0107500), the Open Fund of State Key Laboratory of Tea Plant Biology and Utilization (SKLTOF20170106), the Science and Technology Innovation Project of the Chinese Academy of Agricultural Sciences (CAAS-ASTIP-2014-TRICAAS), the Henan University of Science and Technology Research Start-up Fund for New Faculty (13480058) and the Key Laboratory of Horticultural Crop Growth and Quality Control in Protected Environment of Luoyang City.

Conflicts of Interest: The authors declare no conflict of interest.

References

1. Macfarlane, A.; Macfarlane, I. *The Empire of Tea: The Remarkable History of the Plant that Took over the World*; Overlook Press: New York, NY, USA, 2004.
2. Han, W.; Li, X.; Yan, P.; Zhang, L.; Ahammed, G.J. Tea cultivation under changing climatic conditions. In *Global Tea Science: Current Status and Future Needs*; Burleigh Dodds Science Publishing: Cambridge, UK, 2018; pp. 455–472.
3. Mancini, E.; Beglinger, C.; Drewe, J.; Zanchi, D.; Lang, U.E.; Borgwardt, S. Green tea effects on cognition, mood and human brain function: A systematic review. *Phytomed. Int. J. Phytother. Phytopharm.* **2017**, *34*, 26–37. [CrossRef] [PubMed]
4. Unno, K.; Noda, S.; Kawasaki, Y.; Yamada, H.; Morita, A.; Iguchi, K.; Nakamura, Y. Reduced stress and improved sleep quality caused by green tea are associated with a reduced caffeine content. *Nutrients* **2017**, *9*, 777. [CrossRef] [PubMed]
5. George, V.C.; Dellaire, G.; Rupasinghe, H.P.V. Plant flavonoids in cancer chemoprevention: Role in genome stability. *J. Nutr. Biochem.* **2017**, *45*, 1–14. [CrossRef] [PubMed]
6. Wang, T.-y.; Li, Q.; Bi, K.-S. Bioactive flavonoids in medicinal plants: Structure, activity and biological fate. *Asian J. Pharm. Sci.* **2018**, *13*, 12–23. [CrossRef]
7. Li, X.; Ahammed, G.J.; Li, Z.-X.; Zhang, L.; Wei, J.-P.; Shen, C.; Yan, P.; Zhang, L.-P.; Han, W.-Y. Brassinosteroids improve quality of summer tea (*Camellia sinensis* L.) by balancing biosynthesis of polyphenols and amino acids. *Front. Plant Sci.* **2016**, *7*, 1304. [CrossRef] [PubMed]
8. Li, X.; Zhang, L.; Ahammed, G.J.; Li, Z.-X.; Wei, J.-P.; Shen, C.; Yan, P.; Zhang, L.-P.; Han, W.-Y. Nitric oxide mediates brassinosteroid-induced flavonoid biosynthesis in *Camellia sinensis* L. *J. Plant Physiol.* **2017**, *214*, 145–151. [CrossRef]
9. Agati, G.; Azzarello, E.; Pollastri, S.; Tattini, M. Flavonoids as antioxidants in plants: Location and functional significance. *Plant Sci.* **2012**, *196*, 67–76. [CrossRef] [PubMed]
10. Zhao, J. Flavonoid transport mechanisms: How to go, and with whom. *Trends Plant Sci.* **2015**, *20*, 576–585. [CrossRef] [PubMed]
11. Xu, W.; Dubos, C.; Lepiniec, L. Transcriptional control of flavonoid biosynthesis by MYB-bHLH-WDR complexes. *Trends Plant Sci.* **2015**, *20*, 176–185. [CrossRef] [PubMed]
12. Li, X.; Zhang, L.; Ahammed, G.J.; Li, Z.-X.; Wei, J.-P.; Shen, C.; Yan, P.; Zhang, L.-P.; Han, W.-Y. Stimulation in primary and secondary metabolism by elevated carbon dioxide alters green tea quality in *Camellia sinensis* L. *Sci. Rep.* **2017**, *7*, 7937. [CrossRef] [PubMed]

13. Singh, K.; Kumar, S.; Rani, A.; Gulati, A.; Ahuja, P.S. Phenylalanine ammonia-lyase (PAL) and cinnamate 4-hydroxylase (C4H) and catechins (flavan-3-ols) accumulation in tea. *Funct. Integr. Genom.* **2009**, *9*, 125–134. [CrossRef] [PubMed]

14. Julkunen-Tiitto, R.; Nenadis, N.; Neugart, S.; Robson, M.; Agati, G.; Vepsäläinen, J.; Zipoli, G.; Nybakken, L.; Winkler, B.; Jansen, M.A.K. Assessing the response of plant flavonoids to UV radiation: An overview of appropriate techniques. *Phytochem. Rev.* **2015**, *14*, 273–297. [CrossRef]

15. Lee, K.; Hwang, O.J.; Reiter, R.J.; Back, K. Flavonoids inhibit both rice and sheep serotonin N-acetyltransferases and reduce melatonin levels in plants. *J. Pineal Res.* **2018**, *65*, e12512. [CrossRef] [PubMed]

16. Di Ferdinando, M.; Brunetti, C.; Fini, A.; Tattini, M. Flavonoids as Antioxidants in Plants Under Abiotic Stresses. In *Abiotic Stress Responses in Plants: Metabolism, Productivity and Sustainability*; Ahmad, P., Prasad, M.N.V., Eds.; Springer New York: New York, NY, USA, 2012; pp. 159–179.

17. Ahammed, G.J.; Li, X.; Zhou, J.; Zhou, Y.-H.; Yu, J.-Q. Role of Hormones in Plant Adaptation to Heat Stress. In *Plant Hormones under Challenging Environmental Factors*; Springer: Heidelberg, Germany, 2016; pp. 1–21.

18. Rani, A.; Singh, K.; Ahuja, P.S.; Kumar, S. Molecular regulation of catechins biosynthesis in tea [*Camellia sinensis* (L.) O. Kuntze]. *Gene* **2012**, *495*, 205–210. [CrossRef] [PubMed]

19. Radojičić, A.; Li, X.; Zhang, Y. Salicylic Acid: A Double-Edged Sword for Programed Cell Death in Plants. *Front. Plant Sci.* **2018**, *9*, 1133. [CrossRef]

20. Gondor, O.K.; Janda, T.; Soos, V.; Pal, M.; Majlath, I.; Adak, M.K.; Balazs, E.; Szalai, G. Salicylic Acid Induction of Flavonoid Biosynthesis Pathways in Wheat Varies by Treatment. *Front. Plant Sci.* **2016**, *7*, 1447. [CrossRef] [PubMed]

21. Tounekti, T.; Joubert, E.; Hernández, I.; Munné-Bosch, S. Improving the Polyphenol Content of Tea. *Crit. Rev. Plant Sci.* **2013**, *32*, 192–215. [CrossRef]

22. Peng, Z.; Han, C.; Yuan, L.; Zhang, K.; Huang, H.; Ren, C. Brassinosteroid enhances jasmonate-induced anthocyanin accumulation in Arabidopsis seedlings. *J. Integr. Plant Biol.* **2011**, *53*, 632–640. [CrossRef] [PubMed]

23. Agati, G.; Brunetti, C.; Di Ferdinando, M.; Ferrini, F.; Pollastri, S.; Tattini, M. Functional roles of flavonoids in photoprotection: New evidence, lessons from the past. *Plant Physiol. Biochem.* **2013**, *72*, 35–45. [CrossRef]

24. Ni, J.; Dong, L.; Jiang, Z.; Yang, X.; Sun, Z.; Li, J.; Wu, Y.; Xu, M. Salicylic acid-induced flavonoid accumulation in *Ginkgo biloba* leaves is dependent on red and far-red light. *Ind. Crop. Prod.* **2018**, *118*, 102–110. [CrossRef]

25. Wang, Z.; Yu, Q.; Shen, W.; El Mohtar, C.A.; Zhao, X.; Gmitter, F.G., Jr. Functional study of CHS gene family members in citrus revealed a novel CHS gene affecting the production of flavonoids. *BMC Plant Biol.* **2018**, *18*, 189. [CrossRef] [PubMed]

26. Donati, L.; Ferretti, L.; Frallicciardi, J.; Rosciani, R.; Valletta, A.; Pasqua, G. Stilbene biosynthesis and gene expression in response to methyl jasmonate and continuous light treatment in *Vitis vinifera* cv. Malvasia del Lazio and *Vitis rupestris* Du Lot cell cultures. *Physiol. Plant* **2018**, *8*. [CrossRef] [PubMed]

27. Arce-Rodriguez, M.L.; Ochoa-Alejoa, N. An R2R3-MYB Transcription Factor Regulates Capsaicinoid Biosynthesis. *Plant Physiol.* **2017**, *174*, 1359–1370. [CrossRef] [PubMed]

28. Lin, J.-Y.; Tang, C.-Y. Determination of total phenolic and flavonoid contents in selected fruits and vegetables, as well as their stimulatory effects on mouse splenocyte proliferation. *Food Chem.* **2007**, *101*, 140–147. [CrossRef]

29. Zheng, H.Z.; Cui, C.L.; Zhang, Y.T.; Wang, D.; Jing, Y.; Kim, K.Y. Active changes of lignification-related enzymes in pepper response to *Glomus intraradices* and/or *Phytophthora capsici*. *J. Zhejiang Univ. Sci. B* **2005**, *6*, 778–786. [CrossRef] [PubMed]

30. Livak, K.J.; Schmittgen, T.D. Analysis of relative gene expression data using real-time quantitative PCR and the 2−ΔΔCT method. *Methods* **2001**, *25*, 402–408. [CrossRef] [PubMed]

Sample Availability: Samples of the compounds are available from the authors.

 *molecules*

Review

The Role of Salicylic Acid in Plants Exposed to Heavy Metals

Anket Sharma [1,*,†], Gagan Preet Singh Sidhu [2,†], Fabrizio Araniti [3,*,†], Aditi Shreeya Bali [4], Babar Shahzad [5], Durgesh Kumar Tripathi [6], Marian Brestic [7,8], Milan Skalicky [8] and Marco Landi [9,10,11,*]

[1] State Key Laboratory of Subtropical Silviculture, Zhejiang A&F University, Hangzhou 311300, China
[2] Department of Environment Education, Government College of Commerce and Business Administration, Chandigarh 160047, India; gagan1986sidhu@gmail.com
[3] Dipartimento AGRARIA, Università Mediterranea di Reggio Calabria, Località Feo di Vito, SNC I-89124 Reggio Calabria, RC, Italy
[4] Mehr Chand Mahajan D.A.V. College for Women, Chandigarh 160036, India; shreeyaaditi02@gmail.com
[5] School of Land and Food, University of Tasmania, Hobart, TAS 7005, Australia; babar.shahzad@utas.edu.au
[6] Amity Institute of Organic Agriculture, Amity University Uttar Pradesh, Noida 201313, India; dktripathiau@gmail.com
[7] Department of Plant Physiology, Faculty of Agrobiology and Food Resources, Slovak University of Agriculture, 94976 Nitra, Slovakia; marian.brestic@uniag.sk
[8] Department of Botany and Plant Physiology, Faculty of Agrobiology, Food and Natural Resources, Czech University of Life Sciences, 16500 Prague, Czech Republic; skalicky@af.czu.cz
[9] Department of Agriculture, Food and Environment, University of Pisa, I-56124 Pisa, Italy
[10] CIRSEC, Centre for Climatic Change Impact, University of Pisa, Via del Borghetto 80, I-56124 Pisa, Italy
[11] Interdepartmental Research Center Nutrafood "Nutraceuticals and Food for Health", University of Pisa, I-56124 Pisa, Italy
* Correspondence: anketsharma@gmail.com (A.S.); fabrizio.araniti@unirc.it (F.A.); marco.landi@unipi.it (M.L.)
† Authors contributed equally.

Academic Editor: Francesca Giampieri
Received: 11 December 2019; Accepted: 25 January 2020; Published: 26 January 2020

Abstract: Salicylic acid (SA) is a very simple phenolic compound (a $C_7H_6O_3$ compound composed of an aromatic ring, one carboxylic and a hydroxyl group) and this simplicity contrasts with its high versatility and the involvement of SA in several plant processes either in optimal conditions or in plants facing environmental cues, including heavy metal (HM) stress. Nowadays, a huge body of evidence has unveiled that SA plays a pivotal role as plant growth regulator and influences intra- and inter-plant communication attributable to its methyl ester form, methyl salicylate, which is highly volatile. Under stress, including HM stress, SA interacts with other plant hormones (e.g., auxins, abscisic acid, gibberellin) and promotes the stimulation of antioxidant compounds and enzymes thereby alerting HM-treated plants and helping in counteracting HM stress. The present literature survey reviews recent literature concerning the roles of SA in plants suffering from HM stress with the aim of providing a comprehensive picture about SA and HM, in order to orientate the direction of future research on this topic.

Keywords: metal toxicity; *ortho*-hydroxybenzoic acid; plant hormone; metal pollution; polyphenols; signaling compound

---

1. Introduction

Salicylic acid (SA) (from Latin *Salix*, willow tree), also known as *ortho*-hydroxybenzoic acid, is a phenolic derivative widely distributed in the plant kingdom and is known as a regulator of several physiological and biochemical processes such as thermogenesis, plant signaling or plant defense, and response to biotic and abiotic stress [1,2].

From a chemical point of view, SA belongs to a large group of plant phenolics, and SA can be isolated in plants in both free and conjugated form. In particular, the conjugated form proceeds from the methylation, hydroxylation, and/or glucosylation of the aromatic ring [3,4].

Salicin, one of the natural SA derivatives, was first isolated from the bark of the willow tree (*Salix* sp.) by Johan Büchner in 1828 [5,6]. Successively, it was discovered that almost all the willow trees including *Salix alba, S. purpurea, S. fragilis,* and *S. daphnoides* were particularly rich in this natural compound, in which the concentration in plants significantly fluctuates during the different seasons (highest content during spring and summer, lowest content during autumn and winter [7]) reaching values of 3 mg/g of fresh biomass in plants of *S. laponum* [8]. The first scientist who was able to identify this natural compound in species different from Salix sp. was the Italian chemist Raffaele Piria in the late 1838, who obtained SA in both flower and buds of the European species *Spiraea ulmaria* successively renamed as *Filipendula ulmaria* (L.) Maxim. The discovery that this molecule was not exclusive to the *Salix* genus has opened the door to the study of its biosynthesis, as well as its biochemical and physiological role in plants and in 1899 the Bayer Company formulated a new drug known today as aspirin [9].

Concerning the biosynthesis of SA, it is known to be produced through the shikimate pathway by two metabolic routes (Figure 1). In the first discovered route, also known as phenylalanine route, occuring in the cytoplasm of the cell, the enzyme phenylalanine ammonia lyase (PAL) converts phenylalanine (Phe) to *trans*-cinnamic acid (t-CA), which gets oxidized to benzoic acid (BA). Subsequently, the enzyme benzoic-acid-2-hydroxylase (BA2H) catalyzes the hydroxylation of BA aromatic ring and leads to SA formation. The enzymatic conversion of BA into SA by BA2H requires the presence of hydrogen peroxide (H_2O_2) [10–12].

The first evidences for the first route were given by Ellis and Amrchein [13], who observed that feeding *Gaultheria procumbens* plants with labeled 14C-benzoic acid or 14C-cinnamic acid resulted in the production of labeled SA. Successively, Yalpani et al. [14] and Silverman et al. [15], working on rice and tobacco, proposed that the side chain of *trans*-cinnamic acid is decarboxylated to generate BA. Then, BA is hydroxylated at the C2 position forming SA. Anyway, recent results indicated that benzoyl glucose, a conjugated form of BA, is more likely to be the direct precursor of SA [12,14].

The second route is called isochorismate (IC) pathway and occurs in the chloroplast [16–18]. In plants, chorismate is transformed to isochorismate and then to SA, a reaction which is catalyzed by two enzymes isochorismate synthase (ICS) and isochorismate pyruvate lyase (IPL). Recent studies carried on *Arabidopsis thaliana* demonstrated that the ~90% of defense-related SA is produced from isochorismate generated by the plastid-localized isochorismate synthase1, whereas ~10% is derived from the cytosolic PAL pathway [1,17].

From the physiological point of view, it is known that SA plays a pivotal role in the regulation of plant growth, development, in defense from biotic and abiotic stress, and in plant immune responses [4,19–23].

For several years, SA was believed to be just one of the several phenolic compounds synthetized by plants with relatively low importance [5,16]. In 1974, after more than a hundred of years from its discovery, it was provided the first evidence that SA could play a role as plant hormone, when Clealand and Ajami [24] observed that SA was a mobile signaling molecule localized in the phloem inducing flowering in different plant species.

However, the final evidence that SA was a plant hormone was only provided several years later by Raskin et al. [25], who described its role during the thermogenesis in *Sauromatum guttatum*.

From that moment, an exponential increase of manuscripts focused on SA (acting alone or in concert with other plant hormones) as a plant growth regulator, signaling molecule, as well as plant elicitor protecting plants from biotic and abiotic stress, was observed [22,23,26–31].

Recently, it has also been demonstrated that SA could play a pivotal role in protecting plants from environmental stress, including heavy metals (HM). In fact, several recent manuscripts reported that SA can alleviate HM toxicity influencing both their uptake and/or accumulation in plant organs [32–38], as well as scavenging of reactive oxygen species (ROS) and/or decreasing their accumulation and/or enhancing the antioxidant defense system [39–42], protecting membrane stability and integrity [43], interacting with plant hormones [44], upregulating heme oxygenase [45], and improving the performance of the photosynthetic machinery [42,46,47].

Focusing on these aspects, the present review provides a comprehensive assemblage concerning SA roles in plant defense from HM stress, with the aim to provide a clear view of SA and HM to orientate the direction of future research on this topic.

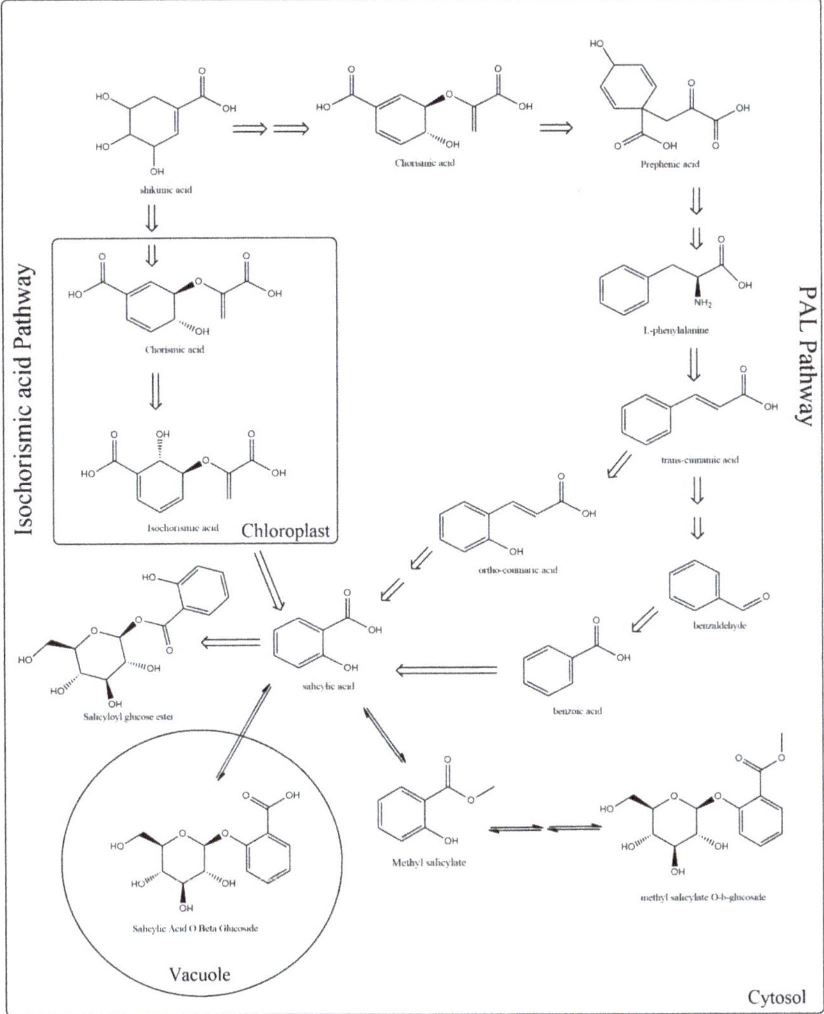

Figure 1. Metabolic pathways involved in the biosynthesis of salicylic acid (SA). Plants use two pathways for SA production, the phenylalanine ammonia-lyase (PAL) (which is divided into two sub-pathways, benzoic acid, and o-coumaric acid) and the isochorismate. In both routes, shikimate serves as a precursor.

2. HM Stress and Its Impacts on Plants

Metals and metalloids with atomic density more than 6 g cm^{-3} are defined as (HM). Both, essential elements, micronutrients that are required in low concentration (e.g., Cu, Cr, Co, and Zn), and nonessential metals such as Pb, Cd, Hg, are incorporated in this group [48,49]. Increased concentration of both essential and nonessential elements is phytotoxic to flora and fauna [50,51]. Heavy metal contamination has become a serious environmental problem worldwide. The increased industrialization, injudicious population growth, and urbanization releases HM that compromise soil and water and pose harms to living biota due to their biomagnification through the food chain [52]. Natural activities such as eruption of volcano and erosion of rocks have contribute in increasing the release of toxic elements to the environment; however, increased human activities such as mining, painting, and refining have enhanced their concentration in the biosphere [53–55].

Soil pollution by HM poses serious concerns to the biotic and abiotic components of the ecosystem [56]. The increased amount of HM in soil leads to greater uptake by plants that can reduce plant growth, biomass, photosynthesis, crop yield, and quality in plant [57]. From a biological point of view, the top soil is the most active zone of soil that accumulates a large amount of toxic metals that poses serious concern to the environment [49,58,59].

The increased level of HM accumulation in plant organs negatively affects the cell metabolism in plants [60]. The different physiological activities in plants such as protein metabolism, photosynthesis, respiration, and morphogenesis are naturally affected by a high concentration of toxic compounds, such as HM [53,54,61,62]. For instance, Rascio et al. [63] documented a decreased root growth and altered morphogenesis in rice seedlings upon treatment with Cd. Many plant species such as *Brassica napus*, *Helianthus annuus*, *Thalaspi caerulescens*, *Vigna radiata* showed inhibition in photosynthesis in response to Cd treatment [64–68]. Recently, Tandon and Srivastava [69] investigated the Pb effect on the morphology and metabolism of *Sesamum indicum* and found that the increasing concentration of metal affected the growth of the plant. Further, the plant showed severe symptoms of chlorosis, necrosis and reduced chlorophyll, and protein content at higher doses of Pb [69].

The major outcome of metal toxicity is the peaked production of ROS due to impairment of photosynthetic process by HM [70]. ROS such as hydroxyl, superoxide, and hydrogen peroxide are produced as by-product during electron transport in photosynthesis and respiration pathways [71]. Under physiological conditions, ROS play a multitude of signaling roles in plants, as well as in other organisms and they take part in a finely-tuned and well-orchestrated regulatory network [72,73]. ROS are indeed integrated into a complex regulatory system in plants which encompasses ROS, plant hormones (e.g., ethylene (ET) and abscisic acid (ABA)), signaling molecules (e.g., salicylic acid (SA) and jasmonic acid (JA)), and secondary messengers (e.g., Ca^{2+}) [74,75]. However, when ROS production exceeds the physiological levels, their accumulation can lead to oxidative stress in the cells, that cause lipids peroxidation, macromolecular degradation, membrane disruption, DNA breakage, and ion leakage in plants [70,74,75]. For instance, Kaur et al. [76] explored Pb-induced ultrastructural changes in roots of wheat and concluded that Pb inhibited root growth, caused ROS generation, and disrupted mitochondrial and nuclear integrity in the tested plant.

The enhanced generation of ROS in the plant cell is controlled by a complex network of antioxidant machinery that maintains ROS homeostasis in the cell [77]. Plants have a finely-tuned and well-orchestrated defense system that includes enzymatic antioxidants such as catalase (CAT), superoxide dismutase (SOD), ascorbate peroxidase (APX), glutathione peroxidase (GPX) and glutathione reductase (GR), and nonenzymatic antioxidants such as ascorbic acid, glutathione, alkaloids, phenol compounds, and α-tocopherol for scavenging excessive ROS [49,61]. Moreover, phytohormones such as auxins, gibberellins, cytokinins, abscisic acid, ethylene, brassinosteroids, jasmonic acid, and SA take part in the defensive mechanism of plants against HM stress.

3. Physiological Roles of SA in Plants Under HM Stress

Concerning the physiological role in plants, SA is known to play a pivotal role in regulating plant morphology, development, flowering, and stomatal closure [78,79]. SA also affects seedling germination, cell growth, and nodulation in legumes [80]. Khan et al. [81] reported increased leaf area and dry weight production in corn and soybean in response to SA. Furthermore, Hussein et al. [82] reported pot studies that documented improved growth, leaf number, dry biomass, and stem diameter in wheat plants when leaves were sprayed with SA. The rate of transpiration and stomatal index of plants increased in response to supplementation of SA [81]. The pigment concentration in wheat seeds significantly enhanced upon exposure to a low concentration (10^{-5} M) of SA. However, foliar application of SA reduced transpiration rate in test plants, *Phaseolus vulgaris* and *Commelina communis* which might be due to the SA-evoked stomatal closure [83–87]. Moreover, SA has been reported to increase the shelf life of cut flowers of rose and defer senescence by controlling water level in rose plants [86].

Plant growth regulators or phytohormones especially, gibberellins, auxin, cytokinins, ethylene, brassinosteroids, and also SA play a key role in providing HM tolerance in plants [83]. SA, a phenolic plant hormone, regulates photosynthesis, respiration, and antioxidant defense mechanism in plants under different abiotic stress such as high temperature, salinity, and HM [78,88,89]. SA pretreatment provides protection from various metals such as Pb, Hg, Cd, in different plants [90–92].

Supplementation of SA in combination with plant growth promoting bacteria reduces Cr-induced oxidative damage in maize by enhancing activities of antioxidant and nonantioxidant enzymes [93,94]. Earlier, Song et al. [95] reported SA mediated enhancement in the activities of CAT and SOD enzymes in barley leaves under Zn, Cu, and Mn stress. Further, carbohydrate metabolism in Cr-treated maize plants improved upon exposure to SA [94]. Alleviation of Cd toxicity was reported in mustard plants in response to exogenous treatment of SA [93]. Recently, SA treatment mitigated Cd stress in *Brassica juncea* plants and enhanced growth and photosynthesis in plants. Moreover, supplementation of SA reduced reactive oxygen species levels by strengthening the antioxidant defense system in plants and provides stability to the plant membrane [96]. The exogenous application of SA upregulates the antioxidant system, improves growth and yield, and results in lowering of oxidative damage under Pb stress in *B. campestris* [97].

A schematization of the protective role exerted by SA in HM-stressed plants is reported in Figure 2, whereas a literature survey on the effect of different HM on plant metabolism is reported in Table 1.

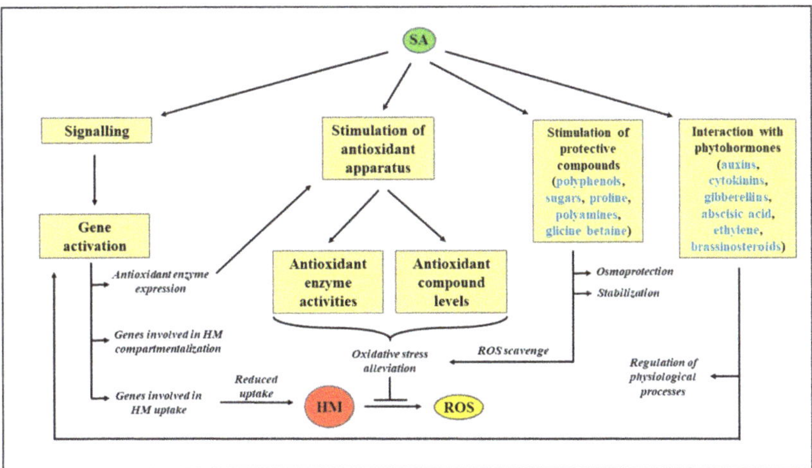

Figure 2. Schematization of the protective role exerted by SA in HM-stressed plants. HM: Heavy metals; ROS: Reactive oxygen species; SA: Salicylic acid.

Table 1. Salicylic acid (SA) effect on different heavy metals (HM) stressed plants.

HM	Species	Effects of SA in plant metabolism	References
Cd	*Lemna minor* L.	Induced a reduction of Cd uptake, the maintenance of ionic homeostasis, improvement of PAL activity, activation of ROS scavenger and of the heat shock proteins.	[98]
	Oryza sativa L.	SA in association with NO reduced Cd uptake and accumulation, as well as ROS accumulation and malondialdehyde production through the maintenance of ascorbate and glutathione levels, and redox status. Improved the activities of antioxidant enzymes such as superoxide dismutase (SOD), catalase (CAT), glutathione S-transferase, and mono dehydroascorbate reductase.	[41]
	Brassica juncea L. Czern	Stimulating the stomatal activity and pore size, alleviated the inhibitory effect of Cd on photosynthesis. The Cd-generated oxidative burst was reduced via enhanced antioxidant activity (CAT and SOD) promoted by SA.	[96]
	Nymphaea tetragona	SA pretreatment decreased Cd concentration and increased the contents of glutathione, nonprotein thiol and phytochelatins.	[99]
	Solanum tuberosum	Cd stress increased endogenous SA level, relative water content, chlorophyll, and proline. Reduced lipid peroxidation, H_2O_2 and O_2^-. SA stimulated enzymatic antioxidants.	[100]
	Triticum aestivum L.	Induced a transient upregulation of protein kinases (SIPK).	[101]
	Mentha piperita	Improved photosynthesis by enhancing activity of RuBisCo and carbonic anhydrase. Reduced the oxidative stress by mitigating the production of free radicals by the maintenance of reduced glutathione pool and free radical scavenging enzymes. Furthermore, restored essential oils production previously affected by Cd.	[102]
Pb	*Brassica juncea* L. Czern	Co-application of 24-epibrassinolide and SA mitigates the negative effects of Pb, by lowering Pb metal uptake and enhancing the heavy metal tolerance index, antioxidative capacities, organic acid levels, phenolic content, water content, and relative water content.	[37]
	Zea mays L.	Improved nitrate reductase activity, glutathione content, and regulated the amino acids metabolism.	[103]
	Triticum aestivum L.	Suppressed chlorophyll degradation, electrolyte leakage, and malondialdehyde accumulation. Furthermore, enhanced the production of total soluble carbohydrates, proline, and the activities of SOD, CAT, and peroxidases.	[104]
	Brassica campestris L.	Improved plant growth and yield upregulating, in the antioxidant defense system, both enzymatic and nonenzymatic components.	[93]
	Zea mays L.	In combination with sodium hydrosulfide reduced arginine, proline, and methionine accumulation and increased nitric oxide and glycine betaine content. Moreover, it regulated the expression of *ZmSAMD* and *ZmACS6* genes (genes involved in methionine metabolism).	[105]

Table 1. *Cont.*

HM	Species	Effects of SA in plant metabolism	References
As	*Trigonella foenum-graecum* L.	Enhanced root growth and increased protein content, free amino acids, and soluble sugars in both cotyledons and radicles. Moreover, it enhanced the activity of hydrolytic enzymes (α- and β-amylase).	[106]
	Artemisia annua L.	Increased endogenous SA level, reduced H_2O_2 and O_2^- generation, as well as lipid peroxidation. Reverted biomass and chlorophyll content. Increased artemisinin, and dihydroartemisinic acid level. Upregulated the expression of four key artemisinin biosynthetic pathway genes (*CYP71AV1, ALDH1, ADS*, and *DBR2*).	[107]
	Artemisia annua L.	Upregulated proteins related to energy metabolism, photosynthesis, secondary metabolism, transcriptional regulators, transport proteins, and proteins related to lipid metabolism.	[108]
	Helianthus annuus L.	Alleviated the negative effect of As on growth and decreased oxidative injuries through the increasing of the enzymatic activity of ROS scavengers such as CAT, ascorbate peroxidase (APX), and glutathione peroxidase, whereas the activity of SOD and guaiacol peroxidase activities was reduced.	[109]
	Oryza sativa L.	As enhanced endogenous level of SA and NO level through the enhancement of nitrate reductase activity.	[110]
Cr	*Sorghum bicolor* L.	Increased both APX and hydrogen peroxide content and decreased the peroxidase activity and ascorbic acid content.	[111]
	Brassica napus L.	Increased dry biomass, enhanced plant growth, and strengthened the reactive oxygen scavenging system by improving the activity in Cr-damaged organelles.	[112]
	Oryza sativa L.	Reduced the concentration and translocation of Cr in shoots but not in roots, suggesting a detoxification strategy based on Cr sequestration in roots. Increased growth parameters, membrane stability, and protein content.	[113]
Ni	*Brassica juncea* L. Czern. & Coss.	Restored growth and photosynthesis increasing the activities of enzymes associated with antioxidant systems, especially the glyoxalase system and the ascorbate–glutathione cycle (AsA–GSH) cycle. It had an additive effect on the activities of the ascorbate and glutathione pools, and the AsA–GSH enzymes and restored the content of mineral nutrient.	[114]
	Eleusine coracana L.	Inhibited Ni transport from roots to shoots, increased chlorophyll content, and the photosynthetic rate, increased the uptake of mineral content, reduced H_2O_2 and proline content, and enhanced the activity of antioxidant enzymes (SOD, CAT, APX).	[115]
	Melissa officinalis L.	Decreased Ni transport to the shoots, increased carotenoid content, induced a significant decrease in electrolyte leakage in stressed plants.	[116]
	Alyssum inflatum Náyr.	Mitigated Ni oxidative effects by reducing H_2O_2 concentration. Reversed the detrimental effects of Ni on carotenoid content and reduced the proline content.	[117]

Table 1. *Cont.*

HM	Species	Effects of SA in plant metabolism	References
Co	*Triticum aestivum* L.	Decreased the accumulation of H_2O_2 and MDA and improved the activity of antioxidant enzymes.	[40]
Cu	*Gossypium barbadense* L.	Limited Cu translocation and improved the activities of antioxidant enzymes.	[118]
	Zea mays L.	Lowered Cu and H_2O_2 accumulation in roots. Induced a reduction of MnSODII activity accompanied by a decrease in H_2O_2 concentration.	[119]
	Zea mays L.	Increased the biomass, root and shoot length, number and leaves area.	[119]

3.1. Effect of SA to Photosynthesis in Plants Subjected to HM Stress

The different stressful conditions encountered by plants affect multiple physiological and biochemical mechanisms in plants. Among these, photosynthesis is usually one of the most affected mechanisms by HM (see a schematization of the effect of HM on chloroplast in Figure 2). HM accumulated in various organs of plants and affect the synthesis of photosynthetic pigments, including carotenoids and chlorophylls [53,54]. HM also alter the chloroplast membrane structure and affect electron transport, thus impairing light-dependent reactions of photosynthesis [120]. Moreover, it was found that the negative effect of HM on PSI and PSII depends on exposure time and concentrations [121,122]. Experiments performed by Khan et al. [123] indicated that PSII is more sensitive to HM stress compared to PSI, however, at high concentrations the activity of PSI resulted inhibited as well. Photosynthesis inhibition caused by HM is also attributable to the impairment of stomatal conductance and transpiration rate [124].

Plants are equipped with multiple mechanisms to preserve the photosynthetic machinery from HM-promoted damages. SA is a major photosynthesis regulator which influences chlorophyll content, stomatal conductivity, and photosynthesis-related enzyme activities in plants [125]. It enhances photosynthetic efficiency and improves photosynthetic apparatus under HM stress [34]. Exogenous application of SA (500 µM) enhanced chlorophyll concentration, CO_2 fixation, and activities of phosphoenolpyruvate carboxylase and RuBISCO in *Triticum aestivum* under Cd toxicity [126]. Further, gas exchange parameters and carbonic anhydrase improved in *B. juncea* under Ni [120] and Mn [127] stress after the exposure to 10 µM SA. SA treatment enhanced Chl*a*, Chl*b*, and carotenoid content in barley plants under Pb stress by increasing antioxidant activity in the plants which might be due to blockage of Ca channels that help in translocation of Pb in roots [60]. Recently, Guo et al. [38] studied the role of SA in Cd alleviation and accumulation in tomato plants. The exogenous exposure of SA also increased pigment content and photosynthetic performance in tomato plants [38]. The consistently observed protective role of SA to the photosynthetic apparatus might be due to increased detoxification of ROS species exerted by SA or by the activation of antioxidant apparatus promoted by SA [125].

3.2. Regulation Mechanism of ROS and Enzymatic Antioxidants Promoted by SA Acid under HM Stress

The generation of ROS is one the first response in plants under HM stress. ROS production is either directly due to Haber-Weiss reaction or it is indirectly because of interference in the antioxidant defense system or electron transport chain [128]. ROS (H_2O_2; hydrogen peroxide, $OH^{\cdot}$; hydroxyl radical, and O_2^{-}; superoxide radical) are very harmful to plants since they lead to oxidative degeneration of cell membranes and large macromolecules [129]. Plants possess a powerful antioxidant apparatus to counteract oxidative stress, which includes different enzymes (SOD, CAT, APX, GR) and nonenzymatic antioxidants (e.g., glutathione, ascorbic acid, phenolics, carotenoids) that scavenge and detoxify ROS over-production in plants [130].

Lipid peroxidation is the first oxidative injury in plants due to HM stress and SA have been shown to provide stability against HM-induced oxidative damage by increasing antioxidant machinery in plants [125]. Parashar et al. [127] and Zhang et al. [131] observed the reduction in lipid peroxidation, electrolyte leakage, and superoxide ion in Mn- and Cd-treated *B. juncea* and *Cucumis melo* upon addition of SA. Few experiments suggest that SA can promote free radical scavenging of HM-promoted ROS by regulating antioxidant enzymes and expression of some proteins and molecules such as OsWRKY45 as reported in rice by Chao et al. [132] that lowers H_2O_2 accumulation in plants. This helps in maintaining the balance between ROS generation and membrane integrity, thereby preventing membrane disruption [133]. Recently, Lu et al. [98] and Gu et al. [99] documented activation of antioxidant enzymes including SOD, APX, and other peroxidases in *Lemna minor* and *Nymphaea tetragona* upon supplementation of SA in plants subjected to Cd stress, which were helpful in conferring Cd tolerance in plants.

3.3. Regulation of Osmolytes and Polyphenols by SA under HM Stress

Plants have evolved various mechanisms to counteract HM-triggered ROS production. Different antioxidant metabolites such as proline, glycine betaine, polyamines, sugars, and polyphenols are all involved in maintaining the ROS balance in plants under stressful conditions, including excess of HM. Below, the intimal connections between SA and other antioxidant compounds are described with the attempt to provide a clear and exhaustive picture about the SA-promoted regulation of antioxidant molecules in plants exposed to HM.

3.3.1. Proline

Proline acts as a free radical scavenger, osmo-protectant, and stabilizer of cellular structures [130,134]. The synthesis of proline occurs from glutamate, which is converted to glutamate-semialdehyde, and then spontaneously to pyrroline-5-carboxylate (P5C) with the help of P5C synthase enzyme. Later, the enzyme P5C reductase aids in the reduction of P5C to proline. The stimulation of proline levels under HM stress was observed, for example, in *Olea europaea* [135] and *Phoenix dactylifera* [136]. However, this is not clear whether the accumulation was attributable to enhanced production of enzymes responsible for proline synthesis, the decrease in enzymes related to its oxidation or both. SA is involved in enhancing proline level under HM toxicity [96]. Parashar et al. [127] reported that SA ameliorated the Mn stress through enhanced accumulation of proline in *Brassica juncea* which might be due to the increased activity of enzymes responsible for proline synthesis [137]. Enhanced proline content also maintains water balance in plants to contrast stressful conditions leading to osmotic stress [138] a condition which can occur when plants reduce the stomatal conductance in order to reduce HM uptake. Further, Chen and Dickman [139] proposed that proline is a powerful ROS scavenger and a pivotal component of protein pathway in plants, besides serving as an osmoprotectant [140]. Zanganeh et al. [141] observed however that SA pre-treatment decreased proline accumulation in *Zea mays* under Pb stress that was supported by the findings of Mostafa et al. [142] in rice plants. Therefore, the pattern of proline (activation/decrement) can be species- or metal-specific and also dependent on the dose of HM experienced by the plant species.

3.3.2. Glycine Betaine

Glycine betaine (GB) is a quaternary level ammonium compound found in higher plants under stress conditions and it acts as osmoprotectant or compatible solutes in plants [143], in which it accumulates at cytosolic level. GB is involved in providing protection against drought, salinity [93], drought [143], and HM stress, as well [144]. Exogenous application of GB is very effective in providing tolerance from HM stress [94,145]. The role of SA in regulating the accumulation of GB in plants under metal stress is still unknown. However, few studies reported that exogenous treatment of GB together with SA can help in alleviating HM toxicity [145]. Recently, Aldesuquy et al. [146] opined that GB and SA regulates osmotic pressure and concentration of osmolytes in plants that maintain osmotic balance and helps in ameliorating the adverse effect of drought stress in wheat, thereby suggesting a possible cooperation. It was also reported that the SA induced the rise in GB level which helped the growth of *Rauwolfia serpentina* plants grown under Na excess [147].

3.3.3. Sugars

The term sugars, collectively used for disaccharides (sucrose, trehalose) and fructans, are water-soluble carbohydrates involved in plant stress tolerance. Sucrose, an important product of photosynthesis, is required for growth, development, storage, and signaling in plants [148,149]. Carbohydrates are building blocks of plants that provide energy and act as a signaling molecule during transcriptional, post-transcriptional processes [150]. Accumulation of soluble sugars has been observed in plants under stressful conditions which indicate their role as osmoprotectant and in maintaining cellular balance in plants [151,152]. The exogenous addition of SA enhanced the amount of polysaccharides and sugars in plants and helped in improving their growth [153]. El-tayeb et al. [154] observed that SA provided Cu tolerance in *Helianthus annuus*. The authors reported an increasing level of soluble sugars in plants treated with SA that protects the photosynthetic pigments from Cu toxicity [154]. Similarly, 0.01 M SA enhanced growth and sugar accumulation in tomato plants and provided stress avoidance and tolerance against Na toxicity [155].

3.3.4. Polyamines

Polyamines (PAs) are water-soluble molecules that play an important role in regulating morphological, developmental, and stress responses in plants [156]. PA have the potential to scavenge HM-triggered ROS [157] and regulate plant defense response to HM toxicity [156,158]. Under stressful conditions, PA operate as signaling compounds and control ion homeostasis and ion transportation in plants, thus actively participating in stress tolerance [159,160]. Many reports suggest that SA treatment influence PA content in plants [131,161]. Recently, Tajti et al. [162] studied the role of putrescine and spermidine on wheat under Cd stress and also reported increased levels of SA in those plants; however, the exact mechanism involved in SA-mediated HM stress tolerance and the relationship between PA and SA in plants are still unknown.

3.3.5. Polyphenols

Phenolics are one of the largest groups of secondary metabolites which include a plethora of compounds with simple aromatic rings to very complex molecules, such as tannins and lignans. They originate from phenylalanine by the activity of PAL. Many reports have demonstrated that enhanced production of phenolic compounds under HM stress can protect from oxidative damage [163,164]. The accumulation of phenolics is principally driven by increased expression of enzymes responsible for phenylpropanoid biosynthesis such as phenylalanine ammonia-lyase, chalcone synthase, shikimate dehydrogenase, cinnamyl alcohol dehydrogenase, and polyphenol oxidase [165,166]. Many studies have documented the role of phytohormones in enhancing the level of some classes of polyphenols, such as anthocyanins [167,168]. Dong et al. [169] reported increased concentrations of phenolics, such as caffeic acid due to exogenous treatment by SA. Similarly, peaked activity of PAL was observed in *Matricaria chamomilla* plants under Ni and Cd stress with the application of SA [170].

3.4. Regulation of Cell Signaling by SA under HM Stress

The HM stress tolerance induced by SA is supportive for its role in stress signaling. The mechanism of tolerance not only depends on the concentration and mode of application of SA but also on the overall status of plants [171]. Abiotic stress not only affects growth and development of plants, but also regulates DNA replication machinery. SA application upregulates the topoisomerase gene and chloroplast elongation factor that help in plant adaptation under stressful conditions [172,173]. Moreover, SA is known to induce expression of *TLC1*, a long terminal repeated retrotransposon family in vivo [171]. This family is transcriptionally activated during stressful conditions and its expression by SA suggests their role in SA-mediated signaling pathways [171]. Another mechanism adopted by SA in regulating HM stress plant response is the increased activity of enzymes involved in AsA-GSH pathway [174]. Both AsA and GSH are active redox compounds that maintain cellular redox balance in plants [175]. SA supplementation also increased SOD and POD level in Cannabis sativa and improved Cd-tolerance [34] which might be related to increased concentration of Ca^{2+} (a second messenger) and H_2O_2, that eventually promote the activity of antioxidant enzymes which reduce cellular ROS level in plants [176,177].

3.5. Crosstalk of SA with Other Plant Growth Regulators

SA regulates different plant responses both under optimal and stressful conditions through the crosstalk with other plant growth regulators or plant hormones [81,178]. The interaction of SA with other hormones such as auxin [179], cytokinin [180], gibberellins [181], abscisic acid [182], ethylene [178], and brassinosteroids [87] has been studied under optimum and stressful environments. The possible outcome of interaction of SA with hormones can be either synergistic or antagonistic under stressful conditions. Recently, Tamás et al. [44] studied the SA regulated alleviation of Cd-stress by restriction of Cd-induced auxin-mediated ROS production in barley roots. The authors suggest that SA treatment reversed indole-3-acetic acid (IAA)-induced stress responses in plants suggesting a role of SA in IAA signaling pathway. Similarly, Agtuca et al. [183] reported an opposite role of IAA and SA in roots of maize. The exogenous application of IAA enhanced lateral growth by depriving primary root growth, while SA increased total root biomass [183].

Exposure to various environmental stresses, such as HM, can enhance ethylene production and induce oxidative stress in plants [175]. The increased ethylene production is due to peaked expression of ethylene-related biosynthetic genes or expression of ethylene-responsive genes [184]. Exogenous SA was reported to mitigate Cd stress in wheat [174] by increasing GSH content that resulted in metal detoxification and scavenging ROS induced by HM-triggered ethylene production. SA supplementation promoted increased ABA level in wheat seedlings under Cd stress that was attributed to a de novo ABA biosynthesis [185]. Further, endogenous ABA controlled SA-mediated alteration of the concentration of dehydrin proteins under HM stress that demonstrate protective mechanism of SA in wheat plants [185].

Under abiotic stress conditions, crosstalk between SA and jasmonates play a crucial role in regulation of plant growth [186]. Generally, SA and jasmonic acid (JA) signaling pathways work in an antagonistic manner [187]. The Mitogen-activated protein kinase (MAPK) signaling pathway mediates the antagonistic action between SA and JA cell signaling [188]. However, nonantagonistic interaction between SA and JA are also reported, but an exact mechanism is still unclear and it needs further studies [186]. For example, in maize plants Cu stress induced the biosynthesis of SA, which further induced JA priming and JA induced volatile organic compounds [189,190].

4. Conclusions

Heavy metal stress has been accepted as one of the major threats for plants growing in contaminated areas. In order to deal with the harmful effects of heavy metals, plants have developed several molecular, metabolic, and physiological processes which allow them to avoid stressful factors or cope with them.

Several researches highlighted that SA, when used at low doses, plays a pivotal role in both alleviating and reducing heavy metal stress in plants. An increase in the endogenous level, as well as exogenous application of this plant hormone has been demonstrated to be helpful for plants either in optimal or in stress conditions. In fact, this ubiquitous plant hormone is involved in the regulation of several metabolic processes in plants, regulating the ex novo biosynthesis of secondary metabolites and osmoprotectants involved in the protection from oxidative stress, thereby increasing the activity of ROS scavenger enzymes and/or acting as antioxidants. However, at high concentrations SA can also act as a negative plant growth regulator [171,191,192].

The scientific literature cited in the present review highlights the important role played by SA in protecting plants from heavy metal stress. However, most of the researches available on this topic are mainly focused on the role played by this molecule after an exogenous application, while very few researches, because of the complexity of the cascade effects generated, have unveiled the defense mechanisms triggered by its endogenous stimulation in response to heavy metals. Therefore, there are still several questions which need further investigation. For example, it would be extremely interesting to disentangle the complexity of SA signaling in response to heavy metals, as well as to unveil if exogenous application of SA might directly or indirectly enhance endogenous SA levels. In the meantime, more genomic, transcriptomic, proteomic, and metabolomics studies are necessary to detect SA responsive genes, proteins, and metabolites altered by heavy metal stress. In addition, it is necessary that a molecular dissection deeply understands the crosstalk between SA with other phytohormones and/or metabolites and the feedback processes involved in controlling the endogenous levels of SA in response to heavy metal stress.

Author Contributions: All authors contributed in writing part of the original draft and also reviewed and edited the whole manuscript. All authors have read and agreed to the published version of the manuscript.

Funding: This research received no external funding.

Conflicts of Interest: The authors declare no conflict of interest.

References

1. Chen, Z.; Zheng, Z.; Huang, J.; Lai, Z.; Fan, B. Biosynthesis of salicylic acid in plants. *Plant Sign. Behav.* **2009**, *4*, 493–496. [CrossRef] [PubMed]
2. Wani, A.B.; Chadar, H.; Wani, A.H.; Singh, S.; Upadhyay, N. Salicylic acid to decrease plant stress. *Environ. Chem. Lett.* **2017**, *15*, 101–123. [CrossRef]
3. Lovelock, D.A.; Šola, I.; Marschollek, S.; Donald, C.E.; Rusak, G.; van Pée, K.H.; Ludwig-Müller, J.; Cahill, D.M. Analysis of salicylic acid-dependent pathways in *Arabidopsis thaliana* following infection with *Plasmodiophora brassicae* and the influence of salicylic acid on disease. *Molecul. Plant Pathol.* **2016**, *17*, 1237–1251. [CrossRef]
4. Maruri-López, I.; Aviles-Baltazar, N.Y.; Buchala, A.; Serrano, M. Intra and extracellular journey of the phytohormone salicylic acid. *Front. Plant Sci.* **2019**, *10*, 423. [CrossRef] [PubMed]
5. Raskin, I. Role of salicylic acid in plants. *Annu. Rev. Plant Biol.* **1992**, *43*, 439–463. [CrossRef]
6. Muthulakshmi, S.; Lingakumar, K. Role of salicylic acid (SA) in plants—A review. *Int. J. Appl. Res.* **2017**, *3*, 33–37.
7. Foster, S. *Tyler's Honest Herbal: A Sensible Guide to the Use of Herbs and Related Remedies*; Routledge: New York, NY, USA, 1999.
8. Petrek, J.; Havel, L.; Petrlova, J.; Adam, V.; Potesil, D.; Babula, P.; Kizek, R. Analysis of salicylic acid in willow barks and branches by an electrochemical method. *Russ. J. Plant Physiol.* **2007**, *54*, 553–558. [CrossRef]
9. Arif, H.; Aggarwal, S. *Salicylic Acid (Aspirin)*; StatPearls Publishing LLC.: Tampa, FL, USA; St. Petersburg, Russia, 2019.
10. Shine, M.; Yang, J.W.; El-Habbak, M.; Nagyabhyru, P.; Fu, D.Q.; Navarre, D.; Ghabrial, S.; Kachroo, P.; Kachroo, A. Cooperative functioning between phenylalanine ammonia lyase and isochorismate synthase activities contributes to salicylic acid biosynthesis in soybean. *New Phytol.* **2016**, *212*, 627–636. [CrossRef]
11. Zhang, Y.; Fu, X.; Hao, X.; Zhang, L.; Wang, L.; Qian, H.; Zhao, J. Molecular cloning and promoter analysis of the specific salicylic acid biosynthetic pathway gene phenylalanine ammonia-lyase (AaPAL1) from *Artemisia annua*. *Biotech. Appl. Biochem.* **2016**, *63*, 514–524. [CrossRef]
12. Chong, J.; Pierrel, M.-A.; Atanassova, R.; Werck-Reichhart, D.; Fritig, B.; Saindrenan, P. Free and conjugated benzoic acid in tobacco plants and cell cultures. Induced accumulation upon elicitation of defense responses and role as salicylic acid precursors. *Plant Physiol.* **2001**, *125*, 318–328. [CrossRef]
13. Hayat, S.; Ali, B.; Ahmad, A. Salicylic acid: Biosynthesis, metabolism and physiological role in plants. In *Salicylic Acid: A Plant Hormone*; Springer: Dordrecht, The Netherlands, 2007; pp. 1–14.
14. Yalpani, N.; León, J.; Lawton, M.A.; Raskin, I. Pathway of salicylic acid biosynthesis in healthy and virus-inoculated tobacco. *Plant Physiol.* **1993**, *103*, 315–321. [CrossRef] [PubMed]
15. Silverman, P.; Seskar, M.; Kanter, D.; Schweizer, P.; Metraux, J.-P.; Raskin, I. Salicylic acid in rice (biosynthesis, conjugation, and possible role). *Plant Physiol.* **1995**, *108*, 633–639. [CrossRef] [PubMed]
16. Métraux, J.-P. Recent breakthroughs in the study of salicylic acid biosynthesis. *Trends Plant Sci.* **2002**, *7*, 332–334. [CrossRef]
17. Garcion, C.; Lohmann, A.; Lamodière, E.; Catinot, J.; Buchala, A.; Doermann, P.; Métraux, J.-P. Characterization and biological function of the Isochorismate Synthase2 gene of Arabidopsis. *Plant Physiol.* **2008**, *147*, 1279–1287. [CrossRef] [PubMed]
18. Rekhter, D.; Lüdke, D.; Ding, Y.; Feussner, K.; Zienkiewicz, K.; Lipka, V.; Wiermer, M.; Zhang, Y.; Feussner, I. Isochorismate-derived biosynthesis of the plant stress hormone salicylic acid. *Science* **2019**, *365*, 498–502. [CrossRef] [PubMed]

19. Wei, Y.; Liu, G.; Chang, Y.; He, C.; Shi, H. Heat shock transcription factor 3 regulates plant immune response through modulation of salicylic acid accumulation and signalling in cassava. *Mol. Plant Pathol.* **2018**, *19*, 2209–2220. [CrossRef]

20. Hartmann, M.; Zeier, J. N-Hydroxypipecolic acid and salicylic acid: A metabolic duo for systemic acquired resistance. *Curr. Opin. Plant Biol.* **2019**, *50*, 44–57. [CrossRef]

21. El-Shazoly, R.M.; Metwally, A.A.; Hamada, A.M. Salicylic acid or thiamin increases tolerance to boron toxicity stress in wheat. *J. Plant Nutr.* **2019**, *42*, 702–722. [CrossRef]

22. Luo, J.; Xia, W.; Cao, P.; Xiao, Z.A.; Zhang, Y.; Liu, M.; Zhan, C.; Wang, N. Integrated transcriptome analysis reveals plant hormones jasmonic acid and salicylic acid coordinate growth and defense responses upon fungal infection in poplar. *Biomolecules* **2019**, *9*, 12. [CrossRef]

23. Pasternak, T.; Groot, E.P.; Kazantsev, F.V.; Teale, W.; Omelyanchuk, N.; Kovrizhnykh, V.; Palme, K.; Mironova, V.V. Salicylic acid affects root meristem patterning via auxin distribution in a concentration-dependent manner. *Plant Physiol.* **2019**, *180*, 1725–1739. [CrossRef]

24. Cleland, C.F.; Ajami, A. Identification of the flower-inducing factor isolated from aphid honeydew as being salicylic acid. *Plant Physiol.* **1974**, *54*, 904–906. [CrossRef] [PubMed]

25. Raskin, I.; Skubatz, H.; Tang, W.; Meeuse, B.J. Salicylic acid levels in thermogenic and non-thermogenic plants. *Ann. Bot.* **1990**, *66*, 369–373. [CrossRef]

26. Dempsey, D.M.A.; Klessig, D.F. How does the multifaceted plant hormone salicylic acid combat disease in plants and are similar mechanisms utilized in humans? *BMC Biol.* **2017**, *15*, 23. [CrossRef]

27. Klessig, D.F.; Choi, H.W.; Dempsey, D.M.A. Systemic acquired resistance and salicylic acid: Past, present, and future. *Mol. Plant-Microbe Interact.* **2018**, *31*, 871–888. [CrossRef] [PubMed]

28. Subban, K.; Subramani, R.; Srinivasan, V.P.M.; Johnpaul, M.; Chelliah, J. Salicylic acid as an effective elicitor for improved taxol production in endophytic fungus *Pestalotiopsis microspora*. *PLoS ONE* **2019**, *14*, e0212736. [CrossRef]

29. Tripathi, D.; Raikhy, G.; Kumar, D. Chemical elicitors of systemic acquired resistance—Salicylic acid and its functional analogs. *Curr. Plant Biol.* **2019**, *17*, 48–59. [CrossRef]

30. Nadeem, M.; Ahmad, W.; Zahir, A.; Hano, C.; Abbasi, B.H. Salicylic acid-enhanced biosynthesis of pharmacologically important lignans and neo lignans in cell suspension culture of *Linum ussitatsimum* L. *Eng. Life Sci.* **2019**, *19*, 168–174. [CrossRef]

31. Li, N.; Han, X.; Feng, D.; Yuan, D.; Huang, L.-J. Signaling crosstalk between salicylic acid and ethylene/jasmonate in plant defense: Do we understand what they are whispering? *Int. J. Mol. Sci.* **2019**, *20*, 671. [CrossRef]

32. Safari, F.; Akramian, M.; Salehi-Arjmand, H.; Khadivi, A. Physiological and molecular mechanisms underlying salicylic acid-mitigated mercury toxicity in lemon balm (*Melissa officinalis* L.). *Ecotoxic Environ. Safety* **2019**, *183*, 109542. [CrossRef]

33. Dalvi, A.A.; Bhalerao, S.A. Response of plants towards heavy metal toxicity: An overview of avoidance, tolerance and uptake mechanism. *Ann. Plant. Sci.* **2013**, *2*, 362–368.

34. Shi, G.; Cai, Q.; Liu, Q.; Wu, L. Salicylic acid-mediated alleviation of cadmium toxicity in hemp plants in relation to cadmium uptake, photosynthesis, and antioxidant enzymes. *Acta Physiol. Plant* **2009**, *31*, 969–977. [CrossRef]

35. Wang, C.; Zhang, S.; Wang, P.; Hou, J.; Qian, J.; Ao, Y.; Lu, J.; Li, L. Salicylic acid involved in the regulation of nutrient elements uptake and oxidative stress in *Vallisneria natans* (Lour.) Hara under Pb stress. *Chemosphere* **2011**, *84*, 136–142. [CrossRef] [PubMed]

36. Wei, T.; Lv, X.; Jia, H.; Hua, L.; Xu, H.; Zhou, R.; Zhao, J.; Ren, X.; Guo, J. Effects of salicylic acid, Fe (II) and plant growth-promoting bacteria on Cd accumulation and toxicity alleviation of Cd tolerant and sensitive tomato genotypes. *J. Environ. Manag.* **2018**, *214*, 164–171. [CrossRef] [PubMed]

37. Kohli, S.K.; Handa, N.; Sharma, A.; Gautam, V.; Arora, S.; Bhardwaj, R.; Alyemeni, M.N.; Wijaya, L.; Ahmad, P. Combined effect of 24-epibrassinolide and salicylic acid mitigates lead (Pb) toxicity by modulating various metabolites in *Brassica juncea* L. seedlings. *Protoplasma* **2018**, *255*, 11–24. [CrossRef] [PubMed]

38. Guo, J.; Zhou, R.; Ren, X.; Jia, H.; Hua, L.; Xu, H.; Lv, X.; Zhao, J.; Wei, T. Effects of salicylic acid, Epi-brassinolide and calcium on stress alleviation and Cd accumulation in tomato plants. *Ecotoxic. Environ. Saf.* **2018**, *157*, 491–496. [CrossRef]

39. Malik, Z.A.; Lal, E.P.; Mir, Z.A.; Lone, A.H. Effect of salicyclic acid and indole acetic acid on tomato crop under induced salinity and cadmium stressed environment: A Review. *Int. J. Plant Soil Sci.* **2018**, *26*, 1–6. [CrossRef]

40. Mohamed, H.E.; Hassan, A.M. Role of salicylic acid in alleviating cobalt toxicity in wheat (*Triticum aestivum* L.) seedlings. *J. Agric. Sci.* **2019**, *11*. [CrossRef]

41. Mostofa, M.G.; Rahman, M.; Ansary, M.; Uddin, M.; Fujita, M.; Tran, L.-S.P. Interactive effects of salicylic acid and nitric oxide in enhancing rice tolerance to cadmium stress. *Int. J. Mol. Sci.* **2019**, *20*, 5798. [CrossRef]

42. Wang, Y.-Y.; Wang, Y.; Li, G.-Z.; Hao, L. Salicylic acid-altering Arabidopsis plant response to cadmium exposure: Underlying mechanisms affecting antioxidation and photosynthesis-related processes. *Ecotoxicol. Environ. Safety* **2019**, *169*, 645–653. [CrossRef]

43. Belkadhi, A.; De Haro, A.; Obregon, S.; Chaïbi, W.; Djebali, W. Positive effects of salicylic acid pretreatment on the composition of flax plastidial membrane lipids under cadmium stress. *Environ. Sci. Poll. Res.* **2015**, *22*, 1457–1467. [CrossRef]

44. Tamás, L.; Mistrík, I.; Alemayehu, A.; Zelinová, V.; Bočová, B.; Huttová, J. Salicylic acid alleviates cadmium-induced stress responses through the inhibition of Cd-induced auxin-mediated reactive oxygen species production in barley root tips. *J. Plant Physiol.* **2015**, *173*, 1–8. [CrossRef] [PubMed]

45. Cui, W.; Li, L.; Gao, Z.; Wu, H.; Xie, Y.; Shen, W. Haem oxygenase-1 is involved in salicylic acid-induced alleviation of oxidative stress due to cadmium stress in *Medicago sativa*. *J. Exp. Bot.* **2012**, *63*, 5521–5534. [CrossRef] [PubMed]

46. Tahjib-Ul-Arif, M.; Siddiqui, M.N.; Sohag, A.A.M.; Sakil, M.A.; Rahman, M.M.; Polash, M.A.S.; Mostofa, M.G.; Tran, L.-S.P. Salicylic acid-mediated enhancement of photosynthesis attributes and antioxidant capacity contributes to yield improvement of maize plants under salt stress. *J. Plant Grow. Regul.* **2018**, *37*, 1318–1330. [CrossRef]

47. Yin, Q.-S.; Yuan, X.; Jiang, Y.-G.; Huang, L.-L.; Li, G.-Z.; Hao, L. Salicylic acid-mediated alleviation in NO_2 phytotoxicity correlated to increased expression levels of the genes related to photosynthesis and carbon metabolism in Arabidopsis. *Environ. Exp. Bot.* **2018**, *156*, 141–150. [CrossRef]

48. Adrees, M.; Ali, S.; Rizwan, M.; Ibrahim, M.; Abbas, F.; Farid, M.; Zia-ur-Rehman, M.; Irshad, M.K.; Bharwana, S.A. The effect of excess copper on growth and physiology of important food crops: A review. *Environ. Sci. Poll. Res.* **2015**, *22*, 8148–8162. [CrossRef]

49. Pinto, A.; De Varennes, A.; Fonseca, R.; Teixeira, D.M. Phytoremediation of soils contaminated with heavy metals: Techniques and strategies. In *Phytoremediation*; Springer: Dordrecht, The Netherlands, 2015; pp. 133–155.

50. Kumar, A.; Usmani, Z.; Ahirwal, J.; Rani, P. Phytomanagement of chromium contaminated brown fields. In *Phytomanagement of Polluted Sites*; Elsevier: Amsterdam, The Neitherland, 2019; pp. 447–469.

51. Park, J.H.; Lamb, D.; Paneerselvam, P.; Choppala, G.; Bolan, N.; Chung, J.-W. Role of organic amendments on enhanced bioremediation of heavy metal (loid) contaminated soils. *J. Haz. Mat.* **2011**, *185*, 549–574. [CrossRef]

52. Kumar, V.; Sharma, A.; Kaur, P.; Sidhu, G.P.S.; Bali, A.S.; Bhardwaj, R.; Thukral, A.K.; Cerda, A. Pollution assessment of heavy metals in soils of India and ecological risk assessment: A state-of-the-art. *Chemosphere* **2019**, *216*, 449–462. [CrossRef]

53. Sidhu, G.P.S.; Singh, H.P.; Batish, D.R.; Kohli, R.K. Effect of lead on oxidative status, antioxidative response and metal accumulation in *Coronopus didymus*. *Plant Physiol. Biochem.* **2016**, *105*, 290–296. [CrossRef]

54. Sidhu, G.P.S.; Singh, H.P.; Batish, D.R.; Kohli, R.K. Tolerance and hyperaccumulation of cadmium by a wild, unpalatable herb *Coronopus didymus* (L.) Sm.(*Brassicaceae*). *Ecotox. Environ. Safety* **2017**, *135*, 209–215. [CrossRef]

55. Sidhu, G.P.S.; Singh, H.P.; Batish, D.R.; Kohli, R.K. Appraising the role of environment friendly chelants in alleviating lead by *Coronopus didymus* from Pb-contaminated soils. *Chemosphere* **2017**, *182*, 129–136. [CrossRef]

56. Keesstra, S.; Mol, G.; de Leeuw, J.; Okx, J.; de Cleen, M.; Visser, S. Soil-related sustainable development goals: Four concepts to make land degradation neutrality and restoration work. *Land* **2018**, *7*, 133. [CrossRef]

57. Ramzani, P.M.A.; Iqbal, M.; Kausar, S.; Ali, S.; Rizwan, M.; Virk, Z.A. Effect of different amendments on rice (*Oryza sativa* L.) growth, yield, nutrient uptake and grain quality in Ni-contaminated soil. *Environ. Sci. Poll. Res.* **2016**, *23*, 18585–18595. [CrossRef] [PubMed]

58. Ihedioha, J.; Ukoha, P.; Ekere, N. Ecological and human health risk assessment of heavy metal contamination in soil of a municipal solid waste dump in Uyo, Nigeria. *Environ. Geochem. Health* **2017**, *39*, 497–515. [CrossRef] [PubMed]

59. Van Nevel, L.; Mertens, J.; Staelens, J.; de Schrijver, A.; Tack, F.M.; de Neve, S.; Meers, E.; Verheyen, K. Elevated Cd and Zn uptake by aspen limits the phytostabilization potential compared to five other tree species. *Ecol. Eng.* **2011**, *37*, 1072–1080. [CrossRef]

60. Arshad, T.; Maqbool, N.; Javed, F.; Wahid, A.; Arshad, M.U. Enhancing the defensive mechanism of lead affected barley (*Hordeum vulgare* L.) genotypes by exogenously applied salicylic acid. *J. Agric. Sci.* **2017**, *9*, 139–146. [CrossRef]

61. Foyer, C.H.; Noctor, G. Redox homeostasis and antioxidant signaling: A metabolic interface between stress perception and physiological responses. *Plant Cell* **2005**, *17*, 1866–1875. [CrossRef]

62. Guerra, F.; Gainza, F.; Pérez, R.; Zamudio, F. Phytoremediation of heavy metals using poplars (*Populus* spp.): A glimpse of the plant responses to copper, cadmium and zinc stress. In *Handbook of Phytoremediation*; Nova Science: New York, NY, USA, 2011; pp. 387–413.

63. Rascio, N.; Dalla Vecchia, F.; La Rocca, N.; Barbato, R.; Pagliano, C.; Raviolo, M.; Gonnelli, C.; Gabbrielli, R. Metal accumulation and damage in rice (cv. Vialone nano) seedlings exposed to cadmium. *Environ. Exp. Bot.* **2008**, *62*, 267–278. [CrossRef]

64. Baryla, A.; Carrier, P.; Franck, F.; Coulomb, C.; Sahut, C.; Havaux, M. Leaf chlorosis in oilseed rape plants (*Brassica napus*) grown on cadmium-polluted soil: Causes and consequences for photosynthesis and growth. *Planta* **2001**, *212*, 696–709. [CrossRef]

65. Di Cagno, R.; Guidi, L.; De Gara, L.; Soldatini, G. Combined cadmium and ozone treatments affect photosynthesis and ascorbate-dependent defences in sunflower. *New Phytol.* **2001**, *151*, 627–636. [CrossRef]

66. Küpper, H.; Parameswaran, A.; Leitenmaier, B.; Trtílek, M.; Šetlík, I. Cadmium-induced inhibition of photosynthesis and long-term acclimation to cadmium stress in the hyperaccumulator *Thlaspi caerulescens*. *New Phytol.* **2007**, *175*, 655–674. [CrossRef]

67. TRAN, T.A.; Popova, L.P. Functions and toxicity of cadmium in plants: Recent advances and future prospects. *Turkish J. Bot.* **2013**, *37*, 1–13.

68. Wahid, A.; Ghani, A.; Javed, F. Effect of cadmium on photosynthesis, nutrition and growth of mungbean. *Agr. Sustain. Develop.* **2008**, *28*, 273–280. [CrossRef]

69. Tandon, P.K.; Srivastava, P. Growth and metabolism of sesame (*Sesamum indicum* L.) plants in relation to lead toxicity. *Agricul. Sci.* **2014**, *6*, 91–92.

70. Rascio, N.; Navari-Izzo, F. Heavy metal hyperaccumulating plants: How and why do they do it? And what makes them so interesting? *Plant Sci.* **2011**, *180*, 169–181. [CrossRef] [PubMed]

71. Kadukova, J.; Kavuličova, J. *Phytoremediation of heavy metal contaminated soils—Plant stress assessment*. *Handbook of Phytoremediation*; Nova Science: New York, NY, USA, 2011; pp. 185–222.

72. Baxter, A.; Mittler, R.; Suzuki, N. ROS as key players in plant stress signalling. *J. Exp. Bot.* **2014**, *65*, 1229–1240. [CrossRef]

73. Mittler, R. ROS are good. *Trends Plant Sci.* **2017**, *22*, 11–19. [CrossRef] [PubMed]

74. Cotrozzi, L.; Pellegrini, E.; Guidi, L.; Landi, M.; Lorenzini, G.; Massai, R.; Remorini, D.; Tonelli, M.; Trivellini, A.; Vernieri, P.; et al. Losing the warning signal: Drought compromises the cross-talk of signaling molecules in *Quercus ilex* exposed to ozone. *Front. Plant Sci.* **2017**, *8*, 1020. [CrossRef] [PubMed]

75. Landi, M.; Cotrozzi, L.; Pellegrini, E.; Remorini, D.; Tonelli, M.; Trivellini, A.; Nali, C.; Guidi, L.; Massai, R.; Vernieri, P.; et al. When "thirsty" means "less able to activate the signalling wave trigged by a pulse of ozone": A case of study in two Mediterranean deciduous oak species with different drought sensitivity. *Sci. Total Environ.* **2019**, *657*, 379–390. [CrossRef] [PubMed]

76. Kaur, G.; Singh, H.P.; Batish, D.R.; Kohli, R.K. Lead (Pb)-induced biochemical and ultrastructural changes in wheat (*Triticum aestivum*) roots. *Protoplasma* **2013**, *250*, 53–62. [CrossRef] [PubMed]

77. Das, K.; Roychoudhury, A. Reactive oxygen species (ROS) and response of antioxidants as ROS-scavengers during environmental stress in plants. *Front. Environ. Sci.* **2014**, *2*, 53. [CrossRef]

78. Miura, K.; Tada, Y. Regulation of water, salinity, and cold stress responses by salicylic acid. *Front. Plant Sci.* **2014**, *5*, 4. [CrossRef] [PubMed]

79. Mohsenzadeh, S.; Shahrtash, M.; Mohabatkar, H. Interactive effects of salicylic acid and silicon on some physiological responses of cadmium-stressed maize seedlings. *Iranian J. Sci. Tech. (Sciences)* **2011**, *35*, 57–60.

80. Vlot, A.C.; Dempsey, D.M.A.; Klessig, D.F. Salicylic acid, a multifaceted hormone to combat disease. *Ann. Rev. Phytopathol.* **2009**, *47*, 177–206. [CrossRef]

81. Khan, M.I.R.; Iqbal, N.; Masood, A.; Per, T.S.; Khan, N.A. Salicylic acid alleviates adverse effects of heat stress on photosynthesis through changes in proline production and ethylene formation. *Plant Sign. Behav.* **2013**, *8*, e26374. [CrossRef] [PubMed]

82. Hussein, M.; Balbaa, L.; Gaballah, M. Salicylic acid and salinity effects on growth of maize plants. *Res. J. Agricul. Biol. Sci.* **2007**, *3*, 321–328.

83. Khokon, M.A.R.; Okuma, E.; Hossain, M.A.; Munemasa, S.; Uraji, M.; Nakamura, Y.; Mori, I.C.; Murata, Y. Involvement of extracellular oxidative burst in salicylic acid-induced stomatal closure in Arabidopsis. *Plant Cell Environ.* **2011**, *34*, 434–443. [CrossRef]

84. Larque-Saavedra, A. Stomatal closure in response to acetylsalicylic acid treatment. *Zeitschrift für Pflanzenphysiologie* **1979**, *93*, 371–375. [CrossRef]

85. Larque-Saavedra, A. The antiranspirant effect of acetylsalcylic acid on *Phaseolus vulgaris*. *Physiol. Plant.* **1978**, *43*, 126–128. [CrossRef]

86. Alaey, M.; Babalar, M.; Naderi, R.; Kafi, M. Effect of pre-and postharvest salicylic acid treatment on physio-chemical attributes in relation to vase-life of rose cut flowers. *Postharvest Biol. Technol.* **2011**, *61*, 91–94. [CrossRef]

87. Divi, U.K.; Rahman, T.; Krishna, P. Brassinosteroid-mediated stress tolerance in Arabidopsis shows interactions with abscisic acid, ethylene and salicylic acid pathways. *BMC Plant Biol.* **2010**, *10*, 151. [CrossRef]

88. Wang, Y.; Hu, J.; Qin, G.; Cui, H.; Wang, Q. Salicylic acid analogues with biological activity may induce chilling tolerance of maize (*Zea mays*) seeds. *Botany* **2012**, *90*, 845–855. [CrossRef]

89. Zengin, F. Effects of exogenous salicylic acid on growth characteristics and biochemical content of wheat seeds under arsenic stress. *J. Environ. Biol.* **2015**, *36*, 249.

90. Ghani, A.; Khan, I.; Ahmed, I.; Mustafa, I.; Abd-Ur, R.; Muhammad, N. Amelioration of lead toxicity in *Pisum sativum* (L.) by foliar application of salicylic acid. *J. Environ. Anal. Toxicol* **2015**, *5*, 10–4172.

91. Gondor, O.K.; Pál, M.; Darkó, É.; Janda, T.; Szalai, G. Salicylic acid and sodium salicylate alleviate cadmium toxicity to different extents in maize (*Zea mays* L.). *PLoS ONE* **2016**, *11*, e0160157. [CrossRef] [PubMed]

92. Zhou, Z.S.; Guo, K.; Elbaz, A.A.; Yang, Z.M. Salicylic acid alleviates mercury toxicity by preventing oxidative stress in roots of *Medicago sativa*. *Environ. Exp. Bot.* **2009**, *65*, 27–34. [CrossRef]

93. Ahmad, P.; Nabi, G.; Ashraf, M. Cadmium-induced oxidative damage in mustard [*Brassica juncea* (L.) Czern. & Coss.] plants can be alleviated by salicylic acid. *South Afr. J. Bot.* **2011**, *77*, 36–44.

94. Islam, F.; Yasmeen, T.; Arif, M.S.; Riaz, M.; Shahzad, S.M.; Imran, Q.; Ali, I. Combined ability of chromium (Cr) tolerant plant growth promoting bacteria (PGPB) and salicylic acid (SA) in attenuation of chromium stress in maize plants. *Plant Physiol. Biochem.* **2016**, *108*, 456–467. [CrossRef]

95. Song, W.Y.; Yang, H.C.; Shao, H.B.; Zheng, A.Z.; Brestic, M. The alleviative effects of salicylic acid on the activities of catalase and superoxide dismutase in malting barley (*Hordeum uhulgare* L.) seedling leaves stressed by heavy metals. *CLEAN–Soil, Air, Water* **2014**, *42*, 88–97. [CrossRef]

96. Faraz, A.; Faizan, M.; Sami, F.; Siddiqui, H.; Hayat, S. Supplementation of salicylic acid and citric acid for alleviation of cadmium toxicity to *Brassica juncea*. *J. Plant Growth Regul.* **2019**, 1–15. [CrossRef]

97. Hasanuzzaman, M.; Matin, M.A.; Fardus, J.; Hasanuzzaman, M.; Hossain, M.S.; Parvin, K. Foliar application of salicylic acid improves growth and yield attributes by upregulating the antioxidant defense system in *Brassica campestris* plants grown in lead-amended soils. *Acta Agrobot.* **2019**, *72*. [CrossRef]

98. Lu, Q.; Zhang, T.; Zhang, W.; Su, C.; Yang, Y.; Hu, D.; Xu, Q. Alleviation of cadmium toxicity in *Lemna minor* by exogenous salicylic acid. *Ecotox. Environ. Saf.* **2018**, *147*, 500–508. [CrossRef] [PubMed]

99. Gu, C.-S.; Yang, Y.-H.; Shao, Y.-F.; Wu, K.-W.; Liu, Z.-L. The effects of exogenous salicylic acid on alleviating cadmium toxicity in *Nymphaea tetragona* Georgi. *S. Afr. J. Bot.* **2018**, *114*, 267–271. [CrossRef]

100. Li, Q.; Wang, G.; Wang, Y.; Yang, D.; Guan, C.; Ji, J. Foliar application of salicylic acid alleviate the cadmium toxicity by modulation the reactive oxygen species in potato. *Ecotox. Environ. Saf.* **2019**, *172*, 317–325. [CrossRef] [PubMed]

101. Tajti, J.; Németh, E.; Glatz, G.; Janda, T.; Pál, M. Pattern of changes in salicylic acid-induced protein kinase (SIPK) gene expression and salicylic acid accumulation in wheat under cadmium exposure. *Plant Biol.* **2019**, *21*, 1176–1180. [CrossRef]

102. Ahmad, B.; Jaleel, H.; Sadiq, Y.; Khan, M.M.A.; Shabbir, A. Response of exogenous salicylic acid on cadmium induced photosynthetic damage, antioxidant metabolism and essential oil production in peppermint. *Plant Growth Regul.* **2018**, *86*, 273–286. [CrossRef]

103. Zanganeh, R.; Jamei, R.; Rahmani, F. Role of salicylic acid and hydrogen sulfide in promoting lead stress tolerance and regulating free amino acid composition in *Zea mays* L. *Acta Physiol. Plant.* **2019**, *41*, 94. [CrossRef]

104. Alamri, S.A.D.; Siddiqui, M.H.; Al-Khaishany, M.Y.; Ali, H.M.; Al-Amri, A.; AlRabiah, H.K. Exogenous application of salicylic acid improves tolerance of wheat plants to lead stress. *Adv. Agric. Sci.* **2018**, *6*, 25–35.

105. Zanganeh, R.; Jamei, R.; Rahmani, F. Impacts of seed priming with salicylic acid and sodium hydrosulfide on possible metabolic pathway of two amino acids in maize plant under lead stress. *Mol. Biol. Res. Commun.* **2018**, *7*, 83.

106. Mabrouk, B.; Kâab, S.; Rezgui, M.; Majdoub, N.; da Silva, J.T.; Kâab, L. Salicylic acid alleviates arsenic and zinc toxicity in the process of reserve mobilization in germinating fenugreek (*Trigonella foenum-graecum* L.) seeds. *S. Afr. J. Bot.* **2019**, *124*, 235–243. [CrossRef]

107. Kumari, A.; Pandey, N.; Pandey-Rai, S. Exogenous salicylic acid-mediated modulation of arsenic stress tolerance with enhanced accumulation of secondary metabolites and improved size of glandular trichomes in *Artemisia annua* L. *Protoplasma* **2018**, *255*, 139–152. [CrossRef]

108. Kumari, A.; Pandey-Rai, S. Enhanced arsenic tolerance and secondary metabolism by modulation of gene expression and proteome profile in *Artemisia annua* L. after application of exogenous salicylic acid. *Plant Physiol. Biochem.* **2018**, *132*, 590–602. [CrossRef]

109. Saidi, I.; Yousfi, N.; Borgi, M.A. Salicylic acid improves the antioxidant ability against arsenic-induced oxidative stress in sunflower (*Helianthus annuus*) seedling. *J. Plant Nutr.* **2017**, *40*, 2326–2335. [CrossRef]

110. Singh, A.P.; Dixit, G.; Kumar, A.; Mishra, S.; Kumar, N.; Dixit, S.; Singh, P.K.; Dwivedi, S.; Trivedi, P.K.; Pandey, V. A protective role for nitric oxide and salicylic acid for arsenite phytotoxicity in rice (*Oryza sativa* L.). *Plant Physiol. Biochem.* **2017**, *115*, 163–173. [CrossRef] [PubMed]

111. Sihag, S.; Brar, B.; Joshi, U. Salicylic acid induces amelioration of chromium toxicity and affects antioxidant enzyme activity in *Sorghum bicolor* L. *Int. J. Phytorem.* **2019**, *21*, 293–304. [CrossRef]

112. Gill, R.A.; Zhang, N.; Ali, B.; Farooq, M.A.; Xu, J.; Gill, M.B.; Mao, B.; Zhou, W. Role of exogenous salicylic acid in regulating physio-morphic and molecular changes under chromium toxicity in black-and yellow-seeded *Brassica napus* L. *Environ. Sci. Poll. Res.* **2016**, *23*, 20483–20496. [CrossRef]

113. Huda, A.N.; Swaraz, A.; Reza, M.A.; Haque, M.A.; Kabir, A.H. Remediation of chromium toxicity through exogenous salicylic acid in rice (*Oryza sativa* L.). *Water Air Soil Poll.* **2016**, *227*, 278. [CrossRef]

114. Zaid, A.; Mohammad, F.; Wani, S.H.; Siddique, K.M. Salicylic acid enhances nickel stress tolerance by up-regulating antioxidant defense and glyoxalase systems in mustard plants. *Ecotoxicol. Environ. Saf.* **2019**, *180*, 575–587. [CrossRef]

115. Kotapati, K.V.; Palaka, B.K.; Ampasala, D.R. Alleviation of nickel toxicity in finger millet (*Eleusine coracana* L.) germinating seedlings by exogenous application of salicylic acid and nitric oxide. *Crop J.* **2017**, *5*, 240–250. [CrossRef]

116. Soltani Maivan, E.; Radjabian, T.; Abrishamchi, P.; Talei, D. Physiological and biochemical responses of *Melissa officinalis* L. to nickel stress and the protective role of salicylic acid. *Arch. Agron. Soil Sci.* **2017**, *63*, 330–343. [CrossRef]

117. Karimi, N.; Ghasempour, H.-R. Salicylic acid and jasmonic acid restrains nickel toxicity by ameliorating antioxidant defense system in shoots of metallicolous and non-metallicolous *Alyssum inflatum* Náyr. Populations. *Plant Physiol. Biochem.* **2019**, *135*, 450–459.

118. Mei, L.; Daud, M.; Ullah, N.; Ali, S.; Khan, M.; Malik, Z.; Zhu, S. Pretreatment with salicylic acid and ascorbic acid significantly mitigate oxidative stress induced by copper in cotton genotypes. *Environ. Sci. Poll. Res.* **2015**, *22*, 9922–9931. [CrossRef] [PubMed]

119. Moravcová, Š.; Tůma, J.; Dučaiová, Z.K.; Waligórski, P.; Kula, M.; Saja, D.; Słomka, A.; Bąba, W.; Libik-Konieczny, M. Influence of salicylic acid pretreatment on seeds germination and some defence mechanisms of *Zea mays* plants under copper stress. *Plant Physiol. Biochem.* **2018**, *122*, 19–30. [CrossRef] [PubMed]

120. Ventrella, A.; Catucci, L.; Piletska, E.; Piletsky, S.; Agostiano, A. Interactions between heavy metals and photosynthetic materials studied by optical techniques. *Bioelectrochemistry* **2009**, *77*, 19–25. [CrossRef] [PubMed]

121. Babu, N.G.; Sarma, P.A.; Attitalla, I.H.; Murthy, S. Effect of selected heavy metal ions on the photosynthetic electron transport and energy transfer in the thylakoid membrane of the cyanobacterium, *Spirulina platensis*. *Acad. J. Plant Sci.* **2010**, *3*, 46–49.

122. Chugh, L.K.; Sawhney, S.K. Photosynthetic activities of *Pisum sativum* seedlings grown in presence of cadmium. *Plant Physiol. Biochem.* **1999**, *37*, 297–303. [CrossRef]

123. Khan, N.; Samiullah Singh, S.; Nazar, R. Activities of antioxidative enzymes, sulphur assimilation, photosynthetic activity and growth of wheat (*Triticum aestivum*) cultivars differing in yield potential under cadmium stress. *J. Agron. Crop Sci.* **2007**, *193*, 435–444. [CrossRef]

124. Yusuf, M.; Fariduddin, Q.; Varshney, P.; Ahmad, A. Salicylic acid minimizes nickel and/or salinity-induced toxicity in Indian mustard (*Brassica juncea*) through an improved antioxidant system. *Environ. Sci. Poll. Res.* **2012**, *19*, 8–18. [CrossRef]

125. Rivas-San Vicente, M.; Plasencia, J. Salicylic acid beyond defence: Its role in plant growth and development. *J. Exp. Bot.* **2011**, *62*, 3321–3338. [CrossRef]

126. Moussa, H.; El-Gamal, S.M. Effect of salicylic acid pretreatment on cadmium toxicity in wheat. *Biol. Plant.* **2010**, *54*, 315–320. [CrossRef]

127. Parashar, A.; Yusuf, M.; Fariduddin, Q.; Ahmad, A. Salicylic acid enhances antioxidant system in Brassica juncea grown under different levels of manganese. *Int. J. Biol. Macromol.* **2014**, *70*, 551–558. [CrossRef]

128. Yadav, S. Heavy metals toxicity in plants: An overview on the role of glutathione and phytochelatins in heavy metal stress tolerance of plants. *S. Afr. J. Bot.* **2010**, *76*, 167–179. [CrossRef]

129. Gill, S.S.; Tuteja, N. Reactive oxygen species and antioxidant machinery in abiotic stress tolerance in crop plants. *Plant Physiol. Biochem.* **2010**, *48*, 909–930. [CrossRef] [PubMed]

130. Sharma, P.; Jha, A.B.; Dubey, R.S.; Pessarakli, M. Reactive oxygen species, oxidative damage, and antioxidative defense mechanism in plants under stressful conditions. *J. Bot.* **2012**, *2012*, 217037. [CrossRef]

131. Zhang, Y.; Xu, S.; Yang, S.; Chen, Y. Salicylic acid alleviates cadmium-induced inhibition of growth and photosynthesis through upregulating antioxidant defense system in two melon cultivars (*Cucumis melo* L.). *Protoplasma* **2015**, *252*, 911–924. [CrossRef]

132. Chao, Y.-Y.; Chen, C.-Y.; Huang, W.-D.; Kao, C.H. Salicylic acid-mediated hydrogen peroxide accumulation and protection against Cd toxicity in rice leaves. *Plant Soil* **2010**, *329*, 327–337. [CrossRef]

133. Liang, Y.; Chen, Q.; Liu, Q.; Zhang, W.; Ding, R. Exogenous silicon (Si) increases antioxidant enzyme activity and reduces lipid peroxidation in roots of salt-stressed barley (*Hordeum vulgare* L.). *J. Plant Physiol.* **2003**, *160*, 1157–1164. [CrossRef]

134. Verbruggen, N.; Hermans, C. Proline accumulation in plants: A review. *Amino Acids* **2008**, *35*, 753–759. [CrossRef]

135. Zouari, M.; Ahmed, C.B.; Elloumi, N.; Bellassoued, K.; Delmail, D.; Labrousse, P.; Abdallah, F.B.; Rouina, B.B. Impact of proline application on cadmium accumulation, mineral nutrition and enzymatic antioxidant defense system of *Olea europaea* L. cv Chemlali exposed to cadmium stress. *Ecotoxicol. Environ. Saf.* **2016**, *128*, 195–205. [CrossRef]

136. Zouari, M.; Ahmed, C.B.; Zorrig, W.; Elloumi, N.; Rabhi, M.; Delmail, D.; Rouina, B.B.; Labrousse, P.; Abdallah, F.B. Exogenous proline mediates alleviation of cadmium stress by promoting photosynthetic activity, water status and antioxidative enzymes activities of young date palm (*Phoenix dactylifera* L.). *Ecotoxicol. Environ. Saf.* **2016**, *128*, 100–108. [CrossRef]

137. Misra, N.; Saxena, P. Effect of salicylic acid on proline metabolism in lentil grown under salinity stress. *Plant Sci.* **2009**, *177*, 181–189. [CrossRef]

138. Zengin, F.K.; Munzuroglu, O. Effects of some heavy metals on content of chlorophyll, proline and some antioxidant chemicals in bean (*Phaseolus vulgaris* L.) seedlings. *Acta Biol. Cracov. Bot.* **2005**, *47*, 157–164.

139. Chen, C.; Dickman, M.B. Proline suppresses apoptosis in the fungal pathogen *Colletotrichum trifolii*. *Proc. Natl. Acad. Sci. USA* **2005**, *102*, 3459–3464. [CrossRef] [PubMed]

140. Szabados, L.; Savoure, A. Proline: A multifunctional amino acid. *Trends Plant Sci.* **2010**, *15*, 89–97. [CrossRef] [PubMed]

141. Zanganeh, R.; Jamei, R.; Rahmani, F. Modulation of growth and oxidative stress by seed priming with salicylic acid in *Zea mays* L. under lead stress. *J. Plant Inter.* **2019**, *14*, 369–375. [CrossRef]

142. Mostofa, M.G.; Fujita, M.; Tran, L.-S.P. Nitric oxide mediates hydrogen peroxide-and salicylic acid-induced salt tolerance in rice (*Oryza sativa* L.) seedlings. *Plant Growth Regul.* **2015**, *77*, 265–277. [CrossRef]

143. Raza, S.; Aown, M.; Saleem, M.F.; Jamil, M.; Khan, I.H. Impact of foliar applied glycinebetaine on growth and physiology of wheat (*Triticum aestivum* L.) under drought conditions. *Pak. J. Agric. Sci.* **2014**, *51*, 327–334.

144. Jabeen, N.; Abbas, Z.; Iqbal, M.; Rizwan, M.; Jabbar, A.; Farid, M.; Ali, S.; Ibrahim, M.; Abbas, F. Glycinebetaine mediates chromium tolerance in mung bean through lowering of Cr uptake and improved antioxidant system. *Arch. Agron. Soil Sci.* **2016**, *62*, 648–662. [CrossRef]

145. Cao, F.; Liu, L.; Ibrahim, W.; Cai, Y.; Wu, F. Alleviating effects of exogenous glutathione, glycinebetaine, brassinosteroids and salicylic acid on cadmium toxicity in rice seedlings (*Oryza sativa*). *Agrotechnology* **2013**, *2*, 107–112. [CrossRef]

146. Aldesuquy, H.S.; Abbas, M.A.; Abo-Hamed, S.A.; Elhakem, A.H. Does glycine betaine and salicylic acid ameliorate the negative effect of drought on wheat by regulating osmotic adjustment through solutes accumulation? *J. Stress Physol. Biochem.* **2013**, *9*, 5–22.

147. Misra, N.; Misra, R. Salicylic acid changes plant growth parameters and proline metabolism in *Rauwolfia serpentina* leaves grown under salinity stress. *Am-Eurasian J. Agric. Environ. Sci.* **2012**, *12*, 1601–1609.

148. Keunen, E.; Peshev, D.; Vangronsveld, J.; Van Den Ende, W.; Cuypers, A. Plant sugars are crucial players in the oxidative challenge during abiotic stress: Extending the traditional concept. *Plant Cell Environ.* **2013**, *36*, 1242–1255. [CrossRef] [PubMed]

149. Salerno, G.L.; Curatti, L. Origin of sucrose metabolism in higher plants: When, how and why? *Trends Plant Sci.* **2003**, *8*, 63–69. [CrossRef]

150. Muller, B.; Pantin, F.; Génard, M.; Turc, O.; Freixes, S.; Piques, M.; Gibon, Y. Water deficits uncouple growth from photosynthesis, increase C content, and modify the relationships between C and growth in sink organs. *J. Exp. Bot.* **2011**, *62*, 1715–1729. [CrossRef] [PubMed]

151. Cheng, Y.-J.; Yang, S.-H.; Hsu, C.-S. Synthesis of conjugated polymers for organic solar cell applications. *Chem. Rev.* **2009**, *109*, 5868–5923. [CrossRef] [PubMed]

152. Van den Ende, W.; Peshev, D. Sugars as antioxidants in plants. In *Crop Improvement under Adverse Conditions*; Springer: Dordrecht, The Netherlands, 2013; pp. 285–307.

153. Luo, Y.; Su, Z.; Bi, T.; Cui, X.; Lan, Q. Salicylic acid improves chilling tolerance by affecting antioxidant enzymes and osmoregulators in sacha inchi (*Plukenetia volubilis*). *Braz. J. Bot.* **2014**, *37*, 357–363. [CrossRef]

154. El-Tayeb, M.; El-Enany, A.; Ahmed, N. Salicylic acid-induced adaptive response to copper stress in sunflower (*Helianthus annuus* L.). *Plant Growth Regul.* **2006**, *50*, 191–199. [CrossRef]

155. Wasti, S.; Mimouni, H.; Smiti, S.; Zid, E.; Ben Ahmed, H. Enhanced salt tolerance of tomatoes by exogenous salicylic acid applied through rooting medium. *OMICS* **2012**, *16*, 200–207. [CrossRef]

156. Nahar, K.; Hasanuzzaman, M.; Alam, M.M.; Rahman, A.; Suzuki, T.; Fujita, M. Polyamine and nitric oxide crosstalk: Antagonistic effects on cadmium toxicity in mung bean plants through upregulating the metal detoxification, antioxidant defense and methylglyoxal detoxification systems. *Ecotoxicol. Environ. Saf.* **2016**, *126*, 245–255. [CrossRef]

157. Benavides, M.P.; Groppa, M.D.; Recalde, L.; Verstraeten, S.V. Effects of polyamines on cadmium-and copper-mediated alterations in wheat (*Triticum aestivum* L.) and sunflower (*Helianthus annuus* L.) seedling membrane fluidity. *Arch. Biochem. Biophys.* **2018**, *654*, 27–39. [CrossRef]

158. Groppa, M.D.; Benavides, M.P.; Tomaro, M.L. Polyamine metabolism in sunflower and wheat leaf discs under cadmium or copper stress. *Plant Sci.* **2003**, *164*, 293–299. [CrossRef]

159. Hasanuzzaman, M.; Nahar, K.; Fujita, M. Regulatory role of polyamines in growth, development and abiotic stress tolerance in plants. In *Plant Adaptation to Environmental Change: Significance of Amino Acids and Their Derivatives*; MD University: Haryana, India, 2014; pp. 157–193.

160. Liu, J.-H.; Wang, W.; Wu, H.; Gong, X.; Moriguchi, T. Polyamines function in stress tolerance: From synthesis to regulation. *Front. Plant Sci.* **2015**, *6*, 827. [CrossRef] [PubMed]

161. Szepesi, Á. Interaction between salicylic acid and polyamines and their possible roles in tomato hardening processes. *Acta Biol. Szeged.* **2011**, *55*, 165–166.

162. Tajti, J.; Janda, T.; Majláth, I.; Szalai, G.; Pál, M. Comparative study on the effects of putrescine and spermidine pre-treatment on cadmium stress in wheat. *Ecotoxicol. Environ. Saf.* **2018**, *148*, 546–554. [CrossRef] [PubMed]

163. Kaur, R.; Yadav, P.; Sharma, A.; Thukral, A.K.; Kumar, V.; Kohli, S.K.; Bhardwaj, R. Castasterone and citric acid treatment restores photosynthetic attributes in *Brassica juncea* L. under Cd (II) toxicity. *Ecotoxicol. Environ. Saf.* **2017**, *145*, 466–475. [CrossRef] [PubMed]

164. Kohli, S.K.; Handa, N.; Sharma, A.; Gautam, V.; Arora, S.; Bhardwaj, R.; Wijaya, L.; Alyemeni, M.N.; Ahmad, P. Interaction of 24-epibrassinolide and salicylic acid regulates pigment contents, antioxidative defense responses, and gene expression in *Brassica juncea* L. seedlings under Pb stress. *Environ. Sci. Poll. Res.* **2018**, *25*, 15159–15173. [CrossRef]

165. Chen, S.; Wang, Q.; Lu, H.; Li, J.; Yang, D.; Liu, J.; Yan, C. Phenolic metabolism and related heavy metal tolerance mechanism in *Kandelia obovata* under Cd and Zn stress. *Ecotoxicol. Environ. Saf.* **2019**, *169*, 134–143. [CrossRef]

166. Zafari, S.; Sharifi, M.; Chashmi, N.A.; Mur, L.A. Modulation of Pb-induced stress in Prosopis shoots through an interconnected network of signaling molecules, phenolic compounds and amino acids. *Plant Physiol. Biochem.* **2016**, *99*, 11–20. [CrossRef]

167. Nakamura, M.; Takeuchi, Y.; Miyanaga, K.; Seki, M.; Furusaki, S. High anthocyanin accumulation in the dark by strawberry (*Fragaria ananassa*) callus. *Biotechnol. Lett.* **1999**, *21*, 695–699. [CrossRef]

168. Narayan, M.; Thimmaraju, R.; Bhagyalakshmi, N. Interplay of growth regulators during solid-state and liquid-state batch cultivation of anthocyanin producing cell line of *Daucus carota*. *Process Biochem.* **2005**, *40*, 351–358. [CrossRef]

169. Dong, J.; Wan, G.; Liang, Z. Accumulation of salicylic acid-induced phenolic compounds and raised activities of secondary metabolic and antioxidative enzymes in *Salvia miltiorrhiza* cell culture. *J. Biotechnol.* **2010**, *148*, 99–104. [CrossRef]

170. Kováčik, J.; Gruz, J.; Hedbavny, J.; Klejdus, B.I.; Strnad, M. Cadmium and nickel uptake are differentially modulated by salicylic acid in *Matricaria chamomilla* plants. *J. Agric. Food Chem.* **2009**, *57*, 9848–9855. [CrossRef] [PubMed]

171. Horváth, E.; Szalai, G.; Janda, T. Induction of abiotic stress tolerance by salicylic acid signaling. *J. Plant Growth Regul.* **2007**, *26*, 290–300. [CrossRef]

172. Hettiarachchi, G.H.; Reddy, M.K.; Sopory, S.K.; Chattopadhyay, S. Regulation of TOP2 by various abiotic stresses including cold and salinity in pea and transgenic tobacco plants. *Plant Cell Physiol.* **2005**, *46*, 1154–1160. [CrossRef] [PubMed]

173. Singh, B.; Mishra, R.; Agarwal, P.K.; Goswami, M.; Nair, S.; Sopory, S.; Reddy, M. A pea chloroplast translation elongation factor that is regulated by abiotic factors. *Biochem. Byophys. Res. Commun.* **2004**, *320*, 523–530. [CrossRef]

174. Kovács, V.; Gondor, O.K.; Szalai, G.; Darkó, É.; Majláth, I.; Janda, T.; Pál, M. Synthesis and role of salicylic acid in wheat varieties with different levels of cadmium tolerance. *J. Hazard. Mater.* **2014**, *280*, 12–19. [CrossRef] [PubMed]

175. Khan, M.I.R.; Fatma, M.; Per, T.S.; Anjum, N.A.; Khan, N.A. Salicylic acid-induced abiotic stress tolerance and underlying mechanisms in plants. *Front. Plant. Sci.* **2015**, *6*, 462. [CrossRef]

176. Arfan, M. Exogenous application of salicylic acid through rooting medium modulates ionaccumulation and antioxidant activity in spring wheat under salt stress. *Int. J. Agric. Biol.* **2009**, *11*, 437–442.

177. Song, W.; Zheng, A.; Shao, H.B.; Chu, L.; Brestic, M.; Zhang, Z. The alleviative effect of salicylic acid on the physiological indices of the seedling leaves in six different wheat genotypes under lead stress. *Plant Omics J.* **2012**, *5*, 486–493.

178. Khan, M.I.R.; Asgher, M.; Khan, N.A. Alleviation of salt-induced photosynthesis and growth inhibition by salicylic acid involves glycinebetaine and ethylene in mungbean (*Vigna radiata* L.). *Plant Physiol. Biochem.* **2014**, *80*, 67–74. [CrossRef]

179. Iglesias, M.J.; Terrile, M.C.; Casalongué, C.A. Auxin and salicylic acid signalings counteract the regulation of adaptive responses to stress. *Plant Signal. Behav.* **2011**, *6*, 452–454. [CrossRef]

180. Jiang, C.-J.; Shimono, M.; Sugano, S.; Kojima, M.; Liu, X.; Inoue, H.; Sakakibara, H.; Takatsuji, H. Cytokinins act synergistically with salicylic acid to activate defense gene expression in rice. *Mol. Plant. Microbe. Interact.* **2013**, *26*, 287–296. [CrossRef] [PubMed]

181. Roghayyeh, S.; Saeede, R.; Omid, A.; Mohammad, S. The effect of salicylic acid and gibberellin on seed reserve utilization, germination and enzyme activity of sorghum (*Sorghum bicolor* L.) seeds under drought stress. *J. Stress Physiol. Biochem.* **2014**, *10*, 5–13.

182. Jiang, C.-J.; Shimono, M.; Sugano, S.; Kojima, M.; Yazawa, K.; Yoshida, R.; Inoue, H.; Hayashi, N.; Sakakibara, H.; Takatsuji, H. Abscisic acid interacts antagonistically with salicylic acid signaling pathway in rice–*Magnaporthe grisea* interaction. *Mol. Plant. Microbe. Interact.* **2010**, *23*, 791–798. [CrossRef] [PubMed]

183. Agtuca, B.; Rieger, E.; Hilger, K.; Song, L.; Robert, C.A.; Erb, M.; Karve, A.; Ferrieri, R.A. Carbon-11 reveals opposing roles of auxin and salicylic acid in regulating leaf physiology, leaf metabolism, and resource allocation patterns that impact root growth in *Zea mays*. *J. Plant Growth Regul.* **2014**, *33*, 328–339. [CrossRef]

184. Thao, N.P.; Khan, M.I.R.; Thu, N.B.A.; Hoang, X.L.T.; Asgher, M.; Khan, N.A.; Tran, L.-S.P. Role of ethylene and its cross talk with other signaling molecules in plant responses to heavy metal stress. *Plant Physiol.* **2015**, *169*, 73–84. [CrossRef]

185. Shakirova, F.; Allagulova, C.R.; Maslennikova, D.; Klyuchnikova, E.; Avalbaev, A.; Bezrukova, M. Salicylic acid-induced protection against cadmium toxicity in wheat plants. *Environ. Exp. Bot.* **2016**, *122*, 19–28. [CrossRef]

186. Per, T.S.; Khan, M.I.R.; Anjum, N.A.; Masood, A.; Hussain, S.J.; Khan, N.A. Jasmonates in plants under abiotic stresses: Crosstalk with other phytohormones matters. *Environ. Exp. Bot.* **2018**, *145*, 104–120. [CrossRef]

187. Khan, M.I.R.; Khan, N.A. Salicylic acid and jasmonates: Approaches in abiotic stress tolerance. *J. Plant Biochem. Physiol.* **2013**, *1*, e113. [CrossRef]

188. Petersen, M.; Brodersen, P.; Naested, H.; Andreasson, E.; Lindhart, U.; Johansen, B.; Nielsen, H.B.; Lacy, M.; Austin, M.J.; Parker, J.E.; et al. Arabidopsis MAP kinase 4 negatively regulates systemic acquired resistance. *Cell* **2000**, *103*, 1111–1120. [CrossRef]

189. Engelberth, J.; Viswanathan, S.; Engelberth, M.J. Low concentrations of salicylic acid stimulate insect elicitor responses in *Zea mays* seedlings. *J. Chem. Ecol.* **2011**, *37*, 263–266. [CrossRef]

190. Rostás, M.; Winter, T.R.; Borkowski, L.; Zeier, J. Copper and herbivory lead to priming and synergism in phytohormones and plant volatiles in the absence of salicylate-jasmonate antagonism. *Plant Signal. Behav.* **2013**, *8*, e24264. [CrossRef] [PubMed]

191. Kumar, B.; Ram, H.; Sarlach, R.S. Enhancing seed yield and quality of Egyptian clover (*Trifolium alexandrinum* L.) with foliar application of bio-regulators. *Field Crops Res.* **2013**, *146*, 25–30. [CrossRef]

192. Canakci, S. Effects of salicylic acid on growth, biochemical constituents in pepper (*Capsicum annuum* L.) seedlings. *Pak. J. Biol. Sci.* **2011**, *14*, 300. [CrossRef] [PubMed]

Review

Response of Phenylpropanoid Pathway and the Role of Polyphenols in Plants under Abiotic Stress

Anket Sharma [1,*,†], Babar Shahzad [2,†], Abdul Rehman [3], Renu Bhardwaj [4], Marco Landi [5] and Bingsong Zheng [1,*]

1   State Key Laboratory of Subtropical Silviculture, Zhejiang A&F University, Hangzhou 311300, China
2   School of Land and Food, University of Tasmania, Hobart, TAS 7005, Australia
3   Department of Crop Science and Biotechnology, Dankook University, Chungnam 31116, Korea
4   Plant Stress Physiology Laboratory, Department of Botanical and Environmental Sciences,
    Guru Nanak Dev University, Amritsar 143005, India
5   Department of Agriculture, Food and Environment, University of Pisa, Via del Borghetto, 80-56124 Pisa, Italy
*   Correspondence: anketsharma@gmail.com (A.S.); bszheng@zafu.edu.cn (B.Z.);
    Tel.: +86-(0)571-63730936 (B.Z.)
†   These authors contributed equally to this work.

Received: 1 June 2019; Accepted: 2 July 2019; Published: 4 July 2019

Abstract: Phenolic compounds are an important class of plant secondary metabolites which play crucial physiological roles throughout the plant life cycle. Phenolics are produced under optimal and suboptimal conditions in plants and play key roles in developmental processes like cell division, hormonal regulation, photosynthetic activity, nutrient mineralization, and reproduction. Plants exhibit increased synthesis of polyphenols such as phenolic acids and flavonoids under abiotic stress conditions, which help the plant to cope with environmental constraints. Phenylpropanoid biosynthetic pathway is activated under abiotic stress conditions (drought, heavy metal, salinity, high/low temperature, and ultraviolet radiations) resulting in accumulation of various phenolic compounds which, among other roles, have the potential to scavenge harmful reactive oxygen species. Deepening the research focuses on the phenolic responses to abiotic stress is of great interest for the scientific community. In the present article, we discuss the biochemical and molecular mechanisms related to the activation of phenylpropanoid metabolism and we describe phenolic-mediated stress tolerance in plants. An attempt has been made to provide updated and brand-new information about the response of phenolics under a challenging environment.

Keywords: abiotic stress; anthocyanin; antioxidant; flavonoid; phenolic acid; polyphenol

1. Introduction

Plants face a plethora of biotic and abiotic stresses during their entire life which have negative impact on their growth, development, and productivity [1–3]. Biotic factors include insect pests, fungi, and weeds whereas abiotic stresses include salinity, drought, heavy metals, pesticides, ultraviolet (UV) radiation, as well as heat and cold stress [3–18]. The amplitude of these abiotic stresses has increased severely in recent years principally due to anthropogenic activities [7,19]. Plants, being sessile, are persistently exposed to these factors and require a set of effective mechanisms which can be activated under unfavorable circumstances to sustain their life cycle [20]. According to some reports, the projection of these stresses contributes significantly and affects the growth and productivity by reducing crop yield and overall crop production by 70% and 50%, respectively [21,22]. Thus, it is imperative to reduce the crop productivity losses by improving crop performance through various approaches, including application of plant bio-stimulant products as well as stimulation of plant secondary metabolism [11,23,24]. Plants need to endure different abiotic stresses and polyphenols accumulate in response to these stresses helping plants to acclimatize to unfavorable environments [25,26]. Hence, the concentration of phenols in plant tissue is a good indicator to predict the extent of abiotic stress tolerance in plants which varies greatly in different plant species under an array of external factors.

Phenolic compounds influence the plant growth and development, including seed germination, biomass accumulation, and improved plant metabolism [27–30]. In this regard, we summarized different studies showing a broad spectrum of different effects of abiotic stresses and discussed how endogenous phenol levels can help in mitigating abiotic stress in plants. Moreover, physiological and molecular mechanisms connected to the phenylpropanoid pathway underlying abiotic stress tolerance have extensively discussed. At the end, we explained phenol-mediated stress tolerance and suggestions have been made to further escalate the extent of deep mechanistic studies.

2. Biosynthetic Pathway of Polyphenols in Plants

Phenolics are known to be the largest groups of secondary metabolites in plants varying from simpler aromatic rings to more complex ones, such as lignins. These compounds originated from phenylalanine therefore are also called as phenylpropanoids. Polyphenols are characterized by the presence of large multiples of phenol structural units. The number and characteristics of these phenol structures underlie the unique physical, chemical, and biological properties of particular members of each class. Phenols are indeed divided into several groups such as phenolic acids, flavonoids, stilbenes, and lignans with peculiar properties. Plant phenolics are synthesized biogenetically through a shikimate/phenylpropanoid pathway, whereas a mevalonate pathway generates terpenoids. Both these secondary pathways produce a wide array of monomeric and polymeric structures encompassing a comprehensive array of physiological and biochemical roles in plants. The term "secondary metabolites" refers to the metabolites or phytochemicals synthesized through secondary metabolism. During the biosynthesis of phenolic compounds, erythrose 4-phosphate is combined with phosphoenolpyruvate (PEP) to form phenylalanine. Then phenylalanine ammonia lyase (PAL) catalyzes the conversion of phenylalanine to *trans*-cinnamic acid. Several other phenolic compound such as flavonoids, coumarins, hydrolysable tannins, monolignols, lignans, and lignins are formed through this pathway, formally known as the phenylpropanoid pathway (see complete details in [26,31–33]).

3. Physiological Roles of Phenolics in Plants

Phenolics are widely distributed and are involved in key metabolic and physiological process in plants [34,35]. Phenolics influence different physiological processes related to growth and development in plants including seed germination, cell division, and synthesis of photosynthetic pigments [36]. Phenolic compounds have been exploited for several application including bioremediation, allelochemical, promotion of plant growth, and antioxidants as food additives [37]. In plants, phenolic accumulation is usually a consistent feature of plants under stress, which represents a defensive mechanism to cope with multiple abiotic stresses [31]. Plant phenolics play an important role in several physiological processes to improve the tolerance and adaptability of plants under suboptimal conditions [38–40]. In particular, a large number of secondary metabolites having antioxidant properties belong to this group [41] which can ameliorate plant performance under stress conditions.

Plants interact with their living environment through secondary metabolites. Polyphenols are, for example, involved in signal transduction from the root to the shoot and also help in nutrient mobilization. The roots exudates contain phenolic compounds which alter the physiochemical properties of the rhizosphere. Soil microbes transform phenolics into compounds which help in N mineralization and humus formation [42]. Furthermore, phenolics improve nutrient uptake through chelation of metallic ions, enhanced active absorption sites, and soil porosity with accelerated mobilization of elements like calcium (Ca), magnesium (Mg), potassium (K), zinc (Zn), iron (Fe), and manganese (Mn) [43]. Recently, Rehman, et al. [44] found that Zn application and plant growth promoting rhizobacteria (PGPRs) treatment enhanced the contents of phenolics and organic acids (pyruvic acid, tartaric acid, citric acid, malonic acid, malic acid, succinic acid, oxaloacetic acid, oxalic acid, and methyl malonic acid) in root exudates of wheat, which helped in nutrient mobilization of Zn, N, and Ca and their uptake [44,45]. Phenolic compounds also help in N fixation in legumes. Legumes release several secondary metabolites from roots, principally flavonoid compounds (flavanols and iso-flavonoids) which play crucial role in Nod factors synthesis and in the production of infection tube during nodulation, given that they inhibits auxin transport and facilitate cell division [46].

Plant phenolics, as physiological regulator or chemical messenger, inhibit the IAA catabolism (dihydroxy B-ring flavonoids) or limit the IAA synthesis (monohydroxy B-ring flavonoids) [47]. Flavonoids play a key role in the development of functional pollen. For instance, addition of a very small dose of flavonol aglycones kaempferol or quercetin restored the fertility in mature pollen during pollination [48,49]. Some phenolic compounds (*trans*-cinnamic acid, coumarin, *p*-hydroxybenzoic acid, and benzoic acid) might be potentially phytotoxic if accumulated in high quantity and can inhibit germination and seedling growth [50] due to the disruption of cellular enzyme functioning and impairment of cell division. For instance, some phenolic compounds inhibit the prolyl aminopeptidase and phosphatase enzyme involved in seed germination in legumes [51]. Conversely, high phenolic acid contents have been reported to exert positive effects in seed germinating. In a recent study, Chen et al. [52] found a substantial increase in free (1042%), bound (120%), and total phenolic acid content (741%) in canary grass during germination. The spruce bark extract containing polyphenols stimulated the germination rate in *Lycopersicon esculentum* while inhibited root elongation [53]. Phenolics reduced the thickness and increased the seed tegument porosity which help in water imbibition and boost the germination rate [54]. Polyphenolic extracts of spruce bark intensified the photosynthetic activity and biosynthesis of assimilatory pigment (chlorophyll *a* and *b*) in maize and sunflower [55,56]. Phenolics reduced the energy required for ion transfer by modifying the structure of thylakoids and mitochondrial membranes [57]. As antioxidants, phenolic compounds participate in the scavenging of reactive oxygen species (ROS), catalyzing oxygenation reactions through formation of metallic complexes, and inhibiting the activities of oxidizing enzymes [58].

In conclusion, polyphenols are produced under optimal and (with higher levels) in suboptimal conditions in plants and play crucial role in the development encompassing signal transduction, cell division, hormonal regulation, photosynthetic activity regulation, germination, and reproduction rate. Plants exhibiting increased synthesis of polyphenols under abiotic stresses usually show a better adaptability to limiting environments.

4. Abiotic Stresses and Their Toxic Effects on Plants

In recent times, producing more food and preventing crop losses to meet the demands of ever-increasing human populations has gained unprecedented importance. Nevertheless, a large proportion of arable land face abiotic stresses (drought, salinity, cold, heat, heavy metal toxicity, UV radiation, etc.) which are expected to increase due to climate change and the incidence of these environmental stresses are further fueled by anthropogenic activities. These abiotic stresses cause alteration in physiological and biochemical processes of plants which results in diminished plant growth and poor yield [59]. These stresses bring rapid changes in cellular redox homeostasis with excessive reactive oxygen species (ROS) generation which eventually damage cell organelles and interfere in ROS-promoted signaling pathways [60]. Contrary to over production of ROS, a physiological redox state hampers normal cell functions and affects the plant immune system, suggesting that a threshold level of ROS is necessary for normal plant functioning (Figure 1; Farooq, et al. [61]). Increased ROS generation under abiotic stresses enhanced itself exponentially the production of ROS [62], which result in peroxidation and destabilization of cellular membranes. Recently Rehman, et al. [63] observed that heat stress and Zn deficiency cause reductions in growth (shoot and root biomass, and root length), and consequently impeded nutrient uptake, enhanced lipid peroxidation and impaired photosynthetic performance. In plants, ROS is produced from 1–2% of total O_2 consumed in high active cell organelles like chloroplasts, mitochondria, and peroxisomes (Figure 1; [64]). The most common ROS are singlet oxygen ($^1O_2$), superoxide radicle ($O_2^{\bullet-}$), hydrogen peroxide (H_2O_2), and hydroxyl radicle ($\bullet$OH) [65].

Abiotic stresses disturb the balance between ROS generation and scavenge and accelerate ROS propagation which damages vital macromolecules (nucleic acids, proteins, carbohydrates, and lipids) and eventually leads to cell death. ROS-induced protein damage is caused by oxidation of amino acid residues (e.g., cysteine) for disulphide bond formation, oxidation of arginine, lysine, and threonine residues resulting in irreversible carbonylation in side chains and oxidation of methionine residue to form methionine sulphoxide [66]. ROS production also limits CO_2 fixation in chloroplasts which are the main site for ROS generation in green plants [67]. ROS reacts with chlorophyll during photosynthesis and forms the chlorophyll triplet state which can rapidly generate ($^1O_2$), thus causing damage to photosynthetic complexes (principally PSII) and perturbing the molecular reaction of the photosynthetic pathway [68]. Apart from the chloroplast, the mitochondria also increase ROS production under abiotic stress which influences plant cellular processes [24]. In mitochondria, ~1–5% of O_2 consumed leads to H_2O_2 formation which is subsequently transformed in $\bullet$OH during the Fenton reaction [69]. Furthermore, intensive respiratory/photorespiratory metabolism demands high electron input leading to escalated ROS production which results in protein oxidation [61]. Peroxisomes are also major sites for ROS production, particularly H_2O_2, and have two- and 50-fold higher concentration of H_2O_2 than chloroplasts and mitochondria, respectively [70]. This H_2O_2 is involved in stress induced oxidative damage given that it can freely pass lipid membranes. Under the physiological level, different antioxidant defense mechanism detoxify ROS. However, over production of ROS can overwhelm the defense system, resulting in oxidative stress, cell damage, and cell death (Figure 1) [71].

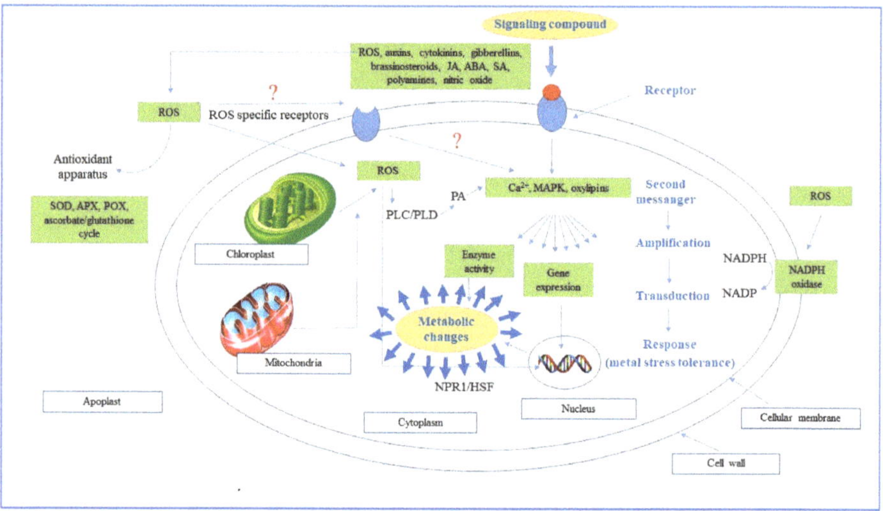

Figure 1. Schematization of signal transmission and transduction in plant cells. Abbreviation: ABA, abscisic acid; APX, ascorbate peroxidase; HSF, redox-sensitive transcription factor; JA, jasmonic acid; MAPK, mitogen-activated protein kinase; NADP, oxidized nicotinamide adenine dinucleotide; NADPH, reduced nicotinamide adenine dinucleotide; NPR1, redox-sensitive transcription factor; OXI1, serine/threonine kinase; PA, phosphatidic acid; PLC/PLD, phospholipases class C and D; POX, peroxidase; ROS, reactive oxygen species; SA, salicylic acid; SOD, superoxide dismutase.

5. Response and Role of Endogenous Phenolics in Plants against Abiotic Stress

In response to abiotic stresses, biosynthesis of secondary metabolites, including polyphenols, is usually increased in plants. Phenolics confer indeed higher tolerance to plants against various stress conditions like heavy metals, salinity, drought, temperature, pesticides, and UV radiations [33,72–77].

Plants growing under stressful environments have the ability to biosynthesize more phenolic compounds in comparison to plants growing under normal conditions [78]. These compounds have antioxidative properties and are capable of scavenging free radicals, resulting in reduction of cell membrane peroxidation [79], hence protecting plant cells from ill effects of oxidative stress. Biosynthesis of phenolics under stressful environments is regulated by the altered activities of various key enzymes of phenolic biosynthetic pathways like PAL and CHS (chalcone synthase). Enhanced performance of enzymes is also accompanied by the up-regulation of the transcript levels of genes encoding key biosynthetic enzymes like *PAL*, *C4H* (cinnamate 4-hydroxylase), *4CL* (4-coumarate: CoA ligase), *CHS*, *CHI* (chalcone isomerase), *F3H* (flavanone3-hydroxylase), *F3′H* (flavonoid 3′-hydroxylase), *F3′5′H* (flavonoid 3′5′-hydroxylase), *DFR* (dihydroflavonol 4-reductase), *FLS* (flavonol synthase), *IFS* (isoflavone synthase), *IFR* (isoflavone reductase), and *UFGT* (UDP flavonoid glycosyltransferase) [74,80–86]. The responses of phenolic compounds under different abiotic stresses have been discussed in individual sections mentioned below.

5.1. Heavy Metal

Metal stress causes oxidative stress to plants by triggering the generation of harmful ROSs and ultimately cause toxicity and retarded growth [11,87,88]. However, enhanced biosynthesis of phenolics in plants under metal stress helps in protecting plants from oxidative stress [72,89,90]. Flavonoids can enhance the metal chelation process which helps in reducing the levels of harmful hydroxyl radical in plant cells [91,92] and this fits well with the observation that the levels of flavonoids in plants have found to be enhanced by metal excess [90,93]. Under metal toxicity, accumulation of specific flavonoids which are involved in aiding to the plant's defense mechanism is also enhanced including anthocyanins and flavonols [72,94–96]. Accumulation of phenolic compounds is due to the up-regulation of the biosynthesis of phenylpropanoid enzymes including phenylalanine ammonia-lyase, chalcone synthase, shikimate dehydrogenase, cinnamyl alcohol dehydrogenase, and polyphenol oxidase [95,97], which in turn, is dependent on the modulation of transcript levels of genes encoding biosynthetic enzymes under metal stress [72,85]. Flavonoids are also known for their scavenging capability of H_2O_2 and are considered to play a crucial role in the phenolic/ascorbate-peroxidase cycle [98,99].

Shikimate dehydrogenase (SKDH) and glucose-6-phosphate dehydrogenase (G6PDH) are two important enzymes which catalyze the biological reaction required for the production of important precursors of phenylpropanoid pathways [100]. Another enzyme cinnamyl alcohol dehydrogenase (CADH) catalyzes biochemical reactions which produce precursors required for synthesis of lignin [101]. Heavy metals stimulate phenylpropanoid the biosynthetic pathway in plants by up-regulating the activities of key biosynthetic enzymes like PAL, SKDH, G6PDH, and CADH [101]. Additionally, polyphenol oxidase (PPO) helps during the process of ROS scavenging, and enhancing a plant's resistance to abiotic stress conditions like heavy metals [100–102]. Table 1 summarizes the impact of metal stress on phenolic composition of plants.

Table 1. Summary table describing the impact of heavy metal stress on the endogenous levels of various phenolic compounds in plants.

Plant Species	Heavy Metal	Response of Endogenous Phenolics and Related Parameters	Reference
	Cu	Increase in contents of total phenols, anthocyanins and other phenolic compounds like catechin, caffeic acid, coumaric acid, kaempferol.	[103]
Brassica juncea	Cr	Increase in total contents of phenols, flavonoids and anthocyanins, accompanied by enhanced expressions of *PAL* and *CHS*.	[72]
	Cr	Increase in anthocyanins accompanied by up-regulation of *CHS* gene.	[93]

Table 1. *Cont.*

	Cd	Increase in the contents of total flavonoids and anthocyanins.	[90]
	Cd	Increase in total contents of flavonoids and anthocyanins, accompanied by enhanced expressions of *PAL* and *CHS*.	[104]
	Cd	Increase in total contents of phenols, polyphenols, flavonoids and anthocyanins.	[105]
	Pb	Increase in total contents of phenols, flavonoids and anthocyanins, accompanied by enhanced expressions of *PAL* and *CHS*.	[106]
	Pb	Increase in total contents of phenols, polyphenols, flavonoids and anthocyanins.	[89]
Fagopyrum esculentum	Al	Increase in total phenolic, flavonoid and anthocyanin contents. Increase in the activity of PAL enzyme.	[77]
Kandelia obovata	Cd and Zn	Enhanced levels of total phenolics accompanied by increased activities of phenol metabolic enzymes like shikimate dehydrogenase, cinnamyl alcohol dehydrogenase and polyphenol oxidase.	[97]
Prosopis farcta	Pb	Increase in total contents of phenols accompanied by enhanced activity of PAL enzyme. Contents of other phenolic compounds were also increased including ferulic acid, cinnamic acid, caffeic acid, daidzein, vitexin, resveratrol, myricetin, quercetin, kaempferol, naringinine, luteolin and diosmin.	[95]
Vitis vinifera	Cu	Enhanced transcript levels of various genes encoding enzymes involved in biosynthesis of phenolics (*PAL, C4H, CHS, F3H, DFR*) and down-regulation of *UFGT* and *ANR*.	[85]
Withania somnifera	Cd	Increase in total contents of flavonoids and phenolics	[101]
Zea mays	Cu, Pb, Cd	Increase in the contents of total phenols and some polyphenols like chlorogenic and vanillic acid.	[96]

PAL (phenylalanine ammonia lyase); CHS (chalcone synthase); CHI (chalcone isomerase); C4H (cinnamate 4-hydroxylase); 4CL (4-coumarate: CoA ligase); F3H (flavanone3-hydroxylase); UFGT (UDP flavonoid glycosyltransferase); IFS (isoflavone synthase); DFR (dihydroflavonol 4-reductase).

5.2. Drought

Phenolic accumulation is very crucial to counteract the negative impacts of drought stress in plants [33]. Transcriptomic and metabolomic studies carried out on *Arabidopsis* plants confirmed that enhanced flavonoid accumulation under drought stress is very helpful to provide resistance [107]. Biosynthesis and accumulation of flavonols were also stimulated in plants under water deficit conditions accompanied by enhanced resistance against drought stress [108,109]. Drought stress also regulated the biosynthetic pathways of phenolic acids and flavonoids, leading to enhanced accumulation of these compounds [82,110,111] which acted as antioxidants and prevented plants from adverse effects of water deficit conditions [112]. For example, contents of flavonoids like kaempferol and quercetin were enhanced in tomato plants accompanied by enhanced drought tolerance [113]. Flavonoid accumulation in cytoplasm can efficiently detoxify harmful H_2O_2 molecules generated as a result of drought stress and, at the end oxidation of flavonoids is followed by ascorbic acid mediated re-conversion of flavonoids into primary metabolites [114]. The main reason for this drought-induced accumulation of phenolic compounds is the modulation of phenylpropanoid biosynthetic pathway. Drought regulates many key genes encoding main enzymes of phenylpropanoid pathway, which results in stimulated biosynthesis of phenolic compounds. The impact of drought stress on accumulation of phenolics and related processes has been summarized in Table 2.

Table 2. Summary table describing the impact of drought stress on the endogenous levels of various phenolic compounds in plants.

Plant Species	Response of Endogenous Phenolics and Related Parameters	Reference
Achillea spp.	Increase in the contents of chlorogenic acid, caffeic acid, rutin, luteolin-7-*O*-glycoside, 1,3-dicaffeoylquinic acid, luteolin, apigenin and kaempferol under 21 days exposure of drought. Enhanced transcript levels of *PAL, CHS, CHI, F3H, F3'H, F3'5'H* and *FLS*.	[82]
	Increase in contents of total phenols and flavonoids.	[115]
Brassica napus	Increase in contents of total phenols, flavonoid and flavonol. Increase in PAL enzyme activity accompanied by enhanced expression of *PAL*.	[110]
Chrysanthemum morifolium	Increase in contents of total phenolics, anthocyanins, chlorogenic acid, luteolin, rutin, ferulic acid, apigenin and quercetin. Enhanced expression of *PAL, CHI*, and *F3H*, particularly in cultivar Taraneh.	[116]
Cucumis sativus	Up-regulation of phenolic metabolites including vanillic acid and 4-hydroxycinnamic acid.	[111]
Fragaria ananassa	Enhanced transcript levels of *PAL, C4H, 4CL, DFR, ANS, FLS* and *UFGT*.	[81]
Lactuca sativa	Increase in the contents of phenolic compounds such as caftaric acid and rutin.	[117]
Larrea spp.	Increase in the contents of polyphenols including flavonoids, proanthocyanidins and flavonols.	[118]
Lotus japonicus	Increase in the contents of kaempferol and quercetine. Up-regulation of the expression of *PAL, C4H, 4CL, CHS, CHI, DFR, IFS* and *IFR*	[119]
Nicotiana tabacum	Increase in PAL enzyme activity and lignin content.	[120]
Ocimum spp.	Increase in content of total phenols	[121]
Thymus vulgaris	Increase in the contents of total flavonoids and polyphenols.	[122]
Triticum aestivum	Increase in content of total phenols	[123]
	Increase in the total contents of phenolics, flavonoids and anthocyanins. Enhanced expression of genes like *CHS, CHI, F3H, FNS, FLS, DFR* and *ANS*.	[84]
Vitis vinifera	Increase in the contents of polyphenols including 4-coumaric acid, caffeic acid, ferulic acid, *cis*-resveratrol-3-O-glucoside, *trans*-resveratrol-3-O-glucoside, catechin, epicatechin, caftaric acid, epicatechin gallate, kaempferol-3-O-glucoside, cyanidin-3-O-glucoside, quercetin-3-O-glucoside and quercetin-3-O glucuronide.	[124]
	Increase in anthocyanin content accompanied by up-regulation of associated biosynthetic genes like *UFGT, CHS* and *F3H*.	[125]

PAL (phenylalanine ammonia lyase); CHS (chalcone synthase); CHI (chalcone isomerase); C4H (cinnamate 4-hydroxylase); 4CL (4-coumarate: CoA ligase); F3H (flavanone3-hydroxylase); F3'H (flavonoid 3'-hydroxylase); F3'5'H (flavonoid 3'5'-hydroxylase); FLS (flavonol synthase); FNS (flavone synthase) UFGT (UDP flavonoid glycosyltransferase); IFS (isoflavone synthase); IFR (isoflavone reductase); DFR (dihydroflavonol 4-reductase); ANS (anthocyanidin synthase).

5.3. Salinity

Salt stress results in generation of ROS like superoxide anions, hydrogen peroxide, and hydroxyl ions [126,127] and require activation of well-orchestrated and finely-tuned plants antioxidant system to contrast ROS propagation [128,129]. Phenolic compounds have powerful antioxidant properties and help in scavenging of harmful ROS in plants under salt stress [130–132]. Moreover, in response to salt stress, phenylpropanoid biosynthetic pathway gets stimulated and results in production of various phenolic compounds which have strong antioxidative potential [131,133,134].

Some genes like *VvbHLH1* are involved in the enhanced production of flavonoids by regulating the genes of the biosynthetic pathways and confer salt tolerance to plants [135,136]. In tobacco plants, *NtCHS1* plays a crucial role in the biosynthesis of flavonoids under salt stress, where accumulation directly favors the scavenging of ROS [130]. Flavone biosynthesis also was enhanced under saline conditions and in *Glycine max*, it was observed that salinity up-regulates the expression of flavone synthase genes, *GmFNSII-1* and *GmFNSII-2* [137]. Some phenolic acids also accumulate in plants under saline conditions including caffeic acid, caftaric acid, cinnamylmalic acid, gallic acid, ferulic acid, and vanillic acid [131,138–140]. Biosynthesis of anthocyanins also was promoted in plants growing under saline conditions [141,142]. A detailed explanation about the effect of salt stress on phenolic composition has been provided in Table 3.

Table 3. Summary table describing the impact of salt stress on the endogenous levels of various phenolic compounds in plants.

Plant Species	Response of Endogenous Phenolics and Related Parameters	Reference
Amaranthus tricolor	Increase in contents of total phenolics, hydroxybenzoic acids (gallic acid, vanilic acid, syringic acid, *p*-hydroxybenzoic acid, ellagic acid), hydroxycinnamic acids (caffeic acid, chlorogenic acid, *p*-coumaric acid, *m*-coumaric acid, ferulic acid, sinapic acid, *trans*-cinnamic acid) and flavonoids (iso-quercetin, hyperoside, rutin)	[140]
Asparagus aethiopicus	Increase in the levels of phenolics like robinin, rutin, apigein, chlorogenic acid and caffeic acid.	[134]
Carthamus tinctorius	Increase in contents of total phenols and flavonoids.	[136]
Chenopodium quinoa	Increase in total polyphenol and flavonoid contents.	[143]
Cynara cardunculus	Increase in contents of phenolic compounds like luteolin-*O*-glucoside, apigenin 6-*c*-glucoside 8-*c*-arabinoside, gallocatechin, leucocyanidin and quercitrin. Decrease in contents of compounds like apigenin, chrysin, genistein, daidzein and ferulic acid	[144]
Fragaria ananassa	Enhanced transcript levels of *PAL, C4H, F3H, DFR* and *FLS*.	[81]
Hordeum vulgare	Increase of total phenolic contents.	[145]
Mentha piperita	Increase of total phenolic contents.	[146]
Ocimum basilicum	Increase in the contents of various phenolic compounds like caffeic acis, caftaric acid, cinnamyl malic acid, feruloyl tartaric acid, quercetin-rutinoside and rosmarinic acid.	[139]
Olea europaea	Increase in contents of total phenolics, kaempf erol and quercetin. Regulation of transcript levels of *PAL, C4H, 4CL, CHS* and *CHI*.	[133]
Salvia mirzayanii	Increase of total phenolic contents.	[132]
Salvia mirzayanii and *Salvia acrosiphon*	Increase in total phenolic content and PAL activity accompanied by enhanced expression of *PAL*.	[147]
Solanum lycopersicon	Increase in total caffeoylquinic acid content	[129]
Solanum villosum	Increase in total phenolic, caffeic acid, and quercetin 3-β-D-glucoside contents. Up-regulation of the expression of *PAL* and *FLS*	[138]
Thymus spp.	Increase in the contents of various phenolic compounds like caffeic acid, gallic acid, trans-2-hydroxycinnamic acid, cinnamic acid, rosmarinic acid, rutin, syringic acid, vanillic acid, apigenin, quercitrin, naringenin and luteolin.	[131]
Triticum aestivum	Increase in contents of total phenols	[123]

PAL (phenylalanine ammonia lyase); CHS (chalcone synthase); CHI (chalcone isomerase); C4H (cinnamate 4-hydroxylase); 4CL (4-coumarate: CoA ligase); F3H (flavanone3-hydroxylase); FLS (flavonol synthase); DFR (dihydroflavonol 4-reductase).

5.4. UV Light

Exposure of UV-B radiations to plants causes damage to their protein structure, causes harmful mutations to DNA and generates harmful ROS. To counteract the negative effects of UV-B exposure, endogenous phenolics accumulated in plant cells and protect cell components by making a shield under epidermal layer. They further reduce DNA damage by preventing dimerization of thymine along with reducing photo-damage of important enzymes like NAD/NADP [33,148]. Moreover, flavonoids also act as light screens due to their capability of absorbing both visible (anthocyanins) and UV radiations (anthocyanins and colorless flavonoids), hence protecting plants from these harmful radiations [26,149]. This fact was supported by various researchers who observed enhanced biosynthesis of flavonoids in plants under UV radiations, accompanied by enhanced UV absorption and plant tolerance to these radiations [98,150] and powerful antioxidant capacity [151]. Moreover, it is also well known that plants growing at high altitude accumulate more phenolics like flavonoids than plants of a temperate region. This enhanced flavonoid accumulation under high light/UV exposure is because of stimulated flavonoid biosynthetic pathways and their corresponding gene transcript levels [33,83,152,153]. The key genes which are up-regulated in plants upon UV exposure include: *CHS* (chalcone synthase); *CHI* (chalcone isomerase); *FLS* (flavonol synthase); *DFR* (dihydroflavonol 4-reductase); *FHT* (flavanone 3β-hydroxylase), *FGT* (flavonoid glycosyltransferases); and *PAL* (phenylalanine ammonia lyase) [154,155]. It is also believed that UV light also utilizes jasmonate dependent/independent pathways to stimulate the biosynthesis of phenols in plants [156]. Additionally, abscisic acid (ABA) is also known to modulate the phenolic biosynthetic pathway in presence of UV light [157]. Table 4 provides a brief summary about impact of UV exposure on the endogenous phenolic composition of plants.

Table 4. Summary table describing the impact of UV light exposure on the endogenous levels of various phenolic compounds in plants.

Plant Species	Response of Endogenous Phenolics and Related Parameters	Reference
Arbutus unedo	Increase in contents of phenolic compounds like theogallin, avicularin and juglanin.	[158]
Brassica oleracea	Increase in contents of gallic acid and sinapic acid.	[159]
Caryopteris mongolica	Increase in contents of flavonoids and anthocyanidins, accompanied by PAL and CHI activity.	[160]
Cuminum cyminum	Increase in contents of total phenolics and anthocyanins, accompanied by enhanced gene expression of *DAHP* and *PAL*.	[153]
Fragaria x ananassa	Increase in contents of kaempferol, ellagic acid and, glucoside derivative of cyaniding, pelargonidin and quercetin. Up-regulation of key genes involved in flavonoid pathway including *CHS, CHI, FHT, DFR, FLS* and *FGT*.	[155]
Kalanchoe pinnata	Increase in contents of total flavonoids and quercitrin.	[161]
Lactuca sativa	Increase in contents of total phenolics, flavonoids and anthocyanins. Contents of phenolic acids were also increased including rosmarinic acid, vanillic acid, *p*-anisic acid, methoxycinnamic acid and chlorogenic acid.	[162]
	Increase in total anthocyanin and phenolic contents. This is accompanied by enhanced activity of PAL enzyme and up-regulation of *PAL* expression.	[163]
Ribes nigrum	Increase in contents of flavonols, anthocyanins, hydroxycinnamic and hydroxybenzoic acids.	[164]
Solanum lycopersicum	Increase in total phenolic content	[165]

Table 4. *Cont.*

Triticum aestivum	After 3 days of UV exposure, increase in contents of total phenolics, ferulic acid, *p*-coumaric acid and vanillic acid, whereas no change in the contents of *p*-hydroxybenzoic acid, syringic acid and sinapic acid. Alterations in the transcript levels of *PAL, C4H, 4CL,* and *COMT*	[83]
Triticum aestivum	Increase in contents of free, bound and total phenolics accompanied by enhanced PAL activity.	[166]
Vigna radiata	Increase in total flavonoid and phenol content, accompanied by enhanced activities of PAL and CHI enzymes.	[154]
Vitis vinifera	Increase in contents of astilbin, quercetin and kaempferol.	[167]
	Increase in contents of phenolic compounds like cyaniding, petunidin, peonidin, malvidin, quercetin, myricetin, kaempferol, procyanidin, gallic acid, protocatechuic acid and vanillic acid.	[157]

CHS (chalcone synthase); CHI (chalcone isomerase); FLS (flavonol synthase); DFR (dihydroflavonol 4-reductase); FHT (flavanone 3β hydroxylase), FGT (flavonoid glycosyltransferases) PAL (phenylalanine ammonia lyase); C4H (cinnamate 4-hydroxylase); 4CL (4-coumarate: CoA ligase); cinnamylalcohol dehydrogenase (CAD); COMT (caffeic acid O-methyltransferase); DAPH (deoxyribonino heptulosinate 7-phosphate synthase).

5.5. Other Abiotic Factors

Other abiotic factors like temperature, nanoparticles, and pesticides also stimulate the endogenous phenolic biosynthesis in plants and help in providing resistance against phytotoxic effects of these abiotic stresses [74,80,153,168–171]. Phenolic biosynthetic pathways also get activated in plants growing under pesticide stress conditions. This leads to more accumulation of phenolic compounds in plants, which confer resistance to survive against pesticide toxicity [73,170]. This stimulated phenolic biosynthesis is due to the activation of key biosynthetic enzymes and up-regulation of key genes of phenylpropanoid branch, including *PAL* and *CHS* [74,80]. Increased accumulation of anthocyanins in plant leaves promote by application of insecticides also helps in recovery of plant photosynthetic efficiency [172]. Similarly, under temperature stress (both heat and chilling), plants synthesize more phenolic compounds such as anthocyanins, flavonoids, flavonols, and phenolic acids, which ultimately protect plant cells [75,129,168,169,173]. In *Festuca trachyphylla* plants growing under heat stress, enhancement in the phenolic compounds was noticed including 4-hydroxybenzoic acid, benzoic acid, caffeic acid, coumaric acid, cinnamic acid, gallic acid, homovanillic acid, ferulic acid, salicylic acid, and vanillic acid [76]. The increased accumulation of these phenolic compounds is accompanied by enhanced tolerance of *F. trachyphylla* plants against high temperature [76]. In carrot, phenolics like coumaric acid, caffeic acid, and anthocyanins are suggested to prevent heat induced oxidative damage by enhancing their accumulation [174]. Some phenolics like salicylic acid also act as stimulant for phenol biosynthesis in plants under high temperature stress. This leads to enhanced accumulation of phenolic compounds which further help in detoxification of ROS and providing heat resistance to plants [175]. Under chilling stress, phenolic compounds like suberin or lignin start accumulating in plant cell walls which helps in enhancing resistance against chilling stress [176]. This enhanced thickness of cell wall due to phenolic accumulation is beneficial for prevention of chilling injury and cell collapse under cold stress [33]. Stimulated phenolic biosynthesis under low temperature stress is due to the enhanced expression of *PAL, CAD* (cinnamylalcohol dehydrogenase), and *HCT* (hydroxycinnamoyl transferase), and increased phenolic levels play crucial role in protection plants against chilling stress [86]. This fact is further supported by the research carried out on peach under chilling stress by Gao et al. [177]. These researchers suggested that 24-epibrassinolide stimulated biosynthesis of phenolics is involved in reduction of heat generated oxidative stress by helping to scavenge of ROS. Table 5 provides a detailed overview about how different abiotic factors affect phenolic metabolism in plants.

Table 5. Summary table describing the impact of various abiotic factors on the endogenous levels of various phenolic compounds in plants.

Plant Species	Abiotic Factor	Response of Endogenous Phenolics and Related Parameters	Reference
Brassica juncea	Insecticide	Increase in total phenol and polyphenol contents.	[73]
	Insecticide	Increase in total phenol, polyphenol and anthocyanin contents accompanied by enhanced expression of *PAL* and *CHS*.	[74]
	Insecticide	Increase in total phenol and anthocyanin contents.	[178]
	Insecticide	Increase in total phenol and anthocyanin contents accompanied by enhanced expression of *PAL* and *CHS*.	[80]
Dracocephalum kotschyi	Silicon dioxide NP	Increase in total phenol, total flavonoid, rosmarinic acid and xantomicrol contents, accompanied by up-regulation of the gene expression of *PAL* and *RAS*.	[179]
Festuca trachyphylla	Heat	Increase in the contents of phenolic compounds like 4-hydroxybenzoic acid, benzoic acid, caffeic acid, coumaric acid, cinnamic acid, gallic acid, homovanillic acid, ferulic acid, salicylic acid and vanillic acid.	[76]
Lens culinaris	Heat	Enhanced levels of total phenolics and flavonoids. Increase in the contents of gallic acid, salicylic acid, chlorogenic acid, ferulic acid and naringenin,	[168]
Nicotiana tabacum	Chilling	Alteration in the contents of various metabolites of phenylalanine metabolic pathway. Enhanced expression of *PAL*, *HCT* and *CAD*.	[86]
Nicotiana langsdorffii	Heat	Increase in the contents of total polyphenols and individual contents of *p*-coumaric acid, chlorogenic acid, cryptochlorogenic acid, neochlorogenic acid and ferulic acid.	[75]
Oryza sativa	Insecticide	Increase in the contents of phenylalanine, p-hydroxybenzoic acid and ferulic acid	[170]
Prunus persica	Chilling	Increase in the activities of enzymes like PAL, C4H, 4CL and CHI. Increase in the contents of phenolic compounds like protocatechuic acid, catechin, cholorogenic acid, neocholorogenic acid, quercetin-3- rutinoside, quercetin-3-glucoside, kaempferol-3- rutinoside	[169]
Solanum lycopersicon	Heat	Increase in total flavonol content	[129]
	Silver NP	Increase in total phenolic content.	[180]
Solanum tuberosum	Zinc NP	Increase in contents of total phenolics and anthocyanins.	[181]
Vigna angularis	Heat	Increase in the contents of anthocyanins and flavonoids.	[173]
Vitis vinifera	Titanium NP	Increase in contents of total phenolics, caftaric acid, quercetin derivatives and kaempferol derivatives.	[171]
Withania somnifera	Copper NP	Increase in contents of total phenolics and flavonoids.	[182]

PAL (phenylalanine ammonia lyase); CHS (chalcone synthase); CHI (chalcone isomerase); C4H (cinnamate 4-hydroxylase); 4CL (4-coumarate: CoA ligase); cinnamylalcohol dehydrogenase (CAD); HCT (hydroxycinnamoyl transferase); COMT (caffeic acid O-methyltransferase); DAPH (deoxyribonino heptulosinate 7-phosphate synthase), RAS (rosmarinic acid synthase); NP (nanoparticles).

6. Conclusions

Phenylpropanoid pathway is likely the most studied pathway of secondary metabolism in planta. In plants growing under challenging environments, accumulation of phenolic compounds usually parallels enhanced plant tolerance as summarized in Figure 2. Abiotic stresses also activate the cell signaling process, resulting in transcriptional up-regulation of phenylpropanoid pathway. The increase in plant's resistance is correlated with the multiple function of polyphenols in plants, principally consisting in their ROS scavenging ability and/or the capacity of some polyphenol classes to protect the plant from excessive light such as UV (flavonoids) and visible light (anthocyanins). In addition, polyphenols might play other key ecological roles under abiotic stress, acting for example as infochemicals for other plants. Aside from the huge body of papers on the matter, further research is needed to deepen, for example, the role of specialized polyphenols as a response to certain abiotic stresses and to describe the intimal mechanisms which shift from primary metabolism to the up-regulation of phenylpropanoid pathway, which is as a cross response to several environmental stressors.

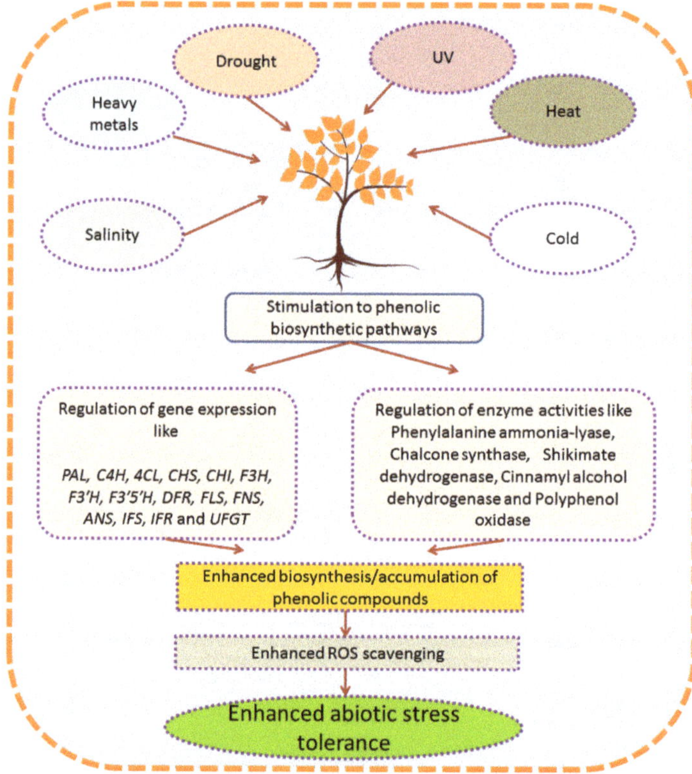

Figure 2. Diagrammatic explanation for response and role of phenolic compounds in plants growing under abiotic stress conditions. ROS (reactive oxygen species); *PAL* (phenylalanine ammonia lyase); *CHS* (chalcone synthase); *CHI* (chalcone isomerase); *C4H* (cinnamate 4-hydroxylase); *4CL* (4-coumarate: CoA ligase); *F3H* (flavanone3-hydroxylase); *F3'H*(flavonoid 3'-hydroxylase); *F3'5'H*(flavonoid 3'5'-hydroxylase); *FLS* (flavonol synthase); *FNS* (flavone synthase) *UFGT* (UDP flavonoid glycosyltransferase); *IFS* (isoflavone synthase); *IFR* (isoflavone reductase); *DFR* (dihydroflavonol 4-reductase); *ANS* (anthocyanidin synthase).

Author Contributions: A.S. and B.Z. drafted the outline of review. A.S., B.S., and A.R. participated in literature collection and writing the initial draft. R.B. and M.L. participated in revision of initial draft.

Funding: We acknowledge the different funding agencies: This study was supported by National Key Research and Development Program of China (2018YFD1000600); Independent research topics of the State Key Laboratory of Subtropical Silviculture (ZY20180208, ZY20180308); Key Research and Development Program of Zhejiang Province (2018C02004); Key Project of Zhejiang Provincial Natural Science Foundation (LZ18C160001); Fruit Innovation Team Project of Zhejiang Province (2016C02052-12); National Undergraduate Innovation and Entrepreneurship Training Project (201610341010); Undergraduate Science and Technology Innovation Plan of Zhejiang Province (2017R412006); Undergraduate Research Training Program in Zhejiang A & F University (102-2013200005, 102-2013200041, 102-2013200042,KX20180047, KX20180043, KX20180065).

References

1. Dresselhaus, T.; Hückelhoven, R. Biotic and Abiotic Stress Responses in Crop Plants. *Agronomy* **2018**, *8*, 267. [CrossRef]
2. Lamaoui, M.; Jemo, M.; Datla, R.; Bekkaoui, F. Heat and Drought Stresses in Crops and Approaches for Their Mitigation. *Front. Plant Sci.* **2018**, *6*, 26. [CrossRef] [PubMed]
3. Mittler, R. Abiotic stress, the field environment and stress combination. *Trends Plant Sci.* **2006**, *11*, 15–19. [CrossRef] [PubMed]
4. Kreps, J.A.; Wu, Y.; Chang, H.-S.; Zhu, T.; Wang, X.; Harper, J.F. Transcriptome changes for *Arabidopsis* in response to salt, osmotic, and cold stress. *Plant Physiol.* **2002**, *130*, 2129–2141. [CrossRef] [PubMed]
5. Mittler, R.; Blumwald, E. Genetic engineering for modern agriculture: Challenges and perspectives. *Annu. Rev. Plant Biol.* **2010**, *61*, 443–462. [CrossRef] [PubMed]
6. Shao, H.-B.; Guo, Q.-J.; Chu, L.-Y.; Zhao, X.-N.; Su, Z.-L.; Hu, Y.-C.; Cheng, J.-F. Understanding molecular mechanism of higher plant plasticity under abiotic stress. *Colloids Surf. B Biointerfaces* **2007**, *54*, 37–45. [CrossRef] [PubMed]
7. Anjum, N.A.; Gill, S.S.; Gill, R. *Plant Adaptation to Environmental Change: Significance of Amino Acids and Their Derivatives*; CABI: Wallingford, UK, 2014.
8. Anjum, S.A.; Ashraf, U.; Tanveer, M.; Khan, I.; Hussain, S.; Shahzad, B.; Zohaib, A.; Abbas, F.; Saleem, M.F.; Ali, I. Drought induced changes in growth, osmolyte accumulation and antioxidant metabolism of three maize hybrids. *Front. Plant Sci.* **2017**, *8*, 19. [CrossRef] [PubMed]
9. Anjum, S.A.; Tanveer, M.; Ashraf, U.; Hussain, S.; Shahzad, B.; Khan, I.; Wang, L. Effect of progressive drought stress on growth, leaf gas exchange, and antioxidant production in two maize cultivars. *Environ. Sci. Pollut. Res.* **2016**, *23*, 17132–17141. [CrossRef]
10. Anjum, S.A.; Tanveer, M.; Hussain, S.; Bao, M.; Wang, L.; Khan, I.; Ullah, E.; Tung, S.A.; Samad, R.A.; Shahzad, B. Cadmium toxicity in Maize (*Zea mays* L.): Consequences on antioxidative systems, reactive oxygen species and cadmium accumulation. *Environ. Sci. Pollut. Res.* **2015**, *22*, 17022–17030. [CrossRef]
11. Shahzad, B.; Tanveer, M.; Che, Z.; Rehman, A.; Cheema, S.A.; Sharma, A.; Song, H.; Ur Rehman, S.; Zhaorong, D. Role of 24-epibrassinolide (EBL) in mediating heavy metal and pesticide induced oxidative stress in plants: A review. *Ecotoxicol. Environ. Saf.* **2018**, *147*, 935–944. [CrossRef]
12. Shahzad, B.; Tanveer, M.; Rehman, A.; Cheema, S.A.; Fahad, S.; Rehman, S.; Sharma, A. Nickel; whether toxic or essential for plants and environment-A review. *Plant Physiol. Biochem.* **2018**, *132*, 641–651. [CrossRef] [PubMed]
13. Sharma, A.; Kumar, V.; Kumar, R.; Shahzad, B.; Thukral, A.K.; Bhardwaj, R. Brassinosteroid-mediated pesticide detoxification in plants: A mini-review. *Cogent Food Agric.* **2018**, *4*, 1436212. [CrossRef]
14. Fahad, S.; Rehman, A.; Shahzad, B.; Tanveer, M.; Saud, S.; Kamran, M.; Ihtisham, M.; Khan, S.U.; Turan, V.; Ur Rahman, M.H. Rice Responses and Tolerance to Metal/Metalloid Toxicity. In *Advances in Rice Research for Abiotic Stress Tolerance*; Elsevier: Amsterdam, The Netherlands, 2019; pp. 299–312.
15. Soares, C.; Carvalho, M.E.; Azevedo, R.A.; Fidalgo, F. Plants facing oxidative challenges—A little help from the antioxidant networks. *Environ. Exp. Bot.* **2019**, *161*, 4–25. [CrossRef]
16. Guo, H.; Feng, X.; Hong, C.; Chen, H.; Zeng, F.; Zheng, B.; Jiang, D. Malate secretion from the root system is an important reason for higher resistance of *Miscanthus sacchariflorus* to cadmium. *Physiol. Planta* **2017**, *159*, 340–353. [CrossRef] [PubMed]

17. Guo, H.; Chen, H.; Hong, C.; Jiang, D.; Zheng, B. Exogenous malic acid alleviates cadmium toxicity in *Miscanthus sacchariflorus* through enhancing photosynthetic capacity and restraining ROS accumulation. *Ecotoxicol. Environ. Saf.* **2017**, *141*, 119–128. [CrossRef] [PubMed]

18. Chen, X.; Qiu, L.; Guo, H.; Wang, Y.; Yuan, H.; Yan, D.; Zheng, B. Spermidine induces physiological and biochemical changes in southern highbush blueberry under drought stress. *Braz. J. Bot.* **2017**, *40*, 841–851. [CrossRef]

19. Khan, M.I.R.; Khan, N.A. Salicylic acid and jasmonates: Approaches in abiotic stress tolerance. *J. Plant Biochem. Physiol.* **2013**, *1*, 4. [CrossRef]

20. Rao, K.M.; Raghavendra, A.; Reddy, K.J. *Physiology and Molecular Biology of Stress Tolerance in Plants*; Springer Science & Business Media: New York, NY, USA, 2006.

21. Kaur, G.; Kumar, S.; Nayyar, H.; Upadhyaya, H. Cold stress injury during the pod-filling phase in chickpea (Cicer arietinum L.): Effects on quantitative and qualitative components of seeds. *J. Agron. Crop Sci.* **2008**, *194*, 457–464.

22. Mantri, N.; Patade, V.; Penna, S.; Ford, R.; Pang, E. Abiotic stress responses in plants: Present and future. In *Abiotic Stress Responses in Plants*; Springer: New York, NY, USA, 2012; pp. 1–19.

23. Shahzad, B.; Cheema, S.; Farooq, M.; Cheema, Z.; Rehman, A.; Abbas, T. Growth Stimulating Influence of Foliage Applied *Brassica* Water Extracts on Morphological and Yield Attributes of Bread Wheat under Different Fertilizer Regimes. *Planta Daninha* **2018**, *36*, e018178331. [CrossRef]

24. Sharma, P.; Jha, A.B.; Dubey, R.S.; Pessarakli, M. Reactive Oxygen Species, Oxidative Damage, and Antioxidative Defense Mechanism in Plants under Stressful Conditions. *J. Bot.* **2012**, *2012*, 26. [CrossRef]

25. Pereira, A. Plant abiotic stress challenges from the changing environment. *Front. Plant Sci.* **2016**, *7*, 1123. [CrossRef] [PubMed]

26. Lattanzio, V. Phenolic Compounds: Introduction. In *Natural Products: Phytochemistry, Botany and Metabolism of Alkaloids, Phenolics and Terpenes*; Ramawat, K.G., Mérillon, J.-M., Eds.; Springer: Berlin/Heidelberg, Germany, 2013; pp. 1543–1580. [CrossRef]

27. Raskin, I. Role of salicylic acid in plants. *Ann. Rev. Plant Biol.* **1992**, *43*, 439–463. [CrossRef]

28. Yalpani, N.; Enyedi, A.J.; León, J.; Raskin, I. Ultraviolet light and ozone stimulate accumulation of salicylic acid, pathogenesis-related proteins and virus resistance in tobacco. *Planta* **1994**, *193*, 372–376. [CrossRef]

29. Senaratna, T.; Touchell, D.; Bunn, E.; Dixon, K. Acetyl salicylic acid (Aspirin) and salicylic acid induce multiple stress tolerance in bean and tomato plants. *Plant Growth Regul.* **2000**, *30*, 157–161. [CrossRef]

30. Nazar, R.; Iqbal, N.; Syeed, S.; Khan, N.A. Salicylic acid alleviates decreases in photosynthesis under salt stress by enhancing nitrogen and sulfur assimilation and antioxidant metabolism differentially in two mungbean cultivars. *J. Plant Physiol.* **2011**, *168*, 807–815. [CrossRef] [PubMed]

31. Cheynier, V.; Comte, G.; Davies, K.M.; Lattanzio, V.; Martens, S. Plant phenolics: Recent advances on their biosynthesis, genetics, and ecophysiology. *Plant Physiol. Biochem.* **2013**, *72*, 1–20. [CrossRef]

32. Saltveit, M.E. Synthesis and metabolism of phenolic compounds. In *Fruit and Vegetable Phytochemicals Chemistry, Nutritional Value, and Stability*; Wiley-Blackwell: Hoboken, NJ, USA, 2010. [CrossRef]

33. Naikoo, M.I.; Dar, M.I.; Raghib, F.; Jaleel, H.; Ahmad, B.; Raina, A.; Khan, F.A.; Naushin, F. Role and Regulation of Plants Phenolics in Abiotic Stress Tolerance: An Overview. In *Plant Signaling Molecules*; Elsevier: Amsterdam, The Netherlands, 2019; pp. 157–168.

34. Boudet, A.M. Evolution and current status of research in phenolic compounds. *Phytochemistry* **2007**, *68*, 2722–2735. [CrossRef]

35. Kumar, V.; Sharma, A.; Kohli, S.K.; Bali, S.; Sharma, M.; Kumar, R.; Bhardwaj, R.; Thukral, A.K. Differential distribution of polyphenols in plants using multivariate techniques. *Biotech. Res. Innov.* **2019**, *3*, 1–21. [CrossRef]

36. Tanase, C.; Bujor, O.-C.; Popa, V.I. Phenolic Natural Compounds and Their Influence on Physiological Processes in Plants. In *Polyphenols in Plants*, 2nd ed.; Watson, R.R., Ed.; Academic Press: Cambridge, MA, USA, 2019; pp. 45–58. [CrossRef]

37. Bujor, O.-C.; Talmaciu, I.A.; Volf, I.; Popa, V.I. Biorefining to recover aromatic compounds with biological properties. *TAPPI J.* **2015**, *14*, 187–193.

38. Andersen, C.P. Source–sink balance and carbon allocation below ground in plants exposed to ozone. *New Phytol.* **2003**, *157*, 213–228. [CrossRef]

39. Lattanzio, V.; Cardinali, A.; Ruta, C.; Fortunato, I.M.; Lattanzio, V.M.T.; Linsalata, V.; Cicco, N. Relationship of secondary metabolism to growth in oregano (*Origanum vulgare* L.) shoot cultures under nutritional stress. *Environ. Exp. Bot.* **2009**, *65*, 54–62. [CrossRef]

40. Dixon, R.A.; Paiva, N.L. Stress-induced phenylpropanoid metabolism. *Plant Cell* **1995**, *7*, 1085. [CrossRef] [PubMed]

41. Oszmanski, J. Polyphenols as antioxidants in food. *Przem. Spo.* **1995**, *3*, 94–96.

42. Halvorson, J.J.; Gonzalez, J.M.; Hagerman, A.E.; Smith, J.L. Sorption of tannin and related phenolic compounds and effects on soluble-N in soil. *Soil Biol. Biochem.* **2009**, *41*, 2002–2010. [CrossRef]

43. Seneviratne, G.; Jayasinghearachchi, H.S. Mycelial colonization by bradyrhizobia and azorhizobia. *J. Biosci.* **2003**, *28*, 243–247. [CrossRef] [PubMed]

44. Rehman, A.; Farooq, M.; Naveed, M.; Nawaz, A.; Shahzad, B. Seed priming of Zn with endophytic bacteria improves the productivity and grain biofortification of bread wheat. *Eur. J. Agron.* **2018**, *94*, 98–107. [CrossRef]

45. Rehman, A.; Farooq, M.; Naveed, M.; Ozturk, L.; Nawaz, A. *Pseudomonas* aided zinc application improves the productivity and biofortification of bread wheat. *Crop Pasture Sci.* **2018**, *69*, 659–672. [CrossRef]

46. Zhang, J.; Subramanian, S.; Stacey, G.; Yu, O. Flavones and flavonols play distinct critical roles during nodulation of *Medicago truncatula* by *Sinorhizobium meliloti*. *Plant J.* **2009**, *57*, 171–183. [CrossRef]

47. Mathesius, U. Flavonoids induced in cells undergoing nodule organogenesis in white clover are regulators of auxin breakdown by peroxidase. *J. Exp. Bot.* **2001**, *52*, 419–426. [CrossRef]

48. Van Der Meer, I.M.; Stam, M.E.; Van Tunen, A.J.; Mol, J.N.; Stuitje, A.R. Antisense inhibition of flavonoid biosynthesis in petunia anthers results in male sterility. *Plant Cell* **1992**, *4*, 253–262. [CrossRef]

49. Taylor, L.P.; Grotewold, E. Flavonoids as developmental regulators. *Curr. Opin. Plant Biol.* **2005**, *8*, 317–323. [CrossRef] [PubMed]

50. Baleroni, C.R.S.; Ferrarese, M.L.L.; Souza, N.E.; Ferrarese-Filho, O. Lipid Accumulation during Canola Seed Germination in Response to Cinnamic Acid Derivatives. *Biol. Planta* **2000**, *43*, 313–316. [CrossRef]

51. Shankar, S.; Girish, R.; Karthik, N.; Rajendran, R.; Mahendran, V. Allelopathic effects of phenolics and terpenoids extracted from Gimelina arborea on germination of Black gram (*Vigna mungo*) and Green gram (*Vigna radiata*). *Allelopath. J.* **2009**, *23*, 323–332.

52. Chen, Z.; Yu, L.; Wang, X.; Gu, Z.; Beta, T. Changes of phenolic profiles and antioxidant activity in canaryseed (*Phalaris canariensis* L.) during germination. *Food Chem.* **2016**, *194*, 608–618. [CrossRef] [PubMed]

53. Balas, A.; Popa, V. Bioactive compounds extracted from *Picea abies* bark. In Proceedings of the 10th European Workshop on Lignocellulosics and Pulp, Stockholm, Sweden, 25–28 August 2008; pp. 345–356.

54. Tobe, K.; Zhang, L.; Qiu, G.Y.; Shimizu, H.; Omasa, K. Characteristics of seed germination in five non-halophytic Chinese desert shrub species. *J. Arid. Environ.* **2001**, *47*, 191–201. [CrossRef]

55. Tanase, C.; Boz, I.; Stingu, A.; Volf, I.; Popa, V.I. Physiological and biochemical responses induced by spruce bark aqueous extract and deuterium depleted water with synergistic action in sunflower (*Helianthus annuus* L.) plants. *Ind. Crops Prod.* **2014**, *60*, 160–167. [CrossRef]

56. Tanase, C.; Bara, C.I.; Popa, V.I. Cytogenetical effect of some polyphenol compounds separated from industrial by-products on. *Cell. Chem. Technol.* **2015**, *49*, 799–805.

57. Moreland, D.E.; Novitzky, W.P. Effects of Phenolic Acids, Coumarins, and Flavonoids on Isolated Chloroplasts and Mitochondria. In *Allelochemicals: Role in Agriculture and Forestry*; American Chemical Society: Washington, DC, USA, 1987; pp. 247–261. [CrossRef]

58. Amarowicz, R.; Weidner, S. Biological activity of grapevine phenolic compounds. In *Grapevine Molecular Physiology & Biotechnology*; Springer: New York, NY, USA, 2009; pp. 389–405.

59. Wani, S.H.; Sah, S. Biotechnology and abiotic stress tolerance in rice. *J. Rice Res.* **2014**, *2*, e105. [CrossRef]

60. Noctor, G.; Reichheld, J.-P.; Foyer, C.H. ROS-related redox regulation and signaling in plants. *Semin. Cell Dev. Biol.* **2018**, *80*, 3–12. [CrossRef]

61. Farooq, M.A.; Niazi, A.K.; Akhtar, J.; Farooq, M.; Souri, Z.; Karimi, N.; Rengel, Z. Acquiring control: The evolution of ROS-Induced oxidative stress and redox signaling pathways in plant stress responses. *Plant Physiol. Biochem.* **2019**, *141*, 353–369. [CrossRef]

62. Suzuki, N.; Koussevitzky, S.; Mittler, R.; Miller, G. ROS and redox signalling in the response of plants to abiotic stress. *Plant Cell Environ.* **2012**, *35*, 259–270. [CrossRef] [PubMed]

63. Rehman, A.; Farooq, M.; Asif, M.; Ozturk, L. Supra-optimal growth temperature exacerbates adverse effects of low Zn supply in wheat. *J. Plant Nutr. Soil Sci.* **2019**. [CrossRef]

64. Corpas, F.J.; Barroso, J.B.; Palma, J.M.; Rodriguez-Ruiz, M. Plant peroxisomes: A nitro-oxidative cocktail. *Redox Biol.* **2017**, *11*, 535–542. [CrossRef]

65. Gill, S.S.; Tuteja, N. Reactive oxygen species and antioxidant machinery in abiotic stress tolerance in crop plants. *Plant Physiol. Biochem.* **2010**, *48*, 909–930. [CrossRef] [PubMed]

66. Kristensen, B.K.; Askerlund, P.; Bykova, N.V.; Egsgaard, H.; Moller, I.M. Identification of oxidised proteins in the matrix of rice leaf mitochondria by immunoprecipitation and two-dimensional liquid chromatography-tandem mass spectrometry. *Phytochemistry* **2004**, *65*, 1839–1851. [CrossRef] [PubMed]

67. Asada, K. Production and Scavenging of Reactive Oxygen Species in Chloroplasts and Their Functions. *Plant Physiol.* **2006**, *141*, 391–396. [CrossRef] [PubMed]

68. Buchert, F.; Forreiter, C. Singlet oxygen inhibits ATPase and proton translocation activity of the thylakoid ATP synthase CF1CFo. *FEBS Lett.* **2010**, *584*, 147–152. [CrossRef] [PubMed]

69. Popov, V.N.; Simonian, R.A.; Skulachev, V.P.; Starkov, A.A. Inhibition of the alternative oxidase stimulates H_2O_2 production in plant mitochondria. *FEBS Lett.* **1997**, *415*, 87–90. [CrossRef]

70. Foyer, C.H.; Bloom, A.J.; Queval, G.; Noctor, G. Photorespiratory metabolism: Genes, mutants, energetics, and redox signaling. *Annu. Rev. Plant Biol.* **2009**, *60*, 455–484. [CrossRef]

71. Halliwell, B. Reactive species and antioxidants. Redox biology is a fundamental theme of aerobic life. *Plant Physiol.* **2006**, *141*, 312–322. [CrossRef]

72. Handa, N.; Kohli, S.K.; Sharma, A.; Thukral, A.K.; Bhardwaj, R.; Abd_Allah, E.F.; Alqarawi, A.A.; Ahmad, P. Selenium modulates dynamics of antioxidative defence expression, photosynthetic attributes and secondary metabolites to mitigate chromium toxicity in *Brassica juncea* L. plants. *Environ. Exp. Bot.* **2019**, *161*, 180–192. [CrossRef]

73. Sharma, A.; Kumar, V.; Thukral, A.K.; Bhardwaj, R. Epibrassinolide-imidacloprid interaction enhances non-enzymatic antioxidants in *Brassica juncea* L. *Ind. J. Plant Physiol.* **2016**, *21*, 70–75. [CrossRef]

74. Sharma, A.; Thakur, S.; Kumar, V.; Kanwar, M.K.; Kesavan, A.K.; Thukral, A.K.; Bhardwaj, R.; Alam, P.; Ahmad, P. Pre-sowing Seed Treatment with 24-Epibrassinolide Ameliorates Pesticide Stress in *Brassica juncea* L. through the Modulation of Stress Markers. *Front. Plant Sci.* **2016**, *7*, 1569. [CrossRef] [PubMed]

75. Ancillotti, C.; Bogani, P.; Biricolti, S.; Calistri, E.; Checchini, L.; Ciofi, L.; Gonnelli, C.; Del Bubba, M. Changes in polyphenol and sugar concentrations in wild type and genetically modified *Nicotiana langsdorffii* Weinmann in response to water and heat stress. *Plant Physiol. Biochem.* **2015**, *97*, 52–61. [CrossRef] [PubMed]

76. Wang, J.; Yuan, B.; Huang, B. Differential Heat-Induced Changes in Phenolic Acids Associated with Genotypic Variations in Heat Tolerance for Hard Fescue. *Crop Sci.* **2019**, *59*, 667–674. [CrossRef]

77. Smirnov, O.E.; Kosyan, A.M.; Kosyk, O.I.; Taran, N.Y. Response of phenolic metabolism induced by aluminium toxicity in *Fagopyrum esculentum* moench. plants. *Ukr. Biochem. J.* **2015**, *87*, 129–135. [CrossRef] [PubMed]

78. Selmar, D. Potential of salt and drought stress to increase pharmaceutical significant secondary compounds in plants. *Landbauforschung Volkenrode* **2008**, *58*, 139.

79. Schroeter, H.; Boyd, C.; Spencer, J.P.; Williams, R.J.; Cadenas, E.; Rice-Evans, C. MAPK signaling in neurodegeneration: Influences of flavonoids and of nitric oxide. *Neurobiol. Aging* **2002**, *23*, 861–880. [CrossRef]

80. Sharma, A.; Yuan, H.; Kumar, V.; Ramakrishnan, M.; Kohli, S.K.; Kaur, R.; Thukral, A.K.; Bhardwaj, R.; Zheng, B. Castasterone attenuates insecticide induced phytotoxicity in mustard. *Ecotoxicol. Environ. Saf.* **2019**, *179*, 50–61. [CrossRef]

81. Perin, E.C.; Da Silva Messias, R.; Borowski, J.M.; Crizel, R.L.; Schott, I.B.; Carvalho, I.R.; Rombaldi, C.V.; Galli, V. ABA-dependent salt and drought stress improve strawberry fruit quality. *Food Chem.* **2019**, *271*, 516–526. [CrossRef]

82. Gharibi, S.; Sayed Tabatabaei, B.E.; Saeidi, G.; Talebi, M.; Matkowski, A. The effect of drought stress on polyphenolic compounds and expression of flavonoid biosynthesis related genes in *Achillea pachycephala* Rech.f. *Phytochemistry* **2019**, *162*, 90–98. [CrossRef]

83. Chen, Z.; Ma, Y.; Yang, R.; Gu, Z.; Wang, P. Effects of exogenous Ca2+ on phenolic accumulation and physiological changes in germinated wheat (*Triticum aestivum* L.) under UV-B radiation. *Food Chem.* **2019**, *288*, 368–376. [CrossRef]

84. Ma, D.; Sun, D.; Wang, C.; Li, Y.; Guo, T. Expression of flavonoid biosynthesis genes and accumulation of flavonoid in wheat leaves in response to drought stress. *Plant Physiol. Biochem.* **2014**, *80*, 60–66. [CrossRef] [PubMed]

85. Leng, X.; Jia, H.; Sun, X.; Shangguan, L.; Mu, Q.; Wang, B.; Fang, J. Comparative transcriptome analysis of grapevine in response to copper stress. *Sci. Rep.* **2015**, *5*, 17749. [CrossRef]

86. Zhou, P.; Li, Q.; Liu, G.; Xu, N.; Yang, Y.; Zeng, W.; Chen, A.; Wang, S. Integrated analysis of transcriptomic and metabolomic data reveals critical metabolic pathways involved in polyphenol biosynthesis in *Nicotiana tabacum* under chilling stress. *Funct. Plant Biol.* **2018**, *46*, 30–43. [CrossRef]

87. Pandey, N.; Sharma, C.P. Effect of heavy metals Co^{2+}, Ni^{2+} and Cd^{2+} on growth and metabolism of cabbage. *Plant Sci.* **2002**, *163*, 753–758. [CrossRef]

88. Villiers, F.; Ducruix, C.; Hugouvieux, V.; Jarno, N.; Ezan, E.; Garin, J.; Junot, C.; Bourguignon, J. Investigating the plant response to cadmium exposure by proteomic and metabolomic approaches. *Proteomics* **2011**, *11*, 1650–1663. [CrossRef] [PubMed]

89. Kohli, S.K.; Handa, N.; Sharma, A.; Gautam, V.; Arora, S.; Bhardwaj, R.; Wijaya, L.; Alyemeni, M.N.; Ahmad, P. Interaction of 24-epibrassinolide and salicylic acid regulates pigment contents, antioxidative defense responses, and gene expression in *Brassica juncea* L. seedlings under Pb stress. *Environ. Sci. Pollut. Res.* **2018**, *25*, 15159–15173. [CrossRef]

90. Kaur, R.; Yadav, P.; Sharma, A.; Kumar Thukral, A.; Kumar, V.; Kaur Kohli, S.; Bhardwaj, R. Castasterone and citric acid treatment restores photosynthetic attributes in *Brassica juncea* L. under Cd(II) toxicity. *Ecotoxicol. Environ. Saf.* **2017**, *145*, 466–475. [CrossRef]

91. Mira, L.; Fernandez, M.T.; Santos, M.; Rocha, R.; Florencio, M.H.; Jennings, K.R. Interactions of flavonoids with iron and copper ions: A mechanism for their antioxidant activity. *Free Radic. Res.* **2002**, *36*, 1199–1208. [CrossRef]

92. Williams, R.J.; Spencer, J.P.; Rice-Evans, C. Flavonoids: Antioxidants or signalling molecules? *Free. Radic. Biol. Med.* **2004**, *36*, 838–849. [CrossRef] [PubMed]

93. Handa, N.; Kohli, S.K.; Sharma, A.; Thukral, A.K.; Bhardwaj, R.; Alyemeni, M.N.; Wijaya, L.; Ahmad, P. Selenium ameliorates chromium toxicity through modifications in pigment system, antioxidative capacity, osmotic system, and metal chelators in *Brassica juncea* seedlings. *S. Afr. J. Bot.* **2018**, *119*, 1–10. [CrossRef]

94. Trejo-Tapia, G.; Jimenez-Aparicio, A.; Rodriguez-Monroy, M.; De Jesus-Sanchez, A.; Gutierrez-Lopez, G. Influence of cobalt and other microelements on the production of betalains and the growth of suspension cultures of *Beta vulgaris*. *Plant Cell Tiss. Org. Cult.* **2001**, *67*, 19–23. [CrossRef]

95. Zafari, S.; Sharifi, M.; Ahmadian Chashmi, N.; Mur, L.A. Modulation of Pb-induced stress in Prosopis shoots through an interconnected network of signaling molecules, phenolic compounds and amino acids. *Plant Physiol. Biochem.* **2016**, *99*, 11–20. [CrossRef] [PubMed]

96. Kısa, D.; Elmastaş, M.; Öztürk, L.; Kayır, Ö. Responses of the phenolic compounds of *Zea mays* under heavy metal stress. *Appl. Biol. Chem.* **2016**, *59*, 813–820. [CrossRef]

97. Chen, S.; Wang, Q.; Lu, H.; Li, J.; Yang, D.; Liu, J.; Yan, C. Phenolic metabolism and related heavy metal tolerance mechanism in *Kandelia Obovata* under Cd and Zn stress. *Ecotoxicol. Environ. Saf.* **2019**, *169*, 134–143. [CrossRef]

98. Michalak, A. Phenolic compounds and their antioxidant activity in plants growing under heavy metal stress. *Pol. J. Environ. Stud.* **2006**, *15*, 523–530.

99. Keilig, K.; Ludwig-Mueller, J. Effect of flavonoids on heavy metal tolerance in *Arabidopsis thaliana* seedlings. *Bot. Stud.* **2009**, *50*, 311–318.

100. Kováčik, J.; Klejdus, B.; Hedbavny, J.; Štork, F.; Bačkor, M. Comparison of cadmium and copper effect on phenolic metabolism, mineral nutrients and stress-related parameters in *Matricaria chamomilla* plants. *Plant Soil* **2009**, *320*, 231. [CrossRef]

101. Mishra, B.; Sangwan, N.S. Amelioration of cadmium stress in *Withania somnifera* by ROS management: Active participation of primary and secondary metabolism. *Plant Growth Regul.* **2019**, *87*, 403–412. [CrossRef]

102. Mishra, B.; Sangwan, R.S.; Mishra, S.; Jadaun, J.S.; Sabir, F.; Sangwan, N.S. Effect of cadmium stress on inductive enzymatic and nonenzymatic responses of ROS and sugar metabolism in multiple shoot cultures of Ashwagandha (*Withania somnifera* Dunal). *Protoplasma* **2014**, *251*, 1031–1045. [CrossRef] [PubMed]

103. Poonam, R.K.; Bhardwaj, R.; Sirhindi, G. Castasterone regulated polyphenolic metabolism and photosynthetic system in *Brassica juncea* plants under copper stress. *J. Pharmacogn. Phytochem.* **2015**, *4*, 282–289.

104. Kaur, P.; Bali, S.; Sharma, A.; Vig, A.P.; Bhardwaj, R. Effect of earthworms on growth, photosynthetic efficiency and metal uptake in *Brassica juncea* L. plants grown in cadmium-polluted soils. *Environ. Sci. Pollut. Res.* **2017**, *24*, 13452–13465. [CrossRef] [PubMed]

105. Kaur, P.; Bali, S.; Sharma, A.; Vig, A.P.; Bhardwaj, R. Role of earthworms in phytoremediation of cadmium (Cd) by modulating the antioxidative potential of *Brassica juncea* L. *Appl. Soil Ecol.* **2018**, *124*, 306–316. [CrossRef]

106. Kohli, S.K.; Handa, N.; Sharma, A.; Kumar, V.; Kaur, P.; Bhardwaj, R. Synergistic effect of 24-epibrassinolide and salicylic acid on photosynthetic efficiency and gene expression in *Brassica juncea* L. under Pb stress. *Turk. J. Biol.* **2017**, *41*, 943–953. [CrossRef] [PubMed]

107. Nakabayashi, R.; Yonekura-Sakakibara, K.; Urano, K.; Suzuki, M.; Yamada, Y.; Nishizawa, T.; Matsuda, F.; Kojima, M.; Sakakibara, H.; Shinozaki, K.; et al. Enhancement of oxidative and drought tolerance in *Arabidopsis* by overaccumulation of antioxidant flavonoids. *Plant J.* **2014**, *77*, 367–379. [CrossRef] [PubMed]

108. Kirakosyan, A.; Seymour, E.; Kaufman, P.B.; Warber, S.; Bolling, S.; Chang, S.C. Antioxidant capacity of polyphenolic extracts from leaves of *Crataegus laevigata* and *Crataegus monogyna* (Hawthorn) subjected to drought and cold stress. *J. Agric. Food Chem.* **2003**, *51*, 3973–3976. [CrossRef]

109. Ballizany, W.L.; Hofmann, R.W.; Jahufer, M.Z.Z.; Barrett, B.A. Multivariate associations of flavonoid and biomass accumulation in white clover (*Trifolium repens*) under drought. *Funct. Plant Biol.* **2012**, *39*, 167–177. [CrossRef]

110. Rezayian, M.; Niknam, V.; Ebrahimzadeh, H. Differential responses of phenolic compounds of Brassica napus under drought stress. *Iran. J. Plant Physiol.* **2018**, *8*, 2417–2425.

111. Li, M.; Li, Y.; Zhang, W.; Li, S.; Gao, Y.; Ai, X.; Zhang, D.; Liu, B.; Li, Q. Metabolomics analysis reveals that elevated atmospheric CO_2 alleviates drought stress in cucumber seedling leaves. *Anal. Biochem.* **2018**, *559*, 71–85. [CrossRef]

112. Nichols, S.N.; Hofmann, R.W.; Williams, W.M. Physiological drought resistance and accumulation of leaf phenolics in white clover interspecific hybrids. *Environ. Exp. Bot.* **2015**, *119*, 40–47. [CrossRef]

113. Sanchez-Rodriguez, E.; Moreno, D.A.; Ferreres, F.; Rubio-Wilhelmi Mdel, M.; Ruiz, J.M. Differential responses of five cherry tomato varieties to water stress: Changes on phenolic metabolites and related enzymes. *Phytochemistry* **2011**, *72*, 723–729. [CrossRef] [PubMed]

114. Hernandez, I.; Alegre, L.; Van Breusegem, F.; Munne-Bosch, S. How relevant are flavonoids as antioxidants in plants? *Trends Plant Sci.* **2009**, *14*, 125–132. [CrossRef] [PubMed]

115. Gharibi, S.; Tabatabaei, B.E.; Saeidi, G.; Goli, S.A. Effect of Drought Stress on Total Phenolic, Lipid Peroxidation, and Antioxidant Activity of Achillea Species. *Appl. Biochem. Biotechnol.* **2016**, *178*, 796–809. [CrossRef] [PubMed]

116. Hodaei, M.; Rahimmalek, M.; Arzani, A.; Talebi, M. The effect of water stress on phytochemical accumulation, bioactive compounds and expression of key genes involved in flavonoid biosynthesis in *Chrysanthemum morifolium* L. *Ind. Crops Prod.* **2018**, *120*, 295–304. [CrossRef]

117. Galieni, A.; Di Mattia, C.; De Gregorio, M.; Speca, S.; Mastrocola, D.; Pisante, M.; Stagnari, F. Effects of nutrient deficiency and abiotic environmental stresses on yield, phenolic compounds and antiradical activity in lettuce (*Lactuca sativa* L.). *Sci. Hortic.* **2015**, *187*, 93–101. [CrossRef]

118. Varela, M.C.; Arslan, I.; Reginato, M.A.; Cenzano, A.M.; Luna, M.V. Phenolic compounds as indicators of drought resistance in shrubs from Patagonian shrublands (Argentina). *Plant Physiol. Biochem.* **2016**, *104*, 81–91. [CrossRef]

119. Garcia-Calderon, M.; Pons-Ferrer, T.; Mrazova, A.; Pal'ove-Balang, P.; Vilkova, M.; Perez-Delgado, C.M.; Vega, J.M.; Eliasova, A.; Repcak, M.; Marquez, A.J.; et al. Modulation of phenolic metabolism under stress conditions in a *Lotus japonicus* mutant lacking plastidic glutamine synthetase. *Front. Plant Sci.* **2015**, *6*, 760. [CrossRef]

120. Silva, F.L.B.; Vieira, L.G.E.; Ribas, A.F.; Moro, A.L.; Neris, D.M.; Pacheco, A.C. Proline accumulation induces the production of total phenolics in transgenic tobacco plants under water deficit without increasing the G6PDH activity. *Theor. Exp. Plant Physiol.* **2018**, *30*, 251–260. [CrossRef]

121. Ghasemi Pirbalouti, A.; Malekpoor, F.; Salimi, A.; Golparvar, A. Exogenous application of chitosan on biochemical and physiological characteristics, phenolic content and antioxidant activity of two species of basil (*Ocimum ciliatum* and *Ocimum basilicum*) under reduced irrigation. *Sci. Hortic.* **2017**, *217*, 114–122. [CrossRef]

122. Khalil, N.; Fekry, M.; Bishr, M.; El-Zalabani, S.; Salama, O. Foliar spraying of salicylic acid induced accumulation of phenolics, increased radical scavenging activity and modified the composition of the essential oil of water stressed *Thymus vulgaris* L. *Plant Physiol. Biochem.* **2018**, *123*, 65–74. [CrossRef] [PubMed]

123. Kaur, L.; Zhawar, V.K. Phenolic parameters under exogenous ABA, water stress, salt stress in two wheat cultivars varying in drought tolerance. *Ind. J. Plant Physiol.* **2015**, *20*, 151–156. [CrossRef]

124. Griesser, M.; Weingart, G.; Schoedl-Hummel, K.; Neumann, N.; Becker, M.; Varmuza, K.; Liebner, F.; Schuhmacher, R.; Forneck, A. Severe drought stress is affecting selected primary metabolites, polyphenols, and volatile metabolites in grapevine leaves (*Vitis vinifera* cv. Pinot noir). *Plant Physiol. Biochem.* **2015**, *88*, 17–26. [CrossRef] [PubMed]

125. Castellarin, S.D.; Pfeiffer, A.; Sivilotti, P.; Degan, M.; Peterlunger, E.; Di Gaspero, G. Transcriptional regulation of anthocyanin biosynthesis in ripening fruits of grapevine under seasonal water deficit. *Plant Cell Environ.* **2007**, *30*, 1381–1399. [CrossRef] [PubMed]

126. Taïbi, K.; Taïbi, F.; Ait Abderrahim, L.; Ennajah, A.; Belkhodja, M.; Mulet, J.M. Effect of salt stress on growth, chlorophyll content, lipid peroxidation and antioxidant defence systems in *Phaseolus vulgaris* L. *S. Afr. J. Bot.* **2016**, *105*, 306–312. [CrossRef]

127. Acosta-Motos, J.R.; Ortuño, M.F.; Bernal-Vicente, A.; Diaz-Vivancos, P.; Sanchez-Blanco, M.J.; Hernandez, J.A. Plant Responses to Salt Stress: Adaptive Mechanisms. *Agronomy* **2017**, *7*, 18. [CrossRef]

128. De Azevedo Neto, A.D.; Prisco, J.T.; Enéas-Filho, J.; Abreu, C.E.B.D.; Gomes-Filho, E. Effect of salt stress on antioxidative enzymes and lipid peroxidation in leaves and roots of salt-tolerant and salt-sensitive maize genotypes. *Environ. Exp. Bot.* **2006**, *56*, 87–94. [CrossRef]

129. Martinez, V.; Mestre, T.C.; Rubio, F.; Girones-Vilaplana, A.; Moreno, D.A.; Mittler, R.; Rivero, R.M. Accumulation of Flavonols over Hydroxycinnamic Acids Favors Oxidative Damage Protection under Abiotic Stress. *Front. Plant Sci.* **2016**, *7*, 838. [CrossRef]

130. Chen, S.; Wu, F.; Li, Y.; Qian, Y.; Pan, X.; Li, F.; Wang, Y.; Wu, Z.; Fu, C.; Lin, H.; et al. NtMYB4 and NtCHS1 Are Critical Factors in the Regulation of Flavonoid Biosynthesis and Are Involved in Salinity Responsiveness. *Front. Plant Sci.* **2019**, *10*, 178. [CrossRef]

131. Bistgani, Z.E.; Hashemi, M.; DaCosta, M.; Craker, L.; Maggi, F.; Morshedloo, M.R. Effect of salinity stress on the physiological characteristics, phenolic compounds and antioxidant activity of *Thymus vulgaris* L. and *Thymus daenensis* Celak. *Ind. Crops Prod.* **2019**, *135*, 311–320. [CrossRef]

132. Valifard, M.; Mohsenzadeh, S.; Kholdebarin, B.; Rowshan, V. Effects of salt stress on volatile compounds, total phenolic content and antioxidant activities of *Salvia mirzayanii*. *S. Afr. J. Bot.* **2014**, *93*, 92–97. [CrossRef]

133. Rossi, L.; Borghi, M.; Francini, A.; Lin, X.; Xie, D.Y.; Sebastiani, L. Salt stress induces differential regulation of the phenylpropanoid pathway in *Olea europaea* cultivars Frantoio (salt-tolerant) and Leccino (salt-sensitive). *J. Plant Physiol.* **2016**, *204*, 8–15. [CrossRef] [PubMed]

134. Al-Ghamdi, A.A.; Elansary, H.O. Synergetic effects of 5-aminolevulinic acid and *Ascophyllum nodosum* seaweed extracts on *Asparagus* phenolics and stress related genes under saline irrigation. *Plant Physiol. Biochem.* **2018**, *129*, 273–284. [CrossRef] [PubMed]

135. Golkar, P.; Taghizadeh, M. In vitro evaluation of phenolic and osmolite compounds, ionic content, and antioxidant activity in safflower (*Carthamus tinctorius* L.) under salinity stress. *Plant Cell Tiss. Org. Cult.* **2018**, *134*, 357–368. [CrossRef]

136. Wang, F.; Zhu, H.; Chen, D.; Li, Z.; Peng, R.; Yao, Q. A grape bHLH transcription factor gene, VvbHLH1, increases the accumulation of flavonoids and enhances salt and drought tolerance in transgenic *Arabidopsis thaliana*. *Plant Cell Tiss. Org. Cult.* **2016**, *125*, 387–398. [CrossRef]

137. Yan, J.; Wang, B.; Jiang, Y.; Cheng, L.; Wu, T. GmFNSII-controlled soybean flavone metabolism responds to abiotic stresses and regulates plant salt tolerance. *Plant Cell Physiol.* **2014**, *55*, 74–86. [CrossRef] [PubMed]

138. Ben-Abdallah, S.; Zorrig, W.; Amyot, L.; Renaud, J.; Hannoufa, A.; Lachâal, M.; Karray-Bouraoui, N. Potential production of polyphenols, carotenoids and glycoalkaloids in *Solanum villosum* Mill. under salt stress. *Biologia* **2019**, *74*, 309–324. [CrossRef]

139. Scagel, C.F.; Lee, J.; Mitchell, J.N. Salinity from NaCl changes the nutrient and polyphenolic composition of basil leaves. *Ind. Crops Prod.* **2019**, *127*, 119–128. [CrossRef]

140. Sarker, U.; Oba, S. Augmentation of leaf color parameters, pigments, vitamins, phenolic acids, flavonoids and antioxidant activity in selected *Amaranthus tricolor* under salinity stress. *Sci. Rep.* **2018**, *8*, 12349. [CrossRef]

141. Dkhil, B.B.; Denden, M. Effect of salt stress on growth, anthocyanins, membrane permeability and chlorophyll fluorescence of Okra (*Abelmoschus esculentus* L.) seedlings. *Am. J. Plant Physiol.* **2012**, *7*, 174–183. [CrossRef]

142. Eryılmaz, F. The Relationships between Salt Stress and Anthocyanin Content in Higher Plants. *Biotechnol. Biotechnol. Equip.* **2006**, *20*, 47–52. [CrossRef]

143. Aloisi, I.; Parrotta, L.; Ruiz, K.B.; Landi, C.; Bini, L.; Cai, G.; Biondi, S.; Del Duca, S. New Insight into Quinoa Seed Quality under Salinity: Changes in Proteomic and Amino Acid Profiles, Phenolic Content, and Antioxidant Activity of Protein Extracts. *Front. Plant Sci.* **2016**, *7*, 656. [CrossRef] [PubMed]

144. Lucini, L.; Borgognone, D.; Rouphael, Y.; Cardarelli, M.; Bernardi, J.; Colla, G. Mild Potassium Chloride Stress Alters the Mineral Composition, Hormone Network, and Phenolic Profile in Artichoke Leaves. *Front. Plant Sci.* **2016**, *7*, 948. [CrossRef] [PubMed]

145. Ma, Y.; Wang, P.; Gu, Z.; Tao, Y.; Shen, C.; Zhou, Y.; Han, Y.; Yang, R. Ca^{2+} involved in GABA signal transduction for phenolics accumulation in germinated hulless barley under NaCl stress. *Food Chem. X* **2019**, *2*, 100023. [CrossRef]

146. Çoban, Ö.; Göktürk Baydar, N. Brassinosteroid effects on some physical and biochemical properties and secondary metabolite accumulation in peppermint (*Mentha piperita* L.) under salt stress. *Ind. Crops Prod.* **2016**, *86*, 251–258. [CrossRef]

147. Valifard, M.; Mohsenzadeh, S.; Niazi, A.; Moghadam, A. Phenylalanine ammonia lyase isolation and functional analysis of phenylpropanoid pathway under salinity stress in 'Salvia' species. *Aust. J. Crop Sci.* **2015**, *9*, 656–665.

148. Daayf, F.; Lattanzio, V. *Recent Advances in Polyphenol Research*; John Wiley & Sons: Hoboken, NJ, USA, 2009.

149. Landi, M.; Tattini, M.; Gould, K.S. Multiple functional roles of anthocyanins in plant-environment interactions. *Environ. Exp. Bot.* **2015**, *119*, 4–17. [CrossRef]

150. Agati, G.; Tattini, M. Multiple functional roles of flavonoids in photoprotection. *New Phytol.* **2010**, *186*, 786–793. [CrossRef]

151. Agati, G.; Azzarello, E.; Pollastri, S.; Tattini, M. Flavonoids as antioxidants in plants: Location and functional significance. *Plant Sci.* **2012**, *196*, 67–76. [CrossRef]

152. Kolb, C.A.; Kaser, M.A.; Kopecky, J.; Zotz, G.; Riederer, M.; Pfundel, E.E. Effects of natural intensities of visible and ultraviolet radiation on epidermal ultraviolet screening and photosynthesis in grape leaves. *Plant. Physiol.* **2001**, *127*, 863–875. [CrossRef]

153. Ghasemi, S.; Kumleh, H.H.; Kordrostami, M. Changes in the expression of some genes involved in the biosynthesis of secondary metabolites in *Cuminum cyminum* L. under UV stress. *Protoplasma* **2019**, *256*, 279–290. [CrossRef] [PubMed]

154. Goyal, A.; Siddiqui, S.; Upadhyay, N.; Soni, J. Effects of ultraviolet irradiation, pulsed electric field, hot water and ethanol vapours treatment on functional properties of mung bean sprouts. *J. Food Sci. Technol.* **2014**, *51*, 708–714. [CrossRef] [PubMed]

155. Xu, Y.; Charles, M.T.; Luo, Z.; Mimee, B.; Veronneau, P.Y.; Rolland, D.; Roussel, D. Preharvest Ultraviolet C Irradiation Increased the Level of Polyphenol Accumulation and Flavonoid Pathway Gene Expression in Strawberry Fruit. *J. Agric. Food Chem.* **2017**, *65*, 9970–9979. [CrossRef] [PubMed]

156. Demkura, P.V.; Abdala, G.; Baldwin, I.T.; Ballare, C.L. Jasmonate-dependent and -independent pathways mediate specific effects of solar ultraviolet B radiation on leaf phenolics and antiherbivore defense. *Plant Physiol.* **2010**, *152*, 1084–1095. [CrossRef] [PubMed]

157. Berli, F.J.; Fanzone, M.; Piccoli, P.; Bottini, R. Solar UV-B and ABA are involved in phenol metabolism of *Vitis vinifera* L. increasing biosynthesis of berry skin polyphenols. *J. Agric. Food Chem.* **2011**, *59*, 4874–4884. [CrossRef] [PubMed]

158. Nenadis, N.; Llorens, L.; Koufogianni, A.; Diaz, L.; Font, J.; Gonzalez, J.A.; Verdaguer, D. Interactive effects of UV radiation and reduced precipitation on the seasonal leaf phenolic content/composition and the antioxidant activity of naturally growing *Arbutus unedo* plants. *J. Photochem. Photobiol. B* **2015**, *153*, 435–444. [CrossRef] [PubMed]

159. Moreira-Rodriguez, M.; Nair, V.; Benavides, J.; Cisneros-Zevallos, L.; Jacobo-Velazquez, D.A. UVA, UVB Light, and Methyl Jasmonate, Alone or Combined, Redirect the Biosynthesis of Glucosinolates, Phenolics, Carotenoids, and Chlorophylls in Broccoli Sprouts. *Int. J. Mol. Sci.* **2017**, *18*, 2330. [CrossRef]

160. Liu, M.; Cao, B.; Zhou, S.; Liu, Y. Responses of the flavonoid pathway to UV-B radiation stress and the correlation with the lipid antioxidant characteristics in the desert plant *Caryopteris mongolica*. *Acta Ecol. Sin.* **2012**, *32*, 150–155. [CrossRef]

161. Nascimento, L.; Leal-Costa, M.V.; Menezes, E.A.; Lopes, V.R.; Muzitano, M.F.; Costa, S.S.; Tavares, E.S. Ultraviolet-B radiation effects on phenolic profile and flavonoid content of *Kalanchoe pinnata*. *J. Photochem. Photobiol. B* **2015**, *148*, 73–81. [CrossRef]

162. Sytar, O.; Zivcak, M.; Bruckova, K.; Brestic, M.; Hemmerich, I.; Rauh, C.; Simko, I. Shift in accumulation of flavonoids and phenolic acids in lettuce attributable to changes in ultraviolet radiation and temperature. *Sci. Hortic.* **2018**, *239*, 193–204. [CrossRef]

163. Lee, M.J.; Son, J.E.; Oh, M.M. Growth and phenolic compounds of *Lactuca sativa* L. grown in a closed-type plant production system with UV-A, -B, or -C lamp. *J. Sci. Food Agric.* **2014**, *94*, 197–204. [CrossRef]

164. Huyskens-Keil, S.; Eichholz, I.; Kroh, L.; Rohn, S. UV-B induced changes of phenol composition and antioxidant activity in black currant fruit (*Ribes nigrum* L.). *J. App. Bot. Food Qual.* **2012**, *81*, 140–144.

165. Mariz-Ponte, N.; Mendes, R.J.; Sario, S.; De Oliveira, J.F.; Melo, P.; Santos, C. Tomato plants use non-enzymatic antioxidant pathways to cope with moderate UV-A/B irradiation: A contribution to the use of UV-A/B in horticulture. *J. Plant Physiol.* **2018**, *221*, 32–42. [CrossRef]

166. Chen, Z.; Ma, Y.; Weng, Y.; Yang, R.; Gu, Z.; Wang, P. Effects of UV-B radiation on phenolic accumulation, antioxidant activity and physiological changes in wheat (*Triticum aestivum* L.) seedlings. *Food Biosci.* **2019**, *30*, 100409. [CrossRef]

167. Alonso, R.; Berli, F.J.; Fontana, A.; Piccoli, P.; Bottini, R. Malbec grape (*Vitis vinifera* L.) responses to the environment: Berry phenolics as influenced by solar UV-B, water deficit and sprayed abscisic acid. *Plant Physiol. Biochem.* **2016**, *109*, 84–90. [CrossRef] [PubMed]

168. Swieca, M. Elicitation with abiotic stresses improves pro-health constituents, antioxidant potential and nutritional quality of lentil sprouts. *Saudi J. Biol. Sci.* **2015**, *22*, 409–416. [CrossRef]

169. Wang, L.; Shan, T.; Xie, B.; Ling, C.; Shao, S.; Jin, P.; Zheng, Y. Glycine betaine reduces chilling injury in peach fruit by enhancing phenolic and sugar metabolisms. *Food Chem.* **2019**, *272*, 530–538. [CrossRef]

170. Mahdavi, V.; Farimani, M.M.; Fathi, F.; Ghassempour, A. A targeted metabolomics approach toward understanding metabolic variations in rice under pesticide stress. *Anal. Biochem.* **2015**, *478*, 65–72. [CrossRef]

171. Korosi, L.; Bouderias, S.; Csepregi, K.; Bognar, B.; Teszlak, P.; Scarpellini, A.; Castelli, A.; Hideg, E.; Jakab, G. Nanostructured TiO2-induced photocatalytic stress enhances the antioxidant capacity and phenolic content in the leaves of *Vitis vinifera* on a genotype-dependent manner. *J. Photochem. Photobiol. B* **2019**, *190*, 137–145. [CrossRef]

172. Sharma, A.; Kumar, V.; Singh, R.; Thukral, A.K.; Bhardwaj, R. Effect of seed pre-soaking with 24-epibrassinolide on growth and photosynthetic parameters of *Brassica juncea* L. in imidacloprid soil. *Ecotoxicol. Environ. Saf.* **2016**, *133*, 195–201. [CrossRef]

173. Zlotek, U.; Szymanowska, U.; Baraniak, B.; Karas, M. Antioxidant activity of polyphenols of adzuki bean (*Vigna angularis*) germinated in abiotic stress conditions. *Acta Sci. Pol. Technol. Aliment.* **2015**, *14*, 55–63. [CrossRef] [PubMed]

174. Commisso, M.; Toffali, K.; Strazzer, P.; Stocchero, M.; Ceoldo, S.; Baldan, B.; Levi, M.; Guzzo, F. Impact of phenylpropanoid compounds on heat stress tolerance in carrot cell cultures. *Front. Plant Sci.* **2016**, *7*, 1439. [CrossRef] [PubMed]

175. Cingoz, G.S.; Gurel, E. Effects of salicylic acid on thermotolerance and cardenolide accumulation under high temperature stress in *Digitalis trojana* Ivanina. *Plant Physiol. Biochem.* **2016**, *105*, 145–149. [CrossRef] [PubMed]

176. Griffith, M.; Yaish, M.W. Antifreeze proteins in overwintering plants: A tale of two activities. *Trends Plant. Sci.* **2004**, *9*, 399–405. [CrossRef] [PubMed]

177. Gao, H.; Zhang, Z.; Lv, X.; Cheng, N.; Peng, B.; Cao, W. Effect of 24-epibrassinolide on chilling injury of peach fruit in relation to phenolic and proline metabolisms. *Post. Biol. Technol.* **2016**, *111*, 390–397. [CrossRef]

178. Sharma, A.; Kumar, V.; Yuan, H.; Kanwar, M.K.; Bhardwaj, R.; Thukral, A.K.; Zheng, B. Jasmonic Acid Seed Treatment Stimulates Insecticide Detoxification in *Brassica juncea* L. *Front. Plant Sci.* **2018**, *9*, 1609. [CrossRef]

179. Nourozi, E.; Hosseini, B.; Maleki, R.; Mandoulakani, B.A. Pharmaceutical important phenolic compounds overproduction and gene expression analysis in *Dracocephalum kotschyi* hairy roots elicited by SiO2 nanoparticles. *Ind. Crops Prod.* **2019**, *133*, 435–446. [CrossRef]

180. Girilal, M.; Fayaz, A.M.; Elumalai, L.K.; Sathiyaseelan, A.; Gandhiappan, J.; Kalaichelvan, P.T. Comparative Stress Physiology Analysis of Biologically and Chemically Synthesized Silver Nanoparticles on *Solanum lycopersicum* L. *Colloid Interface Sci. Commun.* **2018**, *24*, 1–6. [CrossRef]
181. Raigond, P.; Raigond, B.; Kaundal, B.; Singh, B.; Joshi, A.; Dutt, S. Effect of zinc nanoparticles on antioxidative system of potato plants. *J. Environ. Biol.* **2017**, *38*, 435–439. [CrossRef]
182. Singh, O.S.; Pant, N.C.; Laishram, M.L.; Tewari, R.D.; Joshi, K.; Pandey, C. Effect of CuO nanoparticles on polyphenols content and antioxidant activity in Ashwagandha (*Withania somnifera* L. Dunal). *J. Pharmacogn. Phytochem.* **2018**, *7*, 3433–3439.

MDPI

St. Alban-Anlage 66

4052 Basel

Switzerland

Tel. +41 61 683 77 34

Fax +41 61 302 89 18

www.mdpi.com

Molecules Editorial Office

E-mail: molecules@mdpi.com

www.mdpi.com/journal/molecules

www.ingramcontent.com/pod-product-compliance
Lightning Source LLC
LaVergne TN
LVHW070931170726
843515LV00005B/918